Handbook of Energy Utilization in Agriculture

Editor

David Pimentel
New York State College of Agriculture and Life Sciences
Cornell University
Ithaca, New York

CRC Press, Inc.
Boca Raton, Florida

Library of Congress Cataloging in Publication Data

Main entry under title:

Handbook of energy utilization in agriculture.

Bibliography: p.
Includes index.
1. Agriculture—Energy consumption. I. Pimentel, David, 1925-
S494.5.E5H36 333.7 79-4378
ISBN 0-8493-2661-3

Direct all inquiries to CRC Press, Inc., 2000 N.W. 24th Street, Boca Raton, Florida, 33431.

International Standard Book Number 0-8493-2661-3
Library of Congress Card Number 79-4378
Printed in the United States

PREFACE

The aim of this volume is to collect and present available data, both published and unpublished, on energy use in agriculture and forestry production. Energy analyses for some sciences such as ecology are not new, but their applications to agriculture started in 1973. These analyses have grown rapidly in number and complexity.

This handbook is intended for agriculturalists and others concerned with energy use in crop, livestock, and forestry production. We have assembled the best data available, but look forward to receiving comments and suggestions from researchers in helping us strengthen any future editions.

The data presented in the Handbook focus specifically on the energy input aspects of crop, livestock, and forest production. In these analyses no attempt was made to assign an energy value to manpower although it is an input in all the systems studied. Assigning an appropriate value to the energy cost of labor is difficult because of the various means of measuring the energy cost of labor. These energy costs individually or in various combinations might include: (1) the mechanical and heat energy expended in doing a particular agricultural task; (2) food energy consumed per day per agricultural worker; (3) accessory energy utilized specifically for the worker himself; (4) total energy for the worker as a part of society including the government and the military support systems; (5) total energy for the worker plus his family as a part of society. In the analysis of the energy inputs for crop, livestock, and forest production, the decision by the Advisory Board was not to measure the energy input for labor and to employ process analysis for the accounting of the other specific energy inputs for agricultural production. The process analysis employed accounted for about 90% of energy inputs. To account for all the energy inputs, a combination of process analysis and input/output analysis would be desirable.

For many crops details concerning the various inputs such as machinery are not fully known, therefore "best estimates" had to be made. In most cases, averages of inputs were used. Although these inputs vary from region to region and from farm to farm in the United States, the estimated average input values at least give a base guideline.

We fully recognize the many limitations of energy accounting for crop, livestock, and forest production. We feel, however, that the data assembled by the numerous authors will prove valuable to agriculturalists as they strive to make more efficient use of energy and other resources needed for successful crop, livestock, and forestry production in the U.S. and elsewhere in the world.

The contributing Authors and Advisory Board to this handbook devoted considerable time and effort in assembling this handbook. The successful completion of this handbook is due to their cooperation and hard work.

I am indebted to Ms. Nancy Goodman, Debra Alesbury, and Susan Pohl for their valuable assistance in compiling this volume. I also wish to thank Sherry Derr of CRC for her technical assistance in editing this handbook.

David Pimentel

THE EDITOR

David Pimentel is Professor of Insect Ecology and Agricultural Sciences in the Department of Entomology, the Section of Ecology and Systematics, and the Field of Natural Resources, College of Agriculture and Life Sciences, Cornell University. Dr. Pimentel received his B.S. in 1948 from the University of Massachusetts and his Ph.D. in 1951 from Cornell University. From 1951 to 1954 he was Chief of the Tropical Research Laboratory, U.S. Public Health Service, San Juan, Puerto Rico. From 1954 to 1955 he was post-doctoral research fellow at the University of Chicago; 1960 to 1961 OEEC Fellow at Oxford University (England); and 1961, Research Scholar at Massachusetts Institute of Technology. He was appointed an Assistant Professor of Insect Ecology at Cornell University in 1955 and Associate Professor in 1961. In 1963 he was appointed Professor and Head of the Department of Entomology and Limnology. He served as Head until 1969 when he returned to full time research and teaching.

Nationally he has served on numerous Presidential Committees and National Academy of Sciences Committees and Boards. In the recent World Food and Nutrition Study of the NAS he chaired the Panel on *Interdependencies of Population, Food, Health, Energy, and the Environment*. He is chairman of the Biomass Panel of Energy Research Advisory Board of the U.S. Department of Energy. He is a Fellow of the American Association for the Advancement of Science and Fellow of the Entomological Society of Canada.

Dr. Pimentel has authored nearly 200 scientific publications, written two books, and edited five books.

ADVISORY BOARD

CONTRIBUTORS

M. W. Adams
Department of Crop and Soil Sciences
Michigan State University
East Lansing, Michigan

J. T. Alexander
American Crystal Sugar Company
Moorhead, Minnesota

M. S. Allen
Department of Animal Science
Cornell University
Ithaca, New York

Jose Alvarez
Food and Resource Economics Department
Agricultural Research and Educational Center
University of Florida
Belle Glade, Florida

Max E. Austin
Horticulture Department
University of Georgia
College of Agriculture
Coastal Plain Experiment Station
Tifton, Georgia

P. K. Avlani
Agricultural Engineering
University of California
Davis, California

L. R. Baker
Department of Horticulture
Michigan State University
East Lansing, Michigan

J. Bardach
Resource Systems Institute
East-West Center
Honolulu, Hawaii

J. Clair Batty
Department of Mechanical and Manufacturing Engineering
Utah State University
Logan, Utah

James A. Beutel
Pomology Department
University of California
Davis, California

G. A. Bradley
Horticulture and Forestry
University of Arkansas
Fayetteville, Arkansas

Alvan R. Brick, Jr.
Section of Ecology and Systematics
Department of Entomology
Cornell University
Ithaca, New York

L. W. Briggle
Science and Education Administration
Agricultural Research
U.S. Department of Agriculture
Beltsville, Maryland

Roger Brook
Department of Agricultural Engineering
Michigan State University
East Lansing, Michigan

Robert Bukantis
Department of Entomology
Cornell University
Ithaca, New York

Michael Burgess
Department of Entomology
Cornell University
Ithaca, New York

George C. Burrill
Center for Studies in Food Self-Sufficiency
Vermont Institute of Community Involvement
Burlington, Vermont

Vashek Cervinka
California Department of Food and Agriculture
Sacramento, California

William J. Chancellor
Agricultural Engineering
University of California
Davis, California

Sterling Chick
Department of Entomology
Cornell University
Ithaca, New York

Norman I. Childers
Department of Horticulture and Forestry
Rutgers University
New Brunswick, New Jersey

John J. Combs
Department of Range Sciences
Colorado State University
Fort Collins, Colorado

C. Wayne Cook
Department of Range Sciences
Colorado State University
Fort Collins, Colorado

R. C. Funt
Department of Horticulture
Ohio State University
Columbus, Ohio

G. J. Galletta
Agricultural Research
Science and Education Administration
U.S. Department of Agriculture
Beltsville, Maryland

C. Kerry Gee
Economics, Statistics, and Cooperative Services
U.S. Department of Agriculture
Colorado State University
Fort Collins, Colorado

Norman C. Glaze
Science and Education Administration
U.S. Department of Agriculture
Coastal Plain Station
Tifton, Georgia

Nancy Goodman
Department of Entomology
Cornell University
Ithaca, New York

W. R. Grant
Economics, Statistics and Cooperatives Service
Department of Agricultural Economics
Texas A & M University
College Station, Texas

G. H. Heichel
USDA, SEA, AR
Department of Agronomy and Plant Genetics
University of Minnesota
St. Paul, Minnesota

Lynn A. Horel
Department of Agricultural Economics
University of California
Davis, California

Carl S. Hoveland
Agronomy and Soil Department
Auburn University
Auburn, Alabama

R. Brian How
Department of Agricultural Economics
Cornell University
Ithaca, New York

Hunter Johnson Jr.
Extension Vegetable Specialist
University of California
Riverside, California

Jack Keller
Department of Agricultural and Irrigation Engineering
Utah State University
Logan, Utah

Robert T. Kelly
Department of Horticulture and Forestry
Rutgers University
New Brunswick, New Jersey

Dale E. Kester
Department of Pomology
University of California
Davis, California

Wayne R. Knapp
Department of Agronomy
Cornell University
Ithaca, New York

P. F. Knowles
Department of Agronomy and Range Science
University of California
Davis, California

John Krummel
Environmental Sciences Division
Oak Ridge, National Laboratory
Oak Ridge, Tennessee

F. W. Liu
Department of Pomology
Cornell University
Ithaca, New York

William Lockeretz
Northeast Solar Energy Center
Cambridge, Massachusetts

George C. Martin
Department of Pomology
University of California
Davis, California

N. P. Martin
Department of Agronomy and Plant Genetics
University of Minnesota
St. Paul, Minnesota

Michael Miltner
Department of Fisheries
Louisiana State University
Baton Rouge, Louisiana

Philip A. Minges (Deceased)
Department of Vegetable Crops
Cornell University
Ithaca, New York

J. W. Mishoe
Agricultural Engineering Department
University of Florida
Gainesville, Florida

Douglas B. Monteith
School of Forestry
State University of New York
College of Environmental Science and Forestry
Syracuse, New York

Donald M. Nafus
Science and Education Administration
U.S. Department of Agriculture
Beltsville, Maryland

James R. Nolfi
Center for Studies in Food Self-Sufficiency
Burlington, Vermont

Martin R. Okos
Department of Agricultural Engineering
Purdue University
West Lafayette, Indiana

P. A. Oltenacu
Department of Animal Science
Cornell University
Ithaca, New York

Charles E. Ostrander
Department of Poultry Science
Cornell University
Ithaca, New York

Donald K. Ourecky
Department of Pomology and Viticulture
New York State Agricultural Experiment Station
Geneva, New York

Robert M. Peart
Department of Agricultural Engineering
Purdue University
West Lafayette, Indiana

David Pimentel
New York State College of Agriculture and Life Sciences
Cornell University
Ithaca, New York

Robert M. Pool
Department of Pomology and Viticulture
New York State Agricultural Experiment Station
Geneva, New York

E. L. Proebsting
Irrigated Agriculture Research and Extension Center
Washington State University
Prosser, Washington

David Ramming
Federal Research
Science and Education Administration
U.S. Department of Agriculture
Fresno, California

D. L. Reeves
Plant Science Department
South Dakota State University
Brookings, South Dakota

J. T. Reid (Deceased)
Department of Animal Science
Cornell University
Ithaca, New York

Herman J. Reitz
Institute of Food and Agricultural Sciences
Agricultural Research and Education Center
University of Florida
Lake Alfred, Florida

Ray Ricaud
Department of Agronomy
Center for Agricultural Sciences and Rural Science
Louisiana Agricultural Experiment Station
Louisiana State University
Baton Rouge, Louisiana

Stephen P. Rochereau
International Energy Associates Limited
Washington, D.C. 20037

J. N. Rutger
Science and Education Administration
U.S. Department of Agriculture
Department of Agronomy and Range Science
University of California
Davis, California

Edward J. Ryder
Agricultural Research
Science and Education Administration
U.S. Department of Agriculture
U.S. Agricultural Research Station
Salinas, California

Roger Sandsted
Vegetable Crops Department
Cornell UniversitypIthaca, New York

Ilse H. Schreiner
Department of Entomology
Cornell University
Ithaca, New York

W. O. Scott
Cooperative Extension Service and Home Agronomy
University of Illinois
Urbana, Illinois

Darrell Sparks
Department of Horticulture
University of Georgia
Athens, Georgia

R. L. Stebbins
Department of Horticulture
Horticulture Extension
Oregon State University
Corvallis, Oregon

Elinor C. Terhune
Office of Arid Land Studies
University of Arizona
Tucson, Arizona

Kiyoto Uriu
Department of Pomology
University of California
Davis, California

Gerald M. Ward
Department of Animal Sciences
Colorado State University
Fort Collins, Colorado

S. H. Weaver
Research and Development
The Quaker Oats Company
Chicago, Illinois

Ottilie D. White
Department of Animal Science
Cornell University
Ithaca, New York

TABLE OF CONTENTS

Introduction

INTRODUCTION

David Pimentel

Yields in U.S. agricultural production have increased rapidly since the 1940s. This improvement has been the result of the development of new crop varieties and improved livestock types and an increase in various management inputs. Specifically, these inputs include fertilizers, pesticides, and irrigation, all of which depend primarily upon fossil energy.

Indeed, at present, fossil energy is one of the major resources in agricultural production. Currently the high yields with relatively small inputs of manpower are associated with mechanized agriculture and are due essentially to ample supplies of inexpensive fossil energy. An often quoted statistic is that one U.S. farmer feeds about 50 people.[7] This is, of course, an oversimplification of the facts because the farmer himself depends upon large inputs from petrochemical, machinery industries, and commercial services. A more realistic statement is that about one person in five in the nation's work force is involved in supplying food in the American system.[5]

The total food system utilizes about 17% of the nation's fossil energy for production, processing, and preparation, with about one third employed in each activity. Forestry production and utilization require another 5% of the energy. Although these total 22%, the total is still less than the energy consumed by the American automobile, which burns nearly 25% of the total fossil energy used by the nation.

In agricultural production fossil energy has a twofold use, that is to increase crop, livestock, and forestry yields and also to replace labor. The major inputs that increase crop yields include energy used in fertilizers, pesticides, and irrigation. The major inputs for livestock production are forage and grains raised for feed. The energy used to reduce human labor input in crop production has little or no effect on crop yield except to facilitate the timing of planting and harvesting. For example, when corn is produced by hand about 1200 hr of manpower per hectare (500 hr per acre) are expended. With mechanized systems, corn can be produced with an input of only 12 hr of manpower per hectare. Hence, machinery has reduced the labor input to 1/100th of what it had been.

In general the total fossil energy input in crop production per hectare per growing season averages about 450 ℓ of petroleum equivalents or about 5 million kcal. This fossil energy input, however, represents a relatively small portion of the total energy needed for crop production when the contribution of solar energy is considered. For example, during a growing season, over 5 billion kcal of solar energy reaches 1 ha cornfield; about 60 million kcal (1.2%) of this light energy is converted into corn, and only 19 million kcal (0.4%) is corn grain. Hence, man's fossil energy input of 5 million kcal represents 8% of the total solar energy captured by the plants. Because corn is a relatively efficient crop in capturing solar energy, the ratio of fossil energy input to solar will be somewhat higher than for other crops.

The energy collected by crop plants and converted into food depends on weather conditions. It has been calculated that only a 0.6° C reduction in temperature may shorten the growing season for some crops 2 weeks. This can have significant impact on a crop such as corn. For example, each day that the planting of corn is delayed after May 1 in the Corn Belt reduces the yield about 29 kg/ha (1 bushel/acre/day). Thus, changes in weather such as temperature, rainfall, frost, and hail may significantly influence crop yields. Energy use in some cases can be used to help reduce the risks of loss.[2]

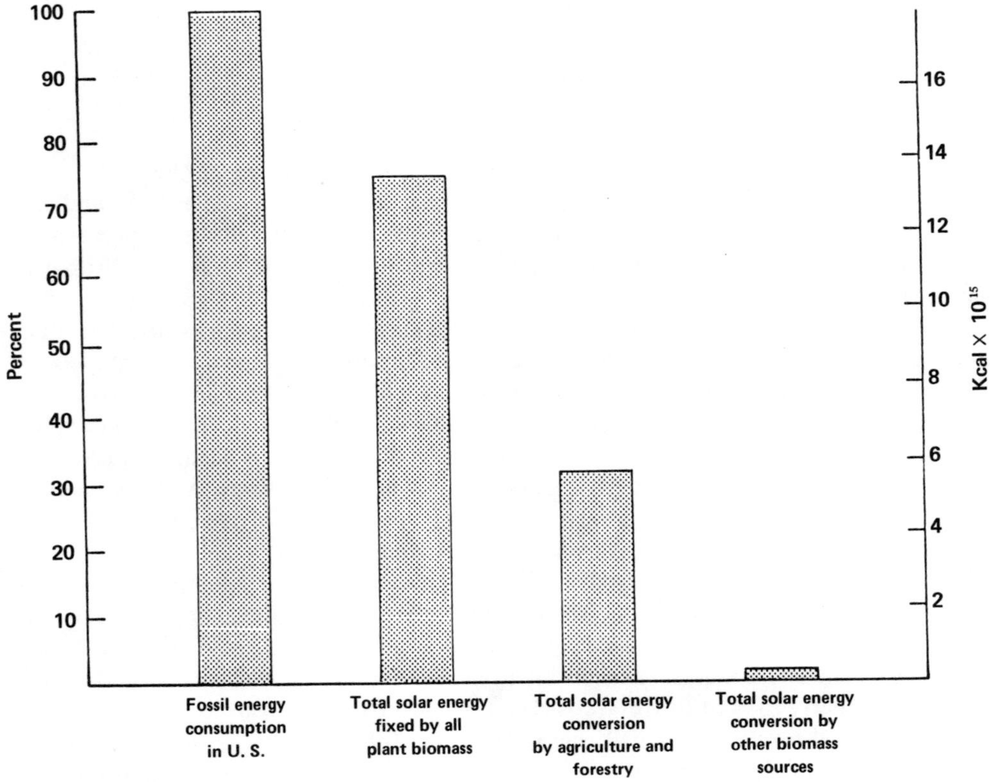

FIGURE 1. Biological solar energy conversion compared with fossil energy consumption in the U.S. All data are calculated for 1 year. (From Pimentel, D., Nafus, D., Vergara, W., Papaj, D., Jacouetta, L., Wulfe, M., Olsvig, L., Frech, K., Loye, M., and Mendoza, E., *Bio Science,* 28, 376, 1978. With permission.)

Although agricultural production in the U.S. depends heavily upon fossil energy, it is also dependent upon solar energy conversion. When agriculture and forestry products are combined and measured by solar energy harvest they represent nearly 6×10^{15} kcal annually (Figure 1). This is roughly equivalent to about one third of total fossil energy use in the U.S. Clearly then, agriculture and forestry are major solar energy conversion systems.

Some optimistic proposals have been presented for utilizing crop remains for biomass energy conversion. This is not recommended because of the undesirable environmental impacts that crop remains removal would have on soil erosion, soil structure, and the soil carbon ratio.[6]

In spite of the fact that food crops utilize solar energy, often the fossil energy input is greater than the energy content of the food harvested. However, measuring the value of the food only in terms of energy (kcal) content may be misleading or not representative of the true value of the food to man. Indeed many fruits, vegetables, and grains are raised for their important nutrients including vitamins, protein, minerals, as well as flavor, and other aesthetic characteristics. Therefore, care must be exercised in evaluating input/output energy ratios for crops because most crops are raised for more than just their food energy.

In this Handbook, fossil energy input and food energy output analyses have been calculated and the amount of protein produced is also listed even though food energy and protein represent only two criteria of the total nutrient value of the food. Fossil

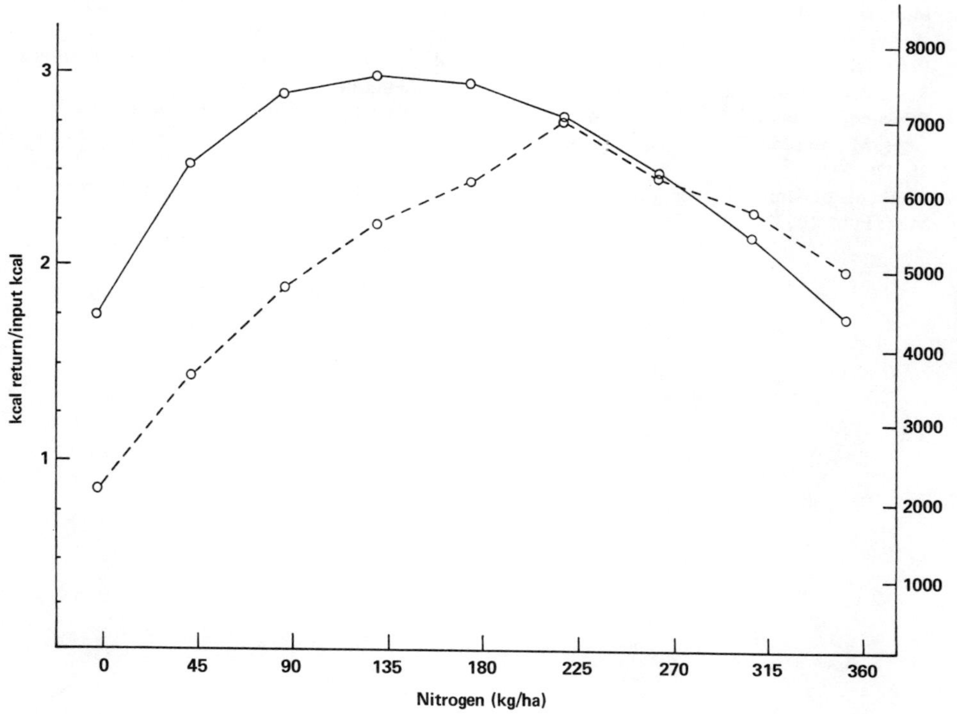

FIGURE 2. Corn yields (kg/ha, o- - -o) with varying amounts of nitrogen (phosphorus = 37 kg/ha) applied per hectare;[4] kcal return per input kcal (o——o).[4,5]

energy input in crop and livestock production was of prime interest here because fossil energy is a finite natural resource that is being rapidly depleted.[3]

In assessing fossil energy inputs and crop yields, it becomes evident that the marginal return in increased production per input kcal decreases as a particular input continues to increase. The yield of corn, for example, continues to increase, but at a decreasing rate, with each additional kilogram of nitrogen fertilizer applied until an input of about 200 kg per hectare is reached. If the quantity of nitrogen applied increases further, corn yields often decline. As a result, the energy output/input also declines with the further increment of nitrogen becoming less favorable (Figure 2).

Obviously determining the best management strategies for crop and livestock production depends upon an analysis of not only the specific environmental conditions for the crop and livestock, but also on energy input data as well as economic cost/benefit data. In other words, the more completely the entire production system is understood, the sounder the agricultural and forestry management strategies.

REFERENCES

1. **Bullard, C. W., Penner, P. S., and Pilati, D. A.,** Energy analysis: handbook for combining process and input-output analysis, Center for Advanced Computation, University of Illinois, Urbana, 1976.
2. **Doering, O. C., III,** Agriculture and energy use in the year 2000, *Am. J. Agric. Econ.*, 59, 1066, 1977.
3. **The National Energy Plan,** Energy Policy and Planning, Executive Office of the President, U.S. Government Printing Office, Washington, D.C., 1977.

4. **Munson, R. D. and Doll, J. P.,** The economics of fertilizer use in crop production, in *Advances in Agronomy,* Vol. 11, Norman, A. G., Ed., Academic Press, New York, 1959, 133.
5. **Pimentel, D., Hurd, L. E., Bellotti, A. C., Forster, M. J., Oka, I. N., Sholes, O. D., and Whitman, R. J.,** Food production and the energy crisis, *Science,* 182, 443, 1973.
6. **Pimentel, D., Nafus, D., Vergara, W., Papaj, D., Jaconetta, L., Wulfe, M., Olsvig, L., Frech, K., Loye, M., and Mendoza, E.,** Biological solar energy conversion and U.S. energy policy, *BioScience,* 28, 376, 1978.
7. **Factbook of United States Agriculture, U.S. Department of Agriculture, Office of Information, Misc. Publ. 1063, 1972.**

Energy Values for Agricultural Inputs

ACCOUNTING FOR ENERGY IN FARM MACHINERY AND BUILDINGS

Otto C. Doering III

There is no precise way to account for the energy used indirectly in agricultural production. This would be the energy that goes into the production of machinery, equipment, buildings and other non-land resources that contribute to food and fiber production over the long term and are normally treated as capital assets. One of the most important of these is farm machinery.

A tremendous amount of virtually unobtainable information would be required to make a precise accounting of the actual energy embodied in a specific stock of farm machinery for any given farming operation. In order to calculate the energy embodied in an engine block one would require information on the blast furnace used for the production of the steel involved, as well as information about the mining, transportation and refining of the iron ore, and other ingredients used in the steel making process. Beyond this, one would have to keep track of the machining and processing of the engine block itself in all intermediate process steps, as well as in final assembly and delivery. Even if such information were available it might not be terribly useful because both production and end use decisions are made on the basis of firm or industry level aggregations of dollar cost information. It is on the basis of economic considerations that energy substitutes or is substituted for other resources rather than on the basis of minimizing energy use per se.

The embodied machinery energy per unit of food or fiber output would be highly variable from year to year and between regions even if the actual machinery energy could be calculated correctly. One source of variability is the degree of intensity with which machinery is used. This is based upon risk considerations, the level of management, labor availability, and a number of other factors. Another source of variability is the actual level of production. This depends upon weather, pests, and many other factors beyond the engineering process and management controls familiar to manufactured products. For any given year, the embodied energy efficiency of a given machinery stock can vary substantially on the basis of all these factors.

Given both the natural and man induced variability from year to year, attempts to achieve maximum energy or dollar efficiency for farm machinery in any one good or even average year might result in less than optimal efficiency over the long run. The experience of the 1978 corn crop may provide a good illustration of this. A wet fall in 1977 prevented much of the fall plowing that often takes place in the Corn Belt. A wet spring in 1978 reduced field working days and the opportunity to complete land preparation and achieve optimal planting dates. However, a record corn crop was planted in the reduced time period because the high farm income of the early to mid 1970s had allowed the extensive replacement of farm machinery with new equipment of increased capacity. Had this "excess" machinery capacity not been available for the unusual 1978 planting conditions the resulting crop would have been smaller and food prices higher. Over the long term, some excess machinery capacity beyond what might be optimal for a single good year could result in improved energy efficiency if the increased productivity from improved timeliness more than offset the extra energy embodied in machinery stock to provide this timeliness. This illustrates that any consideration of machinery efficiency must be made in a dynamic context which includes the interactions of weather, cropping systems, management, etc.

The specific concern here is to provide estimates of the energy embodied in farm machinery and a method of allocating the expenditure of this energy on the production process. What has been developed is an accounting convention. It is recognized that there are a number of different data sources and methods which might have been used. Final choices were often based on a judgment of what was logical and consistent.

Under the system developed here, three categories of energy are calculated separately before being combined to represent the total energy associated with a piece of farm machinery. These are (1) the energy embodied in the materials that go into a piece of machinery; (2) the energy used at the point of manufacture to treat and shape materials and fabricate the piece of machinery; and (3) the energy, both embodied and fabricated, in the repair parts and materials that would be applied to a piece of machinery over its useful life.

The energy embodied in materials was calculated on the basis of industry average values, standard reference values or general usage. The fabrication energy data were collected by a major farm machinery manufacturer. They were gathered by drawing an envelope around the facility that manufactured an item or type of machinery and then counting all energy that went into the facility. Where a class of machinery was involved, this similar equipment ends up with identical fabrication coefficients for individual pieces within the group. The energy in repair parts and materials was calculated from engineering equations that estimated the dollar costs of repairs over the life of a piece of machinery. The dollar costs were used as a proxy for energy in that proportion of repair costs that could be attributed to parts and materials. The total accumulated repair equations are exponential and indicate increasing repairs over the life of the machine. Finally, the embodied fabrication and repair energy is adjusted to an estimated reliable machinery life and combined.

The actual coefficients are given below in the sequence used for calculating machinery energy.

Embodied energy — Generally:

- Tires were accounted for on the basis of 20,500 kcal/kg. This amount is slightly more than that given for the industry.[1]
- Steel is accounted for on the basis of 15,000 kcal/kg which is consistent with Berry[2] for rolled steel and Beaty for poured steel. This is more energy than is ascribed to plain carbon steel in some of the analyses of energy used in automobile production.[3]
- The processing, forging or other treatment or shaping of metals is accounted for in the fabrication energy.
- No attempt is made to give credit for scrappage of machinery and the recycling of metal.
- Tractors and Combines: for tractors and combines, John Deere has computed an embodied energy coefficient that includes the energy value of every bit of material that goes into a 4430 tractor and a 6600 combine.
- The coefficient for the tractor is 11,814 kcal/kg and for the combine is 12,013 kcal/kg. These coefficients are used in the computations of the total embodied energy for all tractors and combines.

Fabrication energy (1976 data) — This was the most recent data available on the energy needed to produce different classes of farm machinery.

Machine	kcal/Ton × 10^6	kcal/kg
Tractors	3.17	3,494
Harvesters, combines, cotton pickers and self-propelled forage harvesters	2.82	3,108

Primary tillage equipment; plows, big discs and chisel plows	1.87	2,061
Large seed planters (primarily corn)	1.87	2,061
Secondary tillage equipment, harrows, cultivators, etc.	1.81	1,995
Sprayers and small grain planters		
Mowers, manure spreaders and handling equipment	1.60	1,764
Balers and all pulled forage equipment	1.36	1,499

The amount of fabrication energy for a given piece of equipment is calculated on the basis of its weight net of rubber tires, with the coefficient appropriate for its class.

Repair parts and materials energy — Repair parts and materials energy were calculated on the basis of total accumulated repair (TAR) equations which represent accumulated repair and maintenance costs as a proportion of the original equipment price up to any point in the life of a piece of equipment. All equipment was assumed to be properly maintained and repaired. Reliable life was estimated to be 82% of total life.[4] The TAR equations were taken to 82% of total life to estimate costs of accumulated repairs and maintenance. Dollar costs were taken as a proxy for energy costs. Because the TAR figures represent total repair and maintenance costs, one third of this amount was taken as representing parts exclusive of labor and other maintenance costs.

The different classes of machinery for the different TAR equations were as follows:[4]

Class 1	Proportion of original equipment cost expended on repairs for each machinery class to 82% of total life
4 Wheel drive tractors	
Crawler tractors	74.25%
Class 2	89.1%
Stationery power units	
2 Wheel drive tractors	
Class 3	45.88%
Self propelled combines	
Self propelled cotton pickers	
Cotton strippers	
Rotary cutters	
Stalk cutters	
Floats and scrapers	
Land planes	
Front end loaders	
Manure spreaders	
Feed trucks	
Pickup trucks	
Balers with engines	
Self propelled forage harvesters	
Class 4	60.69%
Mounted cotton pickers	
Corn pickers	
Flail harvesters	
Potato harvesters	

Beet sugar harvesters
Farm trucks
PTO powered bailers
Ensilage blowers
Pull-type forage harvesters
Self propelled sprayers

Class 5	75.98%

PTO powered combines
Self propelled swathers
Wagon and box
Corn heads
Hay conditioners
Ensilage loaders
Side delivery rakes
Seeding equipment
Mounted sprayers
Feed wagons

Class 6	91.28%

Dry fertilizer equipment
Liquid fertilizer equipment

Class 7	92.58%

Mowers
Tillage equipment

Once repair costs were calculated to 82% of total life (i.e., for the reliable life period) based on the full energy cost of the piece of machinery then the total embodied and fabrication energy was recalculated to reflect the machine being used for 82% of its total life and this amount was added to the repair parts energy.

Examples

Two examples are presented here to exemplify the technique developed above. The computation of the total embodied, fabrication, and repair parts energy for a tractor is the first example. The second example derives the total energy for a primary tillage implement.

The tractor is a two-wheel drive 130 hp tractor that weighs 13,400 lb (or 6078 kg). This example is written up in detail below and then shown in tabular form. The embodied energy for this tractor is computed from the product of the weight and the embodied energy coefficient, 11,814 kcal/kg. The result is 71,805,492 kcal. The fabrication energy is computed by first determining the weight exclusive of tires, i.e., computed to be 0.821 times the total weight. Then this is multiplied by the fabrication energy coefficient, 3494 kcal/kg. The fabrication energy figure is 0.821 × 6078 × 3494 = 17,435,193 kcal. The replacement parts and materials energy over the reliable life is derived by multiplying the total embodied and fabrication energy by the TAR number, 0.891, and by 0.333. The result is 26,477,979 kcal. Then the total embodied and fabrication energy total must be adjusted to the tractor's reliable life, i.e., multiply 89,240,685 × 0.82. The sum of this product and the repair parts energy figure results in the total embodied, fabrication, and repair parts energy figure. The tabular form of this example is shown below.

Tractor:	Embodied energy	kcal
	6,078 kg @ 11,814 kcal/kg	= 71,805,492

Fabrication energy 4,990 kg @ 3,494 kcal/kg	= 17,435,193
Total embodied and fabrication energy	= 89,240,685
Replacement parts and materials over reliable life 89,240,685 × TAR of 0.891 × 0.333	= 26,477,979
Total embodied and fabrication energy adjusted to reliable life 0.82 × 89,240,685	= 73,177,362
Total embodied, fabrication, and repair parts energy	= 99,655,341

The second example for the chisel plow will be shown in tabular form only. The implement under consideration is a 14 foot chisel plow weighing 5,100 lb (2,213 kg). The tires weigh 25 kg. The example is shown below.

Chisel Plow:	Embodied energy	kcal
	Total weight = 2,213 kg 25 kg of tires @ 20,500 kcal/kg	= 512,500
	2,188 kg of steel @ 15,000 kcal/kg	= 32,820,000
	Fabrication energy 2,188 kg @ 2,061 kcal/kg	= 4,509,468
	Total embodied and fabrication energy	= 37,841,968
	Replacement parts and materials over reliable life 37,841,968 × TAR of 0.9258 × 0.333	= 11,666,353
	Total embodied and fabrication energy adjusted to reliable life 0.82 × 37,841,968	= 31,030,414
	Total embodied, fabrication, and repair parts energy	= 42,696,767 kcal

Once the energy for the entire machinery complement has been estimated, then an annual per acre machinery cost can be calculated for each piece of equipment, and this can be summed to give an annual machinery energy cost on a per acre basis for the entire complement. For example, combining the two pieces of machinery listed above:

Tractor: Life of 12 years working 300 acres annually 99,655,341 divided by 12 and by 300 = 27,682

Chisel Plow: Life of 15 years working 300 acres annually 42,696,767 divided by 15 and 300 = 9,488

Therefore, 27,682 + 9,488 or 37,170 kcals is the annual machinery energy cost of this tractor and plow complement of machinery given the life and acreage worked for each machine and implement.

Farm Service Buildings

The energy inputs for farm buildings were calculated by Hannon et al. employing a combination of process analysis and input/output analysis. This is an average for farm residence and service buildings and clearly there are significant differences in the structure and construction materials for various buildings. At the present detailed analyses for the various farm structures have not been calculated. Therefore, the average energy values given below* were employed

Unit energy cost for buildings	kcal/ft²
Residence	139,000
Service building	38,000

1 kcal = 4 BTU
ft² = 0.09 m²

* Courtesy of Bruce Hannon.

ACKNOWLEDGMENTS

This manuscript was reviewed by R. M. Klemme and K. Harling who made a number of corrections and improvements.

REFERENCES

1. **Hall, J. R.,** The Energy Crisis and the Tire Industry, speech to Akron Rubber Group, October 5, 1973, in Rubber and Plastic News.
2. **Berry, R. S.,** The energy cost of automobiles, *Science and Public Affairs,* December, 1973, 58-60.
3. **"Automobile Material Breakdown and Energy for Manufacture"**, in *Auto Products Magazine,* November, 1974, Oak Park, Mich.
4. 1976 Agricultural Engineers Handbook, American Society of Agricultural Engineers, St. Joseph, 1976, 326-329.
6. **Hannon, B., Stein, R., Segal, B., and Serber, D.,** Energy Use in Building Construction, CAC Doc. 228, Center for Advanced Computation, University of Illinois, Urbana, 1977.
7. **Bullard, C. W., Penner, P. S., and Pilati, D. A.,** Energy Analysis: Handbook for Combining Process and Input-Output Analysis, Center for Advanced Computation, University of Illinois, Urbana, 1976.

FUEL AND ENERGY EFFICIENCY

Vashek Cervinka

Table 1
SUMMARY TABLE OF ENERGY VALUES FOR VARIOUS FUELS

Energy Source	Unit	kcal/Unit	Production inputs (kcal)	Total kcal/unit
Gasoline	liter	8,179	1,930	10,109
Diesel	liter	9,235	2,179	11,414
Fuel oil	liter	9,235	2,179	11,414
L.P. gas	liter	6,234	1,471	7,705
Natural gas	m^3	9,885	1,928	11,813
Coal, hard	kg	7,222	563	7,785
Coal, soft	kg	7,260	566	7,826
Hardwood	kg	4,600	345[a]	4,945
Softwood	kg	4,200	315[a]	4,515
Electricity	kWh	859	2,004	2,863

[a] Assumed an inverse energy efficiency of 1.075 for production and harvesting.

Table 2
ENERGY EFFICIENCY[a]

	Energy efficiency (%)	Inverse energy[b] efficiency (%)
Crude oil		
Imported[c]	80.1	1.248
Domestic	81.7	1.224
Natural gas[c]		
Offshore	83.7	1.195
LNG from Alaska	61.0	1.639
Coal		
High-Btu gasification	51.9	1.929
Low-Btu gasification	57.5	1.739
Liquefaction	64.5	1.550
Solvent refined solids	59.5	1.681
Solids	92.7	1.078

Electrical power (including efficiency of input energy)	Coal	Oil[c]	NG[c]	Coal	Oil	NG
Boiler fired (35%)	32.4	28.4	29.3	3.082	3.527	3.412
Gas turbine (29%)	—	—	24.3	—	—	4.115
Combined cycle (36%)	33.4	29.2	30.1	2.997	3.429	3.319
Fuel cell (39%)	—	—	32.6	—	—	3.067

[a] Production/generation and distribution.
[b] Inverse energy efficiency = the total primary energy required to deliver 1 unit of energy of various types to final demand.
[c] Energy for extraction is not included.

Calculated from *Energy Alternatives — A Comparative Analysis,* University of Oklahoma, 1975.

Table 2 (continued)
ENERGY EFFICIENCY[a]

CRUDE OIL

	Domestic		Imported	
	Primary %	Auxiliary 10^9 Btu/ 10^{12} Btu	Primary %	Auxiliary 10^9 Btu/ 10^{12} Btu
Extraction	NA	NA	—	—
Barges	100.0	25.9	—	—
Tankers	—	—	99.9	40.7
Pipeline	100.0	6.2	100.0	6.2
Refining	91.9	55.1	91.9	55.1
Tank trucks	99.9	14.1	99.9	14.1
Total	**91.8**	**101.3**	**91.7**	**116.1**
Efficiency	81.7%		80.1%	

OFFSHORE NATURAL GAS

	Energy	
	Primary %	Auxiliary 10^9 Btu/ 10^{12} Btu
Extraction	NA	NA
Gathering	89.2	0
Processing	96.6	0
Distribution	97.1	0
Total	**83.7**	
Efficiency	83.7%	

ALASKAN NG VIA ALASKAN PIPELINE AND LNG TANKER

	Primary %	Auxiliary 10^9 Btu/10^{12} Btu
Extraction	NA	NA
Gathering	89.2	0
Processing	96.6	0
Transmission	97.1	0
LNG liquefaction	83.0	0
LNG tanker	96.4	24.3
LNG storage	100.0	2.81
LNG vaporization	98.0	.71
Distribution	97.1	0
Total	**63.7**	**27.82**
Efficiency	61.0%	

Table 2 (continued)
ENERGY EFFICIENCY[a]

COAL

Low Btu Gasification

	Primary %	Auxiliary 10^9 Btu/10^{12} Btu
Mining	—	4.46
Transport (conveyor)	100.0	.23
Breaking & sizing	100.0	1.89
Gasification	78.3	96.90
Gathering	89.2	0
Distribution	97.1	0
Total	**67.8**	**103.48**

Efficiency 57.5%

COAL

High Btu Gasification

Mining	—	4.46
Transport (conveyor)	100.0	.23
Breaking & sizing	100.0	1.89
Gasification	60.7	0
Gathering	89.2	0
Distribution	97.1	0
Total	**.526**	**6.58**

Efficiency 51.9%

Liquefaction

Mining	—	4.46
Transport (conveyor)	100.0	.23
Breaking & sizing	100.0	1.89
Liquefaction	65.8	0
Transport (pipeline)	100.0	6.20
Total	**65.8**	**12.78**

Efficiency 64.5%

Solvent Refined Solids

Mining	—	4.46
Transport	100.0	.23
Breaking & sizing	100.0	1.89
Washing	96.8	2.26
Solvent refining	72.7	72.3
Distribution	98.0	12.6
Total	**68.9**	**93.74**

Efficiency 59.5%

Table 2 (continued)
ENERGY EFFICIENCY[a]

Solids

Mining	—	4.46
Transport	100.0	.23
Breaking & sizing	100.0	1.89
Washing	96.8	2.26
Distribution	98.0	12.6
Total	**94.9**	**21.44**

Efficiency 92.7%

Table 3
ENERGY EFFICIENCY

Energy Type	Efficiency (%)	Inverse energy[a] efficiency (%)
Coal mining	.977	1.024
Petroleum refining		
Imported	.828	1.208
Domestic	.924	1.082
Electric utilities	.258	3.870
Gas utilities		
NG imported	.855	1.169
NG domestic	.882	1.134

[a] Inverse energy efficiency = the total primary energy required to deliver 1 unit of energy of various types to final demand.

From Herendeen, R. A., *An Energy Input-Output Matrix for the United States, 1963: User's Guide,* Center for Advanced Computation Doc. No. 69, 1963.

Table 4
ELECTRICAL POWER GENERATION

	Efficiency (%)	Including transmission (%)
Boiler-fired	38	35
Gas turbine	31	29
Combined cycle	39	36
Fuel cell	42	39
Transmission	92	—

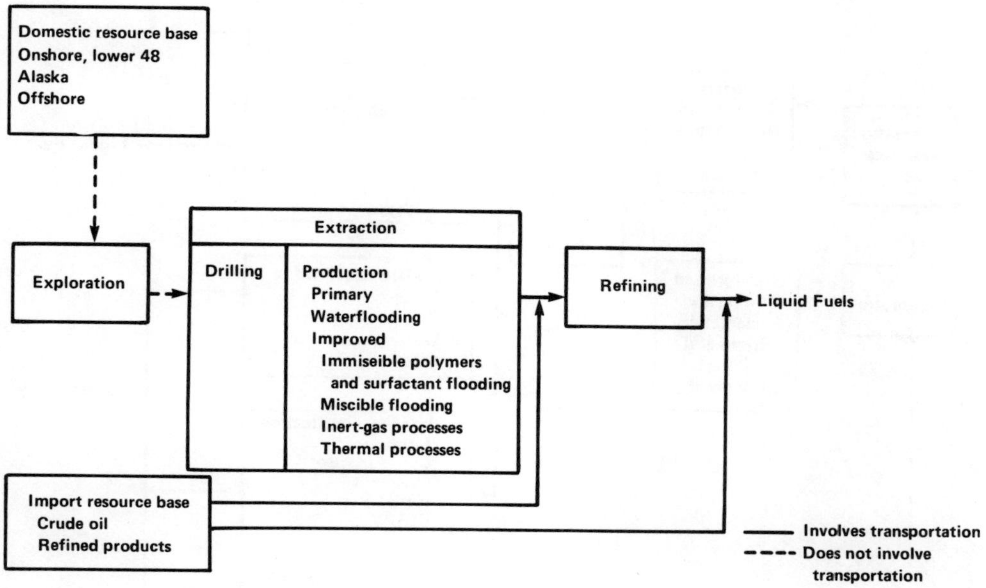

FIGURE 1. Crude oil resource development.

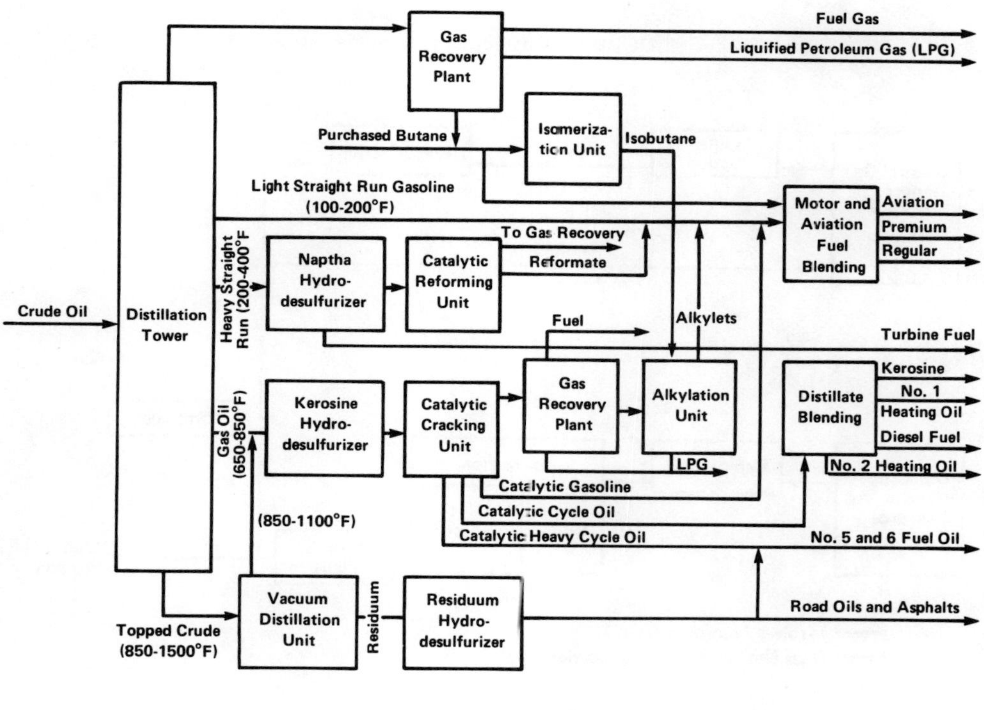

FIGURE 2. Oil refinery.

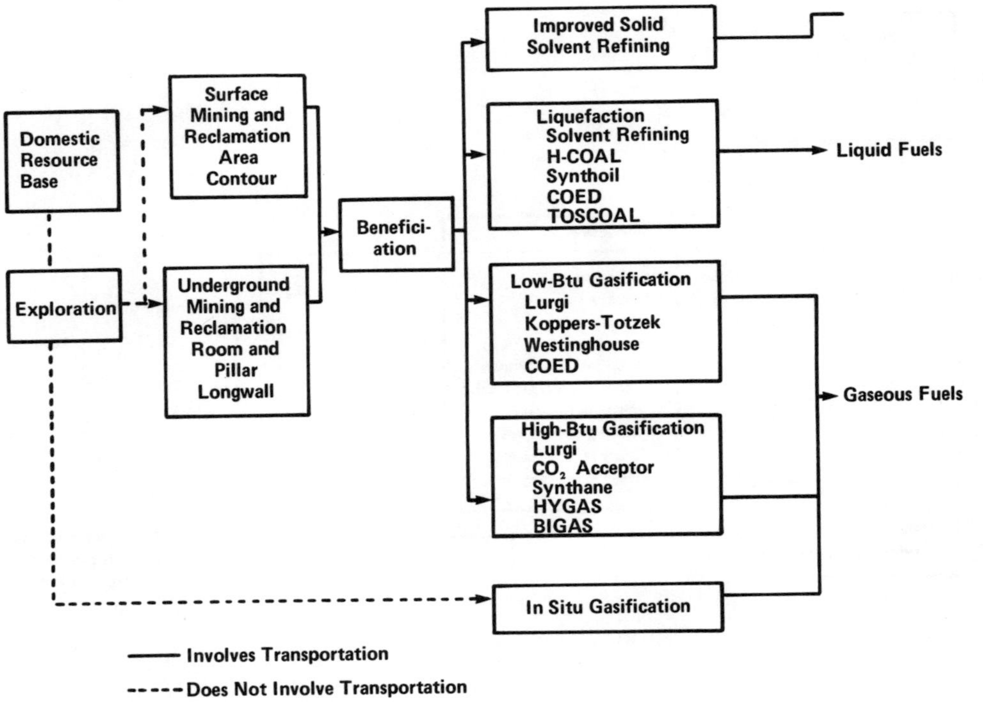

FIGURE 3. Coal resource development.

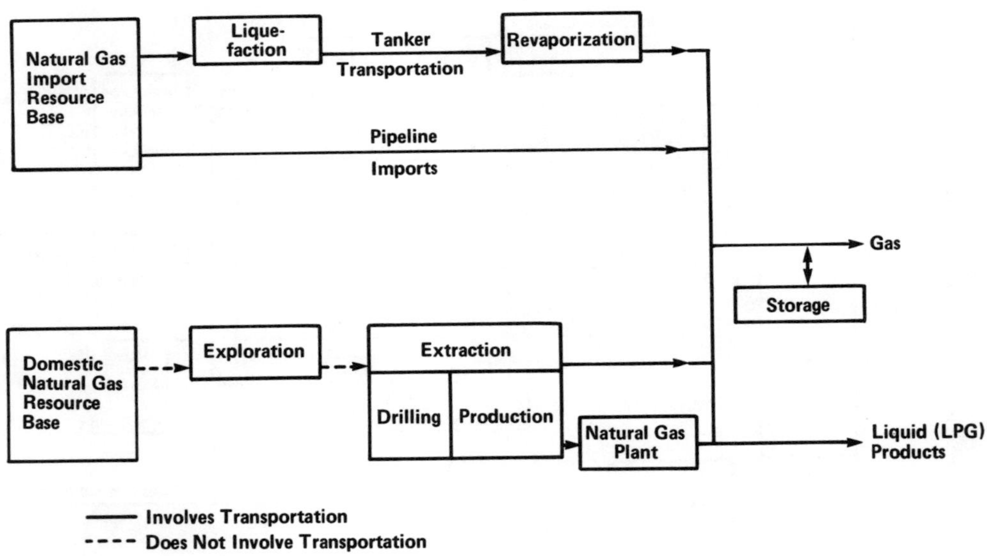

FIGURE 4. Natural gas resource development.

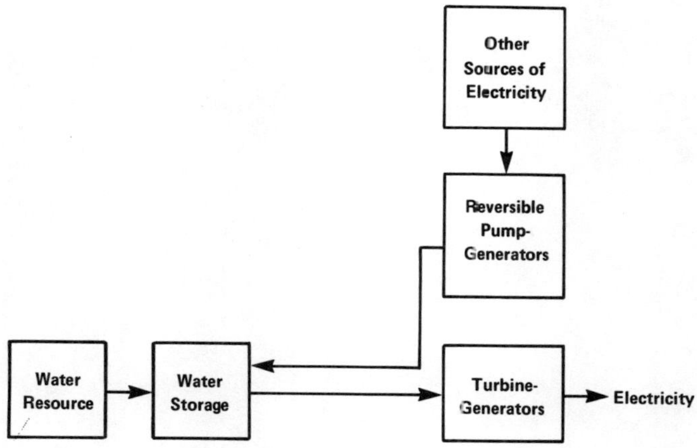

FIGURE 5. Hydroelectric resource development.

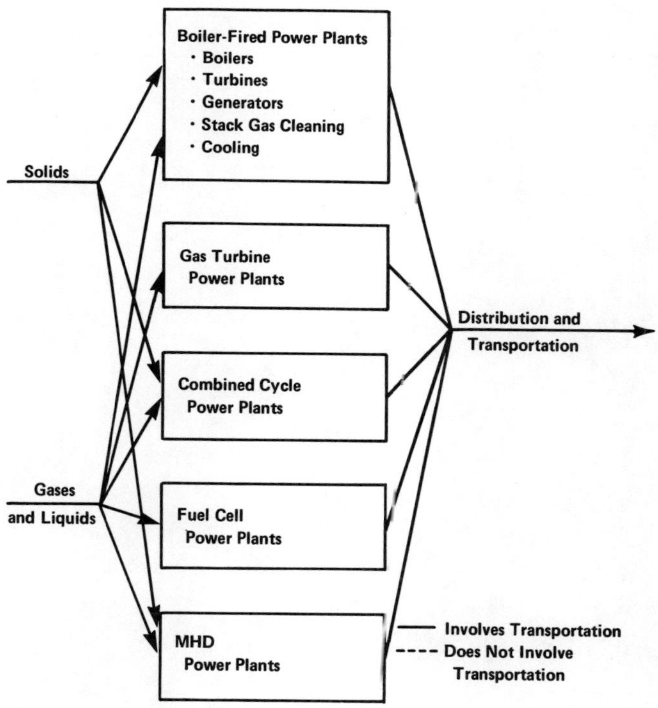

FIGURE 6. Electrical generation system.

ENERGY INPUTS FOR NITROGEN, PHOSPHORUS, AND POTASH FERTILIZERS

William Lockeretz

The following tables give average figures for energy consumed in manufacturing the leading fertilizers used in U.S. agriculture, using standard U.S. processes. Because of variations in the processes used, and the differences in the efficiencies of various plants using a given process, these figures are not completely accurate. It is difficult to give the uncertainties reliably, although some qualitative statements can be made.

For nitrogen, the nutrient which is the leading energy consumer both in terms of energy per kilogram and the total amount of energy used in producing the U.S. supply, the inaccuracies are probably rather small. Virtually all ammonia manufactured in the U.S. (95%) is done by the same method, the Haber-Bosch process using natural gas as a feedstock, and all recently constructed plants have approximately the same efficiency. Other forms of nitrogen fertilizer are synthesized from ammonia, and there are some variations in the methods used. However, the subsequent processing steps use relatively little additional energy compared to ammonia synthesis, so that the resulting uncertainty is not very great.

For phosphate the problem is complicated by the fact that a considerable fraction of the total energy consumption is for extraction of raw materials, both phosphate rock as well as sulfur, which is used to make the sulfuric acid used in converting the phosphate rock to phosphoric acid and eventually into superphosphate. The energy used in this extraction varies with the particular deposits. For potash, essentially all of the energy requirement is for extraction, since potash does not receive subsequent chemical processing. Because of the considerable variation in the depth of potash deposits, the potash data have considerable uncertainties. Fortunately, potash is the least energy-intensive nutrient, so that these uncertainties will not be so significant for crop production budgets that also involve nitrogen and phosphate fertilizers.

The data on energy for transportation are rough approximations, since they are derived from highly aggregated figures on transportation modes and distances. However, transportation accounts for only a small fraction of the total energy consumption, so that this problem is not very serious.

The tables are based on data developed at the Tennessee Valley Authority (TVA). Although figures for various fertilizers are also available from other sources, TVA data were used since they represent the most systematic treatment of this topic. In any case, discrepancies among various published data are not very great. Note, however, that these figures are specifically for U.S. techology. The corresponding numbers for other countries often differ appreciably, because of differences in the hydrocarbon feedstock used in ammonia synthesis, differences in the sources of phosphate rock, sulfur, and potash, and differences in the processes used for converting ammonia to other forms of nitrogen and for converting phosphate rock to superphosphate.

Table 1
ENERGY INPUTS FOR NITROGEN FERTILIZERS (Mcal/kg N)[1,2]

Type	Production				Transportation	Storage and transfer	Total
	Natural gas	Fuel oil	Electricity[a]	Total			
Anhydrous ammonia	11.6	0	0.1	11.7	0.2	0.1	12.0
Urea (prilled or granular)	12.7	0[b]	0.9	13.6	0.4	0.3	14.3
Ammonium nitrate (prilled or granular)	13.4	0[b]	0.5	13.9	0.5	0.3	14.7

[a] Fossil energy requirement (2.52 Mcal/kWh generated).

[b] Assumes the steam used in converting ammonia to urea or ammonium nitrate is generated with natural gas. If fuel oil is used, 1.1 and 1.8 Mcal/kg should be transferred from natural gas to fuel oil for urea and ammonium nitrate, respectively.

Table 2
ENERGY INPUTS FOR PHOSPHATE AND POTASH FERTILIZERS ($Mcal/kg$ of P_2O_5 or K_2O)[1-3]

Type	Production	Transportation of raw materials	Transportation and distribution of final product
Phosphate rock	0.4	—	0.9
Normal superphosphate (0-20-0)	0.6[a]	0.2	1.5
Triple superphosphate (0-46-0)	2.2[a]	0.2	0.6
Muriate of potash (0-0-60)	1.1	—	0.5

[a] Assumes that by-product sulfuric acid is used for normal superphosphate, and Frasch process sulfur for triple superphosphate.

REFERENCES

1. **Blouin, G. M. and Davis, C. H.,** Energy Requirements for the Production and Distribution of Chemical Fertilizers in the United States, paper presented at Southern Regional Educational Board Meeting, Atlanta, Ga., October 2, 1975.
2. **Davis, C. H. and Blouin, G. M.,** Energy consumption in the U.S. chemical fertilizer system from the ground to the ground, in *Agriculture and Energy,* Lockeretz, W., Ed., Academic Press, New York, 1977.
3. **Davis, C. H.,** Energy Requirements for Alternative Methods for Processing Phosphate Fertilizers, paper presented at Tech. Conf. Int. Superphosphate and Compound Manufacturers Assoc. Ltd., Prague, September 1974.

ENERGY USED IN THE UNITED STATES FOR AGRICULTURAL LIMING MATERIALS

Elinor C. Terhune

In 1975 U.S. farms applied to the soil approximately 31×10^6 tons of liming materials (30.8×10^6 tons, National Lime Association data; 31.3×10^6 tons, National Limestone Institute data). This was an increase of about 7×10^6 tons over consumption in 1972 (23.8×10^6 tons, National Lime Association; 25.1×10^6 tons, National Limestone Institute). Nearly all (98.6%) of the liming material used in 1975 consisted of ground, pulverized, and crushed limestone. Burned lime (used in the Northeast and East Central regions) and hydrated lime (used in those regions plus Ohio, Florida, and Washington) accounted for only 0.2% of the agricultural lime. Marl and miscellaneous material (ground mollusc and eggshells, paper and sugar mill refuse, and water-softening sludge) accounted for 1.2% of the total.

The energy value for mining crushed and broken limestone is given as 10×10^9 kWh equivalent for 613×10^6 (short) tons[1] or 15.45 kcal/kg.

$$\frac{10 \times 10^9 \text{ kWh} \times 859.184 \text{ kcal/kWh}}{613 \times 10^6 \text{ tons} \times 907.18 \text{ kg/ton}} = 15.45 \text{ kcal/kg}$$

Data on energy used in manufacturing are available for all plants whose primary output is lime, for 1971 (27.0×10^9 kWh equivalent) and 1974 (28.0×10^9 kWh equivalent).[2] Production of limestone products in 1972 was 13.666×10^6 tons,[2] comprising 9.825×10^6 tons of quicklime, 2.546×10^6 tons of hydrated lime, and 1.295×10^6 tons of dead burnt dolomite. Assuming the energy use increased linearly from 1971 to 1974, 27.3×10^9 kWh equivalent was used in 1972. Lacking a breakdown of energy used specifically to manufacture hydrated and burned lime (the two products of interest, see Table 1), quicklime, hydrated lime, and burned lime have been combined here to give an approximate energy value of 1893 kcal/kg for manufacturing lime products, based on 1972 data:

$$\frac{27.3 \times 10^9 \text{ kWh} \times 859.184 \text{ kcal/kWh}}{13.666 \times 10^6 \text{ tons} \times 907.18 \text{ kg/ton}} = 1893 \text{ kcal/kg}$$

The energy value should be estimated for each lime product, as some are undoubtedly more energy intensive than others, but that cannot be done with the data available here. It is also impossible to tell from these data if the energy value is conservative or liberal because the amount of other products produced by those plants is unknown, and the amount of lime produced as a minor product by other plants is also unknown; the assumption here is that at least these two sources of error approximately cancel each other.

Table 1
ENERGY CONSUMPTION FOR LIMING MATERIALS (kcal/kg)[a]

Type	Mining	Manufacturing	Transportation of raw materials	Transportation and distribution of final product
Crushed and ground limestone	15.45	0	0[b]	300
Burned lime	15.45	1893	200	300
Hydrated lime	15.45	1893	200	300
Marl and refuse from industrial processes	15.45	0[c]		

Note: This table does not include energy for application on the farm.

[a] Calculated from U.S. Bureau of the Census data.
[b] The raw material is the final product, so all transportation costs are counted only for the final product.
[c] The refuse is considered a "free good" because the energy consumed during manufacturing is attributed to the primary product, such as sugar in beet sugar processing mills.

REFERENCES

1. U.S. Bureau of the Census, Census of Mineral Industries, U.S. Department of Commerce, U.S. Government Printing Office, Washington, D.C., 1972.
2. U.S. Bureau of the Census, Census of Manufacturers, U.S. Department of Commerce, U.S. Government Printing Office, Washington, D.C., 1972.
3. Approximate Consumption of Liming Materials on United States Farms During 1972, 44th Consecutive Annual Survey, National Lime Association, Washington, D.C., December 1973.
4. Approximate Consumption of Liming Materials on United States Farms During 1975, 47th Consecutive Annual Survey, National Lime Association, Washington, D.C., December 1976.
5. Report of Tonnage of Agricultural Limestone Used in the United States in 1975 Compared with the Tonnage Reported in 1974 and the Annual Application Needed as Determined by the Respective States, National Limestone Institute, Inc., Fairfax, Va., September 1976.
6. Report of Tonnage of Agricultural Limestone Used in the United States in 1973 Compared with the Tonnage Reported in 1972 and the Annual Application Needed as Determined by the Respective States, National Limestone Institute, Inc., Fairfax, Va., September 1974.

ASSESSING THE FOSSIL ENERGY COSTS OF PROPAGATING AGRICULTURAL CROPS

G. H. Heichel*

INTRODUCTION

Energy analyses of producing agricultural crops dissaggregate and quantify the fossil energy resources spent on the principal crop production inputs. It is now well documented that machinery, direct fuel consumption, and nitrogenous fertilizer manufacture frequently are the most energy demanding of the on- and off-farm inputs to crop culture. Because of the seemingly small quantities of energy involved (supposedly less than 5% of the crop energy budget), the energy investment attributable to propagating crops by seed, bulbs, tubers, cuttings, or root stocks has heretofore largely been ignored.

The few published attempts to disaggregate the energy costs of propagation materials from the remainder of the production inputs have relied on questionable estimates of the fossil energy costs involved. The approximation that fossil energy attributable to producing seed of grain crops can be estimated either by the digestible energy content or by the enthalpy of the seed has been invoked (e.g., Pimentel et al.[5]). No attempts have been made to estimate the fossil fuel costs of the bulbs, tubers, cuttings, or rootstocks used in reproducing other crops.

That the calculation of the energy costs of propagation from the energy content of seeds or other reproductive structures is fraught with difficulty is amply demonstrated by comparing the market price of feed grain with that of agricultural seed. For example, the price of hybrid corn is about 20 times that of Number 2 yellow corn. Because the price of commodities is, within broad limits, proportional to the fossil energy costs of their production, we are intuitively led to the tentative conclusion that agricultural seed purchased by the farmer is energetically more expensive to produce than seed sold for feed or food grain.

The purpose of this chapter is to exemplify and compare methods that can be used to quantify the fossil energy costs of propagating crops in energy analyses of crop production.

METHOD 1

In this method, the fossil energy costs of crop propagation use the estimate that they are a multiple of the enthalpy or of the digestible energy content of the seed. This technique appears in the literature,[5] but its continued use seems inadvisable as subsequent comparisons will reveal. Furthermore, digestible energy values or calorimetric values of the enthalpy of many propagation materials are unavailable.

METHOD 2

This method is applicable only to crops for which the harvested portion is anatomically identical to the organs or tissues used for establishing the crop. The purpose is to eliminate fossil energy costs of propagation as a line in the energy budget. For a

* Plant Physiologist, Science and Education Administration—Agricultural Research, U.S. Department of Agriculture, St. Paul, Minn. Cooperative investigation of the SEA-FR, USDA, and the Minnesota Agriculture Experiment Station (Scientific Journal Series Paper No. 10436).

seed-bearing crop, the quantity of seed used to sow a hectare of land is subtracted from the total yield of the crop, with the output energy of the cropping system being the "net above seed requirements". Specifically, a corn crop yielding 6250 kg/ha grain would be diminished by the 17 kg/ha seed required to establish the crop.

This is a simple and straightforward procedure for grain or other crops in which the material harvested is anatomically identical to the plant material sown for propagation. The obvious deficiency is that it makes the unproven assumption that seed and feed or food grains have similar fossil energy requirements for production. The second deficiency is that the method is useless for crops for which the plant materials sown and harvested are anatomically different. How can alfalfa seed and forage, tomato seed and fruit, sugarbeet seed and beetroots, or rootstocks and grapes or apples be equated?

METHOD 3

This method relies upon the economic costs of propagation materials to estimate the fossil energy costs of producing seed or other organs for reproducing crops. Lacking a detailed input-output analysis for every crop, the technique relies upon knowledge of the energy intensity of a unit of gross national product (kilocalorie per dollar) and the assumption that seeds or other materials of crop reproduction are comparable in energy intensiveness to other goods and services in the economy. At this time, the credibility of the assumption cannot be confirmed with analytical results.

Despite this reservation, a precedent for the technique has been set by Heichel,[3] who aggregated seed fossil energy costs with other variable costs of production in energy analyses of cropping systems. Leach[4] and Hannon et al.[2] have used variations of this approach to estimate fossil energy costs that were not readily obtained from first principles.

Calculations of seed energy costs by this technique contrast markedly with those based upon the enthalpy of seed, as the following examples show:

1. Corn silage production. Seed corn sold for about $40/bushel or approximately $1.57/kg in 1977. The most recent calculation of the energy intensity of a unit of gross national product was 15,800 kcal/dollar for 1973.[2] Using the dollar to energy transformation, 1 kg of corn seed would require 24,830 kcal to grow, process, and distribute. At a normal seeding rate of 17 kg/ha, the energy investment in seed corn for growth of corn silage would be about 422,110 kcal/ha. The impact of this energy investment is apparent from Table 1. Using Pimentel et al.[5] estimate of energy investment in seed as twice the enthalpy, the seed energy cost at 119,680 kcal/ha or 3% of the crop energy budget was calculated. Using Method 3, as outlined, results in a seed energy investment of 422,000 kcal/ha, 253% of the earlier estimates, and fully 10% of the energy budget for producing corn silage. The ratio of output energy to input energy is decreased from 9.1 to 8.5, a 7% decline in overall energetic efficiency attributable to the method of calculating seed energy costs.

2. Alfalfa production, seeding year. Establishing alfalfa using a herbicide instead of a companion crop is becoming increasingly prevalent in the midwest. Certified alfalfa seed sold for $3.90/kg in Minnesota in 1977. Transforming cost to energy at 15,800 kcal/dollar reveals that 1 kg of alfalfa seed would require 61,530 kcal to grow, process, and distribute. At a normal seeding rate of 14 kg/ha, the energy investment in seed for sowing alfalfa would be approximately 868,000 kcal/ha. Estimating fossil energy for seed at twice the enthalpy, the seed energy cost is

Table 1
ENERGETICS OF CORN SILAGE PRODUCTION IN SOUTHEASTERN MINNESOTA USING TWO METHODS OF CALCULATING THE FOSSIL ENERGY COST OF SEED

Item	Quantity/ha	kcal/ha[a]
	Input	
Labor	6.83 hr	
Machinery	1,481 kg	42,050
Liquid fuel	141 ℓ	1,620,790
Nitrogen	114 kg	1,675,800
Phosphorus (P_2O_5)	45 kg	49,500
Potassium (K_2O)	39 kg	62,400
Seed	17 kg	[119,680](422,110)
Pesticides	5.7 kg	138,000
Electricity	—	—
Transportation	1,778 kg	456,870
Total energy input		**[4,165,090](4,467,520)**
	Output	
Crop energy yield (70% H_2O)	40,910 kg	37,922,000
kcal return/kcal input		[9.1] (8.5)

[a] Values in [] estimated by Method 2, doubling the enthalpy of the seed. Values in () derived by estimating seed energy costs by Method 3. See text for details.

about 94,000 kcal/ha (Table 2), or 4.6% of the crop energy budget. Method 3 increases the energy investment in seed more than 800% to 868,000 kcal/ha. This energy investment is 31% of the crop energy budget and exceeds that of any other single component in the budget except liquid fuel. The ratio of output energy to input energy is decreased from 6.9 to 5.0, a 28% decline in overall energetic efficiency attributable to the method of calculating seed energy costs.

METHOD 4

Rather than using the approximations of the previous three methods, the ideal situation would be to calculate a fossil energy budget for producing seed, rootstocks, tubers, bulbs, and other reproductive structures of the major field, fruit and vegetable crops. Because production of propagation materials is largely the activity of commercial seedsmen, horticulturists, and agronomists, the needed information on inputs to production is of proprietary nature and not easily accessible to energy analysts.

In cooperation with Dr. D. E. Brown of Land O'Lakes, Inc., Caldwell, Idaho, information on the production, processing, and distribution of alfalfa seed from two counties has been subjected to a fossil energy analysis using the techniques applied to other crops in this volume. This analysis reveals the complexity of cultural practices bearing upon the energy analysis and the relative energy needs of alfalfa seed production compared with processing and distribution of the seed to the grower. Finally, the analysis provides limited data for comparison with the results of Method 3.

Fossil energy resources equivalent to nearly 27 bbl of crude oil are needed to grow,

Table 2
ENERGETICS OF ESTABLISHING ALFALFA WITH A HERBICIDE IN SOUTHEASTERN MINNESOTA USING TWO METHODS OF CALCULATING THE FOSSIL ENERGY COST OF SEED

Item	Quantity/ha	kcal/ha[a]
	Input	
Labor	6.4 hr	
Machinery	1,465 kg	30,375
Liquid fuel	95 ℓ	1,084,330
Nitrogen	—	—
Phosphorus (P_2O_5)	45 kg	49,500
Potassium (K_2O)	136 kg	217,600
Seed	14 kg	[94,000](868,000)
Pesticides	5.7 kg	138,000
Electricity	—	—
Transportation	1,717 kg	441,269
Total energy input		**[2,055,074](2,829,704)**
	Output	
Crop energy yield (15% H_2O)	5,682 kg	14,262,000
Energy yield/energy input		[6.9] (5.0)

[a] Values in [] estimated by Method 2, doubling the enthalpy of the seed. Values in () derived by estimating seed energy costs by Method 3. See text for details.

process, and distribute the 900 kg of alfalfa seed from 1 ha of cropland. Nearly 40% of this energy is attributable to providing the approximately 30 ha-cm of water for irrigating the crop. About 4.1% of the crop fossil energy budget is invested in chemical defoliants to remove leaves prior to harvesting of the seed. The fossil energy cost of the seed, computed by Method 3, represents about 2.1% of the energy budget, with lesser amounts being used in machinery, fertilizers, and pesticides.

The processing and distribution of alfalfa seed requires more energy resources than does seed production (Table 3). Packaging of the seed requires about 46% of the fossil energy budget which exceeds the energy commitment to irrigation and is only slightly less than all of the energy needs for seed production. The results with alfalfa seed substantiate findings with other commodities that packaging materials are more demanding of fossil energy than the contents of the package.[1] Approximately 44,000 kcal of fossil energy resources per kilogram of seed are presently required by the alfalfa seed industry to propagate and distribute seed in two counties in Idaho (Table 3). This fossil energy requirement is 28% less than the estimate derived by the dollar to energy transformation of Method 3. Lack of quantitative information on the fossil energy investment in the irrigation system and the seed processing facilities would contribute to the energy budget in Table 3 yielding conservative calculations. Another factor resulting in conservative calculations in Table 3 is that the seed yields in Canyon and Owyhee Counties are nearly double the Idaho state average (USDA, 1977). Thus, the fossil energy input per kilogram of seed produced might range to a value double that in Table 3. It is concluded that the transformation of Method 3 yields fossil energy

Table 3
AN ENERGY BUDGET OF PRODUCING, PROCESSING, AND DISTRIBUTING ALFALFA SEED, CANYON AND OWYHEE COUNTIES, IDAHO

Item	Quantity/ha	kcal/ha
Input		
Labor	15.8 hr	
Machinery[a]	1700 kg	33,073
Liquid fuel	68 ℓ	776,150
Electricity	—	—
Nitrogen	—	—
Phosphorus (P_2O_5)	128 kg	140,800
Potassium (K_2O)	—	—
Limestone	—	—
Seed	13.5 kg	837,000
Irrigation water[b]	30 cm	15,685,490
Insecticides	5.4 kg	472,260
Herbicides	21.9 kg	171,660
Defoliants	112 1 + 3.4 kg	1,614,000
Transport to farm	1978 kg	508,346
Transport to processor	16 km	3,330
Seed cleaning	200 kwh	572,600
Packaging	—	17,981,600
Transport to retail	2400 km	461,650
Total fossil energy input		**39,257,959**
Output		
Dry matter yield	900 kg	
Protein yield	not applicable	
Fossil energy input/seed output		43,620 kcal/kg

[a] Exclusive of irrigation system, seed processing.
[b] Energy for pumping H_2O.

values for alfalfa seed that are in acceptable compliance with those computed from detailed energy budgets.

ENERGY COSTS OF FIELD SEED PRODUCTION, PROCESSING, AND DISTRIBUTION, 1977

Because of the paucity of data on the energy needs of propagating the major field crops, the fossil energy costs of producing, processing, and distributing 24 types of field crop seed have been computed using the dollar to energy transformation (Table 4). Knowing the sowing rate for the crop, these values can be applied in calculations of the fossil energy attributable to seed in energy budgets of crop production until more precise values like those of Table 3 become available for specific crops.

SUMMARY

Four methods of calculating the fossil energy costs of the seed used in crop production systems have been exemplified. Two methods that were closely compared revealed

Table 4
ESTIMATED FOSSIL ENERGY COSTS OF FIELD SEED PRODUCTION, PROCESSING AND DISTRIBUTION, 1977

Crop	Retail seed cost[7] $/lb	Retail seed cost[7] $/kg	Energy cost[a] (kcal/kg)
Alfalfa (uncertified)	1.53	3.37	53,246
Alfalfa (certified)	1.77	3.90	61,620
Clover, Red	1.08	2.38	37,604
Clover, Alsike	0.895	1.97	31,126
Clover, Sweet	0.385	0.85	13,430
Clover, Ladino	2.20	4.85	76,630
Lespedeza, Sericea	2.26	4.98	78,684
Timothy	0.60	1.32	20,856
Orchardgrass	0.62	1.37	21,646
Ryegrass, annual	0.35	0.77	12,166
Bluegrass, Kentucky	1.47	3.24	51,192
Fescue, tall	0.475	1.05	16,590
Sudangrass	0.345	0.76	12,008
Seed potatoes	0.056	0.12	1,896
Seed corn, hybrid	0.71	1.57	24,806
Seed wheat, spring	0.087	0.19	3,002
Seed oats	0.116	0.26	4,108
Seed barley	0.096	0.21	3,318
Seed flax	0.196	0.43	6,794
Seed peanuts	0.52	1.15	18,170
Soybean seed	0.217	0.48	7,584
Cottonseed for planting	0.303	0.67	10,586
Grain sorghum, hybrid	0.41	0.90	14,220
Forage sorghum	0.275	0.61	9.638

[a] Computed using the dollar to energy transformation, 15,800 kcal/dollar.[2]

that discrepancies of 7 to 28% in the energetic efficiency of a cropping system are possible. The results suggest that the fossil energy attributable to seed production, processing, and distribution can range from 10 to 31% of the crop energy budget, in contrast with previous estimates of less than 5%. A method of transforming the economic costs of seed or other propagation materials into a fossil energy value for use in crop energy budgets is exemplified and compared with an energy analysis of alfalfa seed production.

ACKNOWLEDGMENT

I thank Dr. D. E. Brown for the alfalfa seed production data.

REFERENCES

1. **Berry, R. S. and Makino, H.,** Energy thrift in packaging and marketing, *Tech. Rev.*, 76, 1, 1974.
2. **Hannon, B. M., Harrington, C., Howell, R. W., and Kirkpatrick, K.,** *The Dollar, Energy, and Employment Costs of Protein Consumption,* Document No. 182, Center for Advanced Computation, University of Illinois, Urbana, 1976.

3. **Heichel, G. H.,** Comparative efficiency of energy use in crop production, *Conn. Agric. Exp. Stn. Bull.*, 739. 26 pp., 1973.
4. **Leach, G.,** *Energy and Food Production,* International Institute for Environment and Development, Washington, D.C., 1975.
5. **Pimentel, D., Lynn, W. R., MacReynolds, W. K., Hewes, M. T., and Rush, S.,** *Proceedings, Workshop on Research Methodologies for Studies of Energy, Food, Man, and Environment, Phase I,* Cornell University, Ithaca, N.Y., 1974.
6. Agricultural Statistics, U.S. Department of Agriculture, Washington, D.C. 1977.
7. U.S. Department of Agriculture, Seed Crops, Crop Reporting Board, Statistical Reporting Service, Washington, D.C., 1977.

ENERGY REQUIREMENTS FOR IRRIGATION

J. Clair Batty and Jack Keller

Table 1 presents estimates of energies to manufacture a limited number of products used in irrigation. These values are representative of the values reported in current technical literature. Also shown in Table 1 are estimates of expected life of these materials installed in a properly designed system.

The estimated fixed annual energy costs shown in Table 1 are explained as follows. Certain of these materials will probably be recycled and the energy to manufacture such components from recycled materials may be considerably less than to manufacture them from raw materials. The following relationship tends to account for the recycling of certain materials.

$$AFEC = \frac{(ERM + ERC)\ (NTR)}{ESL}$$

where: AFEC = Annual fixed energy cost, ERM = Energy input to product manufactured from raw material, ERC = Energy input to product manufactured from recycled materials, NTR = Number of times product is replaced over the expected life of the system, ESL = Expected system life,

As an example, consider aluminum components in an irrigation system. If we assume an expected system life of 40 years, a life of aluminum components of 20 years (so that the aluminum is replaced once over the 40-year evaluation period), ERM = 280 MJ*/kg and ERC = 110 MJ/kg the annual fixed energy cost is computed as

$$AFEC = \frac{(280 + 110)\ (1)}{40} = 9.8\ MJ/kg - year$$

Summarized in Table 2 are values of mass per unit length for various kinds of irrigation pipe and lined ditches or canals. By multiplying the appropriate value of mass per unit length from Table 2 by the appropriate energy per unit mass from Table 1, the manufacturing energy per unit length of pipe may be obtained.

It may be concluded from inspection of Table 6 that:

- For low lifts, the installation energies are significant and should be considered in energy saving analyses.
- Where pumping is necessary, pumping energies dominate the installation energies even for low lifts.
- As the lift increases, systems with lower irrigation efficiency become increasingly energy consumptive.
- Where high lifts are involved, reasonable steps to increase irrigation efficiency are probably justified from an energy conservation point of view.

* 4186 J = 1 kcal

Table 1
MANUFACTURING ENERGIES FOR CERTAIN PRODUCTS USED IN IRRIGATION SYSTEMS (4186 J = 1 kcal)

Product	Energy input[a] raw material (ERM) MJ/kg	Energy input[a] recycled material (ERC) MJ/kg	Expected product life (EPL)	Number of times replaced over 40-year period (NTR)	Annual fixed energy cost 40-year evaluation period MJ/kg-year
Steel	65	30	20 years	1	2.4
Aluminum	280	110	20 years	1	9.8
Brass	160	80	7 years	5	14.0
PVC	120	120[b]	40 years	0	3.0
Polyethylene	160	160[b]	10 years	3	16.0
Asbestos-cement	15	15[b]	40 years	0	0.4
Concrete (ditches)	2	2[b]	15 years	2	0.2
Electric motor driven pumps	84	63	12 years	3	6.8
Diesel engine driven pumps	75	35	12 years	3	4.5
Excavation & fill	10×10^{-3}	10×10^{-3}[b]	40	0	0.25×10^{-3}
Ditching & trenching	15 MJ/m	15 MJ/m[b]	40	0	0.375 MJ/m-year
			20	1	0.75 MJ/m-year
			10	3	1.50 MJ/m-year

[a] These values taken from the general literature. (For summary see Batty, J. C., Hamad, S. N., and Keller, J., Energy Inputs to Irrigation, Journal of the Drainage Division of the American Society of Civil Engineers, Dec. 1975.

[b] These products are not presently recyclable.

Table 2
SUMMARY OF MASS PER UNIT LENGTH OF VARIOUS KINDS OF IRRIGATION PIPING AND CHANNELS

A. Polyethylene Tubing[a]

Nominal size		Inside Diameter	Wall	Mass
in.	(mm)	(mm)	(mm)	(kg/m)
1/16	1.59	1.02	0.51	2.2
1/8	3.18	1.78	0.71	4.5
1/8	3.18	1.52	1.02	7.4
3/16	4.76	3.18	0.76	8.9
7/32	5.56	4.06	0.76	11.9
1/4	6.35	4.32	1.02	17.9
1/4	6.35	3.18	1.52	23.8
9/32	7.14	5.72	0.76	17.9
3/8	9.53	9.53	1.02	38.7
1/2	12.70	13.21	1.52	67.0
1/2	12.70	14.73	1.58	77.4
5/8	15.88	17.78	1.65	113.1
3/4	19.05	20.32	1.78	145.8
1	25.40	25.91	2.28	223.2

[a] Based on data provided by manufacturer.

Table 2 (continued)
SUMMARY OF MASS PER UNIT LENGTH OF VARIOUS KINDS OF IRRIGATION PIPING AND CHANNELS

B. Rigid PVC Pipe[a]

Size		125 psi (862 kPa)		460 psi (1103 kPa)		200 psi (1379 kPa)		Class 315 psi (2172 kPa)		Schedule 40	
(in.)	(mm)	Inside diameter (mm)	Mass (kg/m)	Inside diameter (mm)	Mass (kg/m)	Inside diameter (mm)	Mass (kg/m)	Inside diameter (mm)	Mass (kg/m)	Inside diameter	Mass
2	50.8	56.6	0.51	55.7	0.62	54.6	0.77				
2½	63.5	68.5	0.74	67.4	0.92	66.1	1.13	51.4	1.18	52.6	1.03
3	76.2	83.3	1.10	82.0	1.37	80.4	1.68	62.2	1.72	62.7	1.64
4	101.6	107.2	1.82	105.5	2.26	103.4	2.77	75.7	2.55	78.0	2.14
5	127.0	132.6	2.80	130.4	3.47	127.8	4.24	97.4	4.20	102.3	3.04
6	152.4	158.0	3.94	155.3	4.96	152.2	6.03				
8	203.2	205.6	6.70	202.2	8.33						
10	254.0	256.2	10.43								

[a] Based on data provided by manufacturer.

C. Aluminum Irrigation Pipe[a]

Plain pipe					Pipe in 40-ft lengths with female coupler and band and latch assembly	Pipe in 40-ft lengths with female coupler and with apron and riser outlet and band and latch assembly
Size and Outside Diameter (in.)	Size and Outside Diameter (mm)	Inside Diameter (mm)	Wall (mm)	Mass (kg/m)	Mass (kg/m)	Mass (kg/m)
2	50.8	48.25	1.27	0.53	0.57	0.58
3	76.2	73.66	1.27	0.81	0.85	0.89
4	101.6	99.06	1.27	1.08	1.15	1.20
4	101.6	97.94	1.83	1.55		
5	127.0	124.36	1.32	1.41	1.52	
5	127.0	123.04	1.98	2.10		
6	152.4	149.81	1.30	1.66		
6	152.4	149.45	1.47	1.89	2.03	
7	177.8	174.55	1.63	2.43		
8	203.2	200.61	1.30	2.22		
8	203.2	199.95	1.63	2.78		
8	203.2	199.54	1.83	3.12		
10	254.9	251.41	1.30	2.78		
10	254.9	250.75	1.63	3.48		
10	254.9	249.22	2.39	5.09		
12	304.8	301.55	1.63	4.18		

[a] Based on data provided by manufacturer.

Table 2 (continued)
SUMMARY OF MASS PER UNIT LENGTH OF VARIOUS KINDS OF IRRIGATION PIPING AND CHANNELS

D. Steel[a]

Nominal size		Mill pipe			Fabricated pipe	
(in.)	(mm)	Inside Diameter (mm)	Wall (mm)	Mass (kg/m)	Wall (mm)	Mass (kg/m)
2	50.8	52.50	3.91	5.43		
2½	63.5	62.71	5.16	8.62		
3	76.2	77.93	5.49	11.28		
3	76.2	76.20	6.35	12.92		
3	76.2	74.63	7.14	14.39		
4	101.6	102.26	6.02	16.06	2.67	6.40
4	101.6	101.6	6.35	16.89	3.43	8.18
4	101.6	100.03	7.14	18.86		
5	127.0	128.19	6.55	21.76		
5	127.0	127.03	7.14	23.62		
5	127.0	125.45	7.92	26.07		
6	152.4	154.05	7.11	28.23	3.43	12.50
6	152.4	152.43	7.92	31.36	4.78	17.26
6	152.4	150.80	8.74	34.32	5.56	20.09
8	203.2	206.38	6.35	33.28	4.55	22.17
8	203.2	205.00	7.04	36.76	4.78	23.22
8	203.2	203.24	7.92	41.28	6.07	29.47
10	234.0	258.88	7.09	46.43	4.55	27.83
10	254.0	257.45	7.80	50.96	4.78	29.17
10	254.0	255.57	8.74	56.85	6.07	37.06
12	304.8	307.09	8.38	65.14	4.55	33.48
12	304.8	304.80	9.53	73.75	4.78	35.12
12	304.8	303.23	10.31	79.71	6.07	44.65
14	355.6	336.55	9.53	81.21	4.78	41.07
14	355.6	333.35	11.13	94.31	6.07	52.24
16	406.4	387.35	9.53	93.13		
16	406.4	384.15	11.13	108.22		

[a] Based on data from the Steel Pipe Manual M11, American Water Works Association, 1964.

E. Asbestos — Cement[a]

Pipe size and inside diameter		Mass including couplings (kg/m) pipe classification ASTM C668				
(In.)	(mm)	30	40	50	60	70
18	457.2	69.80	95.24	97.18	118.76	142.12
20	508.0	84.68	115.78	118.31	148.67	175.16
21	533.4	92.71	129.17	128.88	160.72	192.72
24	609.6	118.91	158.64	168.02	209.54	251.35
27	685.8	150.01	205.82	214.74	267.87	320.70
30	762.0	183.19	245.10	263.56	330.23	394.07
33	838.2	220.10	298.38	318.62	398.53	476.96
36	914.4	259.54	347.04	378.74	473.99	568.19

[a] Based on data provided by manufacturer.

Table 2 (continued)
SUMMARY OF MASS PER UNIT LENGTH OF VARIOUS KINDS OF IRRIGATION PIPING AND CHANNELS

F. Mass of Concrete (Conc.) and Excavated (Excav.) Material Associated with Concrete Lined Ditches and Canals

Depth (m) Bottom Width (m)	.25 (m) (kg/m)		.5m (kg/m)		.75m (kg/m)		1.0m (kg/m)		1.5m (kg/m)		2.0m (kg/m)	
	Excav.	Conc.	Excav.	Conc.	Excav.	Conv.	Excav.	Conc.	Excav.	Conc.	Excav.	Conc.
0	110	99	440	198	990	296	1760	395	3960	593	7040	790
0.25	220	134	660	232	1320	331	2200	430	4620	628	7920	825
0.5	330	169	880	267	1650	366	2640	465	5280	663	8800	860
0.75	440	204	1100	302	1980	401	3080	500	5940	697	9680	895
1.0	550	238	1320	337	2310	436	3520	535	6600	732	10560	930
1.5	770	308	1760	407	2970	506	4400	605	7920	802	12320	1000
2.0	990	378	2200	477	3630	576	5280	675	9240	872	14080	1070

Note: The thickness and side slopes of nonreinforced concrete ditch and canal linings are specified by Part 1 Engineering Conservation Practices of the National Engineering Handbook, Soil Conservation Service of the U.S. Department of Agriculture. This table assumes a uniform thickness of 2.5 in (63.5 mm) with a sideslope of 1 to 1. Density of Concrete is taken as 2200 kg/m^3, density of excavation as 1600 kg/m^3.

Table 3
ENERGY REQUIRED TO PUMP 1 ha-cm OF WATER[a]

Total head m	Direct energy (pump efficiency = .70) MJ	Electricity (Motor efficiency = .88) kWh	Diesel[b] (Engine efficiency = 0.30) ℓ	Natural gas[b] engine efficiency = 0.20 m^3
10	14.0	4.4	1.21	1.92
50	70.0	22.1	6.07	9.59
100	140.1	44.2	12.14	19.18
200	280.2	88.4	24.28	38.37
300	420.3	132.6	36.42	57.55

[a] 1 hectare - cm = $100 m^3$ = .9729 acre · in ≈ 1 acre · in.

[b] Heating value of diesel = 38.4 MJ/l; heating value of Natural Gas = 36.5 MJ/m^3. Direct energy calculated as:

$100\ m^3 \times 1000\ kg/m^3 \times 9.806\ 65\ N/kg \times 1 MJ/10^6 N \cdot m \times$ total head (m)/Pump efficiency

Conversion may be made as follows: 4186 J = 1 kcal; 1 kwh = 859 kcal.

Table 4
TYPICAL DESIGN SUMMARIES OF NINE IRRIGATION SYSTEMS FOR A 64.8 HA (160 ACRE) FIELD[a]

	Pumping Unit[b]		Pipe (10^3kg)				Earth Work	
Irrigation System	Power (kW)	Approximate mass (kg)	Polyethylene	PVC	Aluminum	Other equipment (10^3 kg)	Grading (m^3)	Ditching (m)
Surface without IRRS[c]	48	650		1.4			50 000	2400
Surface with IRRS	51	650		2.4	4.5		50 000	2400
Solid set sprinkle	92	800		6.5	34.6	8.6		1140
Permanent sprinkle	92	800		27.6		9.6		44900
Hand-moved sprinkle	92	800		6.5	2.5	0.6		2400
Side roll sprinkle	92	800		6.5	4.3	2.5		2400
Center-pivot sprinkle	75	700		3.8		15.9		460
Traveler sprinkle	133	1000		8.8		7.5		1560
Trickle	70	700	13.05	16.9		0.8		2400

[a] 1 kWh = 859 kcal.

[b] Assumes electric motor driven pumping unit.

[c] Irrigation Runoff Return System.

From Batty, J. C., Hamad, S. N., and Keller, J., Energy inputs to irrigation, Irrigation Drainage Division, American Society of Civil Engineers, December 1975.

Table 5
ANNUAL FIXED ENERGY INPUTS IN MJ/HA-YEAR ASSOCIATED WITH INSTALLATION OF NINE IRRIGATION SYSTEMS DESCRIBED IN TABLE 4: 40-YEAR EVALUATION PERIOD

Irrigation system	Area irrigated hectares	Pumping unit	Piping			Other equipment	Earth work		
			Poly	PVC	Aluminum		Grading	Ditching	Total
Surface without IRRS[a]	63.2	70		66			316	14	466
Surface with IRRS	62.8	70		115	702		318	14	1219
Solid set sprinkle	64.0	85		305	5298	323		7	6018
Perma-nent sprinkle	64.0	85		1294		360		25	1764
Hand-moved sprinkle	64.0	85		304	382	23		14	808
Side-roll sprinkle	64.0	85		304	658	94		14	1155
Center-pi-vot sprinkle	50.6	94		225		754		4	1077
Traveler sprinkler	61.6	110		429		292		10	841
Trickle	64.0	74	3263	792		30		56	4215

[a] IRRS = Irrigation runoff return system.

Table 6
COMPARISON OF ANNUAL INSTALLATION ENERGIES WITH PUMPING ENERGIES REQUIRED TO APPLY 1 HA-M OF NET IRRIGATION AT VARIOUS LIFTS (4186 J = 1 kcal)

			Lift of 0 m		Lift of 50 m		Lift of 100 m	
System	Irrigation efficiency	Installation energy[a] MJ/ha-year	Total head (m)	Pumping energy[b] (MJ/ha-m)	Total head (m)	Pumping energy (MJ/ha-m)	Total head	Pumping energy (ha-m)
Surface without IRRS[c]	.50	466	3	3,184	53	56,250	103	109317
Surface with IRRS	.85	1,219	5	3,122	55	34,337	105	65552
Solid set sprinkle	.80	6,018	53	35,156	103	68,323	153	101489
Permanent sprinkle	.80	1,764	53	35,156	103	68,323	153	101089
Hand-moved sprinkle	.75	808	53	37,500	103	72,878	153	108255
Side-roll sprinkle	.75	1,155	53	37,500	103	72,878	153	108255
Center-pivot sprinkle	.80	1,074	60	39,800	110	72,966	160	106132
Traveler sprinkle	.70	841	95	72,018	145	109,923	195	147828
Trickle	.90	4,215	35	20,637	85	50,118	135	79599

[a] See Table 5.

[b] Pumping Energy Calculated using the following relationship:

$$PE = \frac{10{,}000\ m^3/\text{hectare}\cdot m \times 1{,}000\ kg/m^3 \times 9.80665\ N/kg\ 1MJ/10^6N\cdot m \times \text{Total head (m)}}{\text{(Pump efficiency) (Electric motor efficiency) (Thermal conversion efficiency) (Irrigation efficiency)}} = 530.66\ \text{(Total Head) MJ (Irrigation Efficiency) hectare}\cdot m$$

In the above pump efficiency = 0.70, electric motor efficiency = 0.88, thermal conversion efficiency = 0.30.

[c] IRRS = Irrigation runoff return system.

ENERGY INPUTS FOR THE PRODUCTION, FORMULATION, PACKAGING, AND TRANSPORT OF VARIOUS PESTICIDES

David Pimentel

Pesticides require fossil energy for their production. In the manufacturing process, the active pesticide ingredient uses direct energy inputs of heat and electricity and indirect energy inputs from the fuels that make up the hydrocarbon stocks used in manufacture. The energy inputs for each pesticide range from 13,810 kcal/kg for methyl parathion to 109,520 kcal/kg for paraquat (Table 1). These inputs vary according to the hydrocarbon feed stocks used and the amount of heat and electricity used in the manufacturing process.

Herbicide production averages about 57,000 kcal/kg of energy, ranging from 19,080 to 109,520 kcal/kg (Table 1). Energy requirements of insecticide production average slightly less than that for herbicides or 44,000 kcal/kg, ranging from 13,810 to 108,100 kcal/kg. Fungicides appear to be the most economical, requiring about 22,000 kcal/kg and ranging from 15,250 to 27,380 kcal/kg for production.

Additional energy inputs are required to formulate the pesticide, to package it, and to transport it to the farm for use (Table 2). The least energy intensive means of supplying a pesticide appears to be a wettable powder formulation (Table 2). Miscible oil formulations and granules are both relatively energy intensive. The energy inputs for formulating, packaging and transporting are presented in Table 2. These inputs represent about a third of the total inputs.

Table 1
ENERGY INPUTS FOR THE BASIC PRODUCTION OF VARIOUS PESTICIDES

Pesticides	kcal for production (1 kg active ingredient)	Ref.
Herbicides		
MCPA	30,952	3
Diuron	64,290	3
Atrazine	45,240	3
Trifluralin	35,170	3
Paraquat	109,520	3
2,4-D	24,200	5
2,4,5-T	56,700	3
Chloramben	71,400	3
Dinoseb	19,080	3
Propanil	52,240	3
Propachlor	69,050	3
Dicamba	70,240	3
Glyphosate	108,100	3
Diquat	95,240	3
Insecticides		
DDT	24,200	4
Toxaphene	38,100	3
Methyl parathion	13,810	3
Carbofuran	108,100	3
Carbaryl	36,430	3
Fumigants		
Methyl bromide	15,950	6
Fungicides		
Ferbam	15,250	3
Maneb	23,570	3
Captan	27,380	3
Sulfur	26,620	2
Pesticide average	**49,020**	

[a] The percentage of oil, natural gas, and coal involved in the production of pesticides has been calculated as 42% oil, 38% natural gas, and 20% coal.

Table 2
ENERGY INPUTS (PRODUCTION, FORMULATION, PACKAGING, TRANSPORT) FOR VARIOUS PESTICIDES[a]

Pesticide	kcal Input					% of Energy types		
	Production Active Ingredient[b]	Formulation	Packaging	Transport[c]	Total	Oil	Gas	Coal
Herbicide								
Miscible oil	57,000	33,300[d]	8,500[e]	1,110	99,910	60	23	17
Wettable powder	57,000	2,500[f]	2,600[g]	670	62,770	43	37	20
Granules	57,000	3,600[h]	20,000[i]	6,720	86,600	42	37	21
Insecticide								
Miscible oil	44,000	33,300[d]	8,500[e]	1,110	86,910	61	23	16
Wettable powder	44,000	2,500[f]	2,600[g]	670	61,470	43	37	20
Granules	44,000	3,600[h]	20,000[i]	6,720	74,300	42	37	21
Dust	44,000	3,600[h]	20,000[i]	6,720	74,300	42	37	21
Fungicide								
Miscible oil	22,000	33,300[d]	8,500[e]	1,110	64,910	70	15	15
Wettable powder	22,000	2,500[f]	2,600[g]	670	27,770	42	37	21
Granules	22,000	3,600[h]	20,000[i]	6,720	51,600	41	37	22
Dust	22,000	3,600[h]	20,000[i]	6,720	51,600	41	37	22

[a] All values given for 1 kg active pesticide ingredient.
[b] Average from Table 1.
[c] Assumed 0.53 kcal/kg/km for transport by truck and rail and at a mean distance of 640 km.
[d] Assumes a 30% formulation.
[e] Packaging includes a metal 5 gal container and is based on data of Reference 1.
[f] Assumes a 50% wettable powder.
[g] Assumes pesticide formulation in 1 kg paper packages and is based on data from Reference 1.
[h] Assumes 5% active ingredient in either granules or dust.
[i] Assumes pesticide formulation in 20 kg paper packages and is based on data from Reference 1.

REFERENCES

1. **Berry, R. S. and Makino, H.,** Energy thrift in packaging and marketing, *Technol. Rev.,* 76(4), 1, 1974.
2. Estimated.
3. **Green, M. B.,** Energy in agriculture, *Chem. Ind. (London),* Aug., 641, 1976.
4. **Leach, G. and Slesser, M.,** *Energy Equivalents of Network Inputs to Food Producing Processes,* Strathclyde University, Glasgow, 1973.
5. **Pimentel, D.,** personal observation, 1973.
6. **Gavett, E.,** personal communication, 1977.

ENERGY REQUIREMENTS FOR VARIOUS METHODS OF CROP DRYING*

Robert M. Peart, Roger Brook, and Martin R. Okos

INTRODUCTION

A variety of methods are used for drying grain, and this paper summarizes energy requirements for all of the drying methods in current use as well as some experimental ones. It also points out the importance of capacity, reliability, and quality assurance in the drying method, because energy efficiency cannot be the only criterion used in selecting a drying method. New dryer designs and new energy sources, such as solar and the heat pump, are also discussed.

Tables 1 and 3 give estimates of the energy requirements for eight drying methods, and these are discussed individually. Energy estimations were based on simulation of high-temperature drying and full-scale demonstrations for the low-temperature methods. The tables show values for drying to 14% wet basis from various starting moisture contents. The relationships are not linear because of the wet basis moisture values and also because of varying efficiencies for different amounts of drying. The estimates given in these tables should be viewed as nominal values for typical designs and weather conditions. New designs and methods will no doubt be developed to improve on these energy efficiencies.

CONVENTIONAL HIGH-TEMPERATURE DRYING

This is probably the most common drying method, extensively used in central grain elevators and on farms. Drying air temperature is typically 90°C (194°F), and airflow is in the range of 75 m^3/min/t. The prevalent arrangement is cross-flow, with grain stationary or flowing down in a column 20 to 50 cm thick and airflow perpendicular to this column. Following drying, ambient air is blown through the grain at about the same rate for about 20 min to cool it before storage.

HIGH-TEMPERATURE DRYING WITH DRYERATION

The dryeration process, developed by G. H. Foster,[2] combines slow cooling and removal of the last two percentage points of moisture in a separate operation, after conventional drying to 16% moisture. The corn is removed from the dryer without cooling at a temperature of about 60°C or less and held for 6 hr with no airflow. This removes temperature and moisture gradients within the kernel. Then a low flow (0.5 m^3/min/t) of ambient air carries away warm high-humidity air, removing the final two percentage points of moisture and cooling the grain. This is usually done overnight, and the grain is cooler than if cooled with daytime air. Table 1 shows dryeration gives a 15 to 25% energy saving, mainly because of the heat energy saved in the least efficient part of the drying cycle. Fan energy is also reduced because of the greatly reduced airflow rate during slow cooling.

Other major advantages of dryeration are better corn quality due to slower cooling and higher capacity for a given high-temperature dryer unit, because cooling and the final two percentage points of drying are not done in the dryer.

* Approved as Journal Paper No. 7073 of the Purdue University Agricultural Experimental Station.

Table 1
ENERGY REQUIRED TO DRY CORN TO 14% WET BASIS (kWh/t)

Drying from (% wet basis)	High speed to 14%, 90°C		High speed to 16%, 90°C, dryeration		High speed to 20%, 90°C, bin finish		Batch-in-bin, 60°C
	Fan	Heat	Fan	Heat	Fan	Heat	Fan and heat
30	4.6	365	3.8	310	16.0	240	370
28	4.1	315	3.3	260	15.5	190	315
26	3.6	270	2.8	210	15.0	145	260
24	3.2	225	2.2	170	14.6	100	210
22	2.7	180	1.8	130	14.1	60	165
20	2.2	135	1.4	90	—	—	120

COMBINATION HIGH TEMPERATURE/LOW TEMPERATURE

This method is being tested by researchers and is used by some farmers. It is similar to dryeration, but corn leaves the high-temperature dryer at 20% moisture with considerable drying still to be done. It is then put into a storage bin with a perforated floor for slow, low-temperature drying with natural air at the rate of 1 m^3/min/t. This low airflow is started immediately when hot corn goes into the bin, without the tempering period used in dryeration. The completion of drying down to 14% may take as long as 4 weeks, as it is dependent on ambient air conditions. A low temperature rise may be used in this stage to assure drying down to 14%. The values in Table 1 assume natural air only in the final stage for 30 days.

These values show a considerable energy saving for this method, but the equipment costs are higher because two complete drying systems are required. We believe there is a trend toward this method by farmers, however, due to lower fuel costs. Quality advantages similar to those in the dryeration method are realized.

BATCH-IN-BIN DRYING

This method has been in use for some time, and was developed to improve the drying rate of the older in-bin layer method covered next. In this method, a storage bin with a perforated floor is the dryer, and a fan and heater unit supplies about 20 m^3/min/t at about 60°C (122°F) through a grain depth of about 1 m. Typically, a farmer harvests this 1 m deep batch in 1 day, and the fan and heater operate through the day and night. Early in the morning the heat is turned off and the corn is cooled. Moisture content will vary from over-dry on the bottom, where the drying air enters, to underdry on top. Removing the grain from the bin mixes it adequately and the moisture content will equalize in storage. Then the bin is ready for another batch.

Fuel efficiency for this method is about equal to the conventional high-temperature method. Drying at 90°C is theoretically more efficient than at 60°C, but the lower airflow and greater bed depth of the batch-in-bin method compensate. The values in Table 1 were obtained using about 1400 kcal/kg water evaporated (2500 Btu/lb), an estimate based on simulation and test results.

DESIGNS TO CONSERVE ENERGY IN HIGH-TEMPERATURE DRYERS

The conventional method of high-speed drying is the cross-flow dryer. Traditional

Table 2
PERFORMANCE OF HIGH-TEMPERATURE DRYERS WITH DESIGNS FOR ENERGY SAVING: ENERGY REQUIRED TO DRY CORN TO 14% WET BASIS (kWh/t)

	Cross flow				Concurrent flow	
	Reverse cooling		Recycling			
Drying from (% wet basis)	Fan	Heat	Fan	Heat	Fan	Heat
30	4.8	353.0	4.6	298.2	7.7	287.7
28	4.3	303.1	4.1	255.8	6.5	244.7
26	3.7	256.1	3.5	215.8	5.4	204.1
24	3.2	211.2	3.0	177.6	4.4	165.6
22	2.6	166.9	2.5	140.0	3.5	129.1
20	2.1	123.6	2.0	103.2	2.5	94.3

designs exhaust all the drying and cooling air, thus losing all the energy contained in the air. Changes in the design of cross-flow dryers could reduce total energy consumption of cross-flow dryers.

A cross-flow dryer drying corn to 14% includes a cooling section in the dryer. The energy in the cooling air can be recycled by use of the inlet to the drying air fan. The inlet drying air would increase in temperature as the average grain temperature decreases about 25°C, thus saving 12.5 kWh/t. The fan energy would increase by 15% due to the increased cooling airflow. The energy consumption for this dryer configuration with reverse cooling is illustrated in Table 2.

Further modification of the cross-flow dryer would allow recycling of 50% of the exhausted drying air. The exhaust air from the top half of the drying column is not recycled because it is normally saturated and has thus used its total drying potential. Reuse of half the drying air would reduce energy consumption about 15% in addition to reversed cooling. The recycling also reduces fan energy by 5%. The energy consumption for a cross-flow dryer with recycling of half the drying air, as well as the cooling air, is illustrated in Table 2. The 15% saving by recycling can also be applied to dryer management schemes which dry corn to 16 to 20% and utilize bin cooling.

New dryer types are appearing on the commercial market. One is the concurrent-flow dryer. Typical energy consumption for a concurrent-flow dryer is shown in Table 2. The figures are based on field experience with a drying bed 2 m deep, with an airflow of 32 m^3/min/t.

One new continuous-flow unit has a device to turn the grain column during the drying cycle so that wetter corn that was in the exhaust side of the dryer is moved to the higher temperature area nearer the input air.

IN-BIN DRYING, 20°C

This is often called the layer drying method because the bin is partially filled, dried, and then a second and third layer is added and dried. The temperature is selected to lower the relative humidity of the natural air enough to bring the corn to an equilibrium moisture content of about 14%. Higher temperature will cause overdrying of the grain on the bottom. An airflow of 1 m^3/min/t is typical.

The energy values for this method in Table 3 were obtained using about 1200 kcal/kg of water evaporated (2200 Btu/lb), based on simulations and average test results.

Table 3
ENERGY REQUIRED TO DRY CORN TO 14% WET BASIS (kWh/t)

Drying from (% wet basis)	In-bin, 20°C Fan and heat	In-bin low-temperature ΔT = 4°C Fan and heat	Low-temperature solar, maximum ΔT = 10°C Fan and heat	Low-temperature heat pump, ΔT = 5.6°C Heat	Low-temperature heat pump, ΔT = 5.6°C Fan
30	—[a]	—[a]	—[a]	—[a]	—[a]
28	305	275	240[b]	190	35
26	230	190	120[b]	75	30
24	185	120	80[c]	35	25
22	145	80	50[c]	15	20
20	105	60	35[c]	10	15

[a] Not recommended.
[b] Includes supplemental heat — solar heat not included.
[c] Fan only — solar heat not included.

Table 4
LOW-TEMPERATURE CORN DRYING

Starting moisture content (% wet basis)	Airflow rate (m^3/min/t)
28	5
26	3
24	2
22	1
20	1

Energy requirements for this method are very dependent upon ambient air conditions, so these values are average over several years of weather in central Indiana.

LOW-TEMPERATURE IN-BIN DRYING

This method is similar to the in-bin method just discussed, but a lower constant temperature rise of about 4°C is used, along with higher airflow rates. It is still experimental for the higher moisture contents, but many farmers have had satisfactory results starting at moisture contents of 24% and lower.

All of these slower methods require a higher level of management than the fast methods. Actions taken on a given day do not show results for several days, and changeable weather can mask other inputs. Only one batch is usually dried in one bin each year, so the operator does not collect experience as quickly as in a system where one or more batches are dried per day and the results can be seen immediately.

This method is not recommended for corn higher than 28% moisture content because of the excessive airflow requirements. Airflow requirements were assumed as shown in Table 4, for this method and for the other low-temperature methods.

At the higher moistures, energy requirements are comparable to the in-bin method. The low-temperature method requires more fan energy and less heat energy at 28% and 26%. At the lower moistures with lower airflows, the low-temperature method obtains more drying energy from the drying capability of the natural air, so it is more efficient than the higher temperature method.

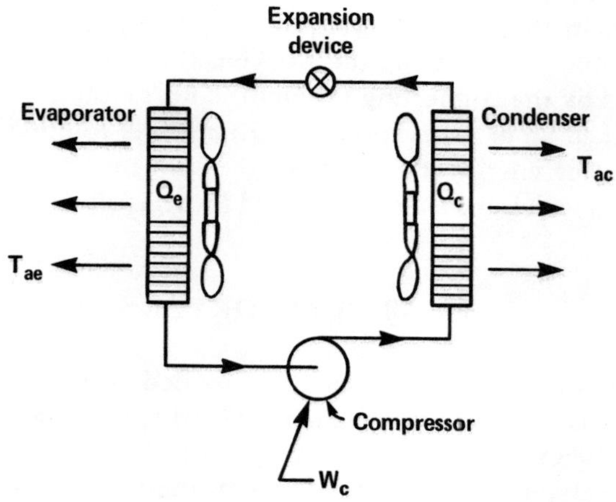

FIGURE 1. Heat pump system.

LOW-TEMPERATURE SOLAR METHOD

This method uses the same airflow rates as those given above, but for starting moistures of 24% or lower no heat other than solar is needed. These energy estimates are for fan energy only for 24% and lower moisture, and they include some supplemental heat and the fan energy for 26% and 28%. The solar energy requirement is that obtainable from an air-heating collector giving a maximum temperature rise of 10°C at noon on a clear day, with the total airflow passing through the collector. This can be obtained with a suspended plate collector with a single fiberglass cover sloped to be perpendicular to the noon sun, with a collector area of 0.8 m^2/t of corn, for an airflow rate of 2 m^3/min/t. Proportionately more or less collector is needed for more or less airflow.

This method is the most dependent on the weather of those discussed, because of the importance of ambient air conditions, as well as solar radiation. These data are based on central Indiana weather conditions for October, November, and December.

DRYING WITH THE HEAT PUMP

An alternative heating method to electrical resistance and gas heating is heat pumping. The electric heat pump consists of two heat exchangers (evaporator and condenser), compressor and expansion device. Figure 1 illustrates the major elements of a heat pump. In addition, fans are employed to move air through the heat exchangers, and controls are required to manage power to the compressor and the fans. The four main components of the heat pump are connected by tubing that contains a working fluid. The fluid is propelled by the action of the compressor.

The heat pump moves energy (heat) from a reservoir of lower temperature to one of higher temperature. In accordance with the second law of thermodynamics, this action cannot be accomplished without the input of work. In the heat pump, the work supplied is in the form of electricity delivered to the compressor.

When the heat pump is used as an air conditioner, it removes heat from a lici g space and transfers it to the warmer ambient air outside the space. When the device is used to heat a living space, heat is removed from a cool ambient exterior and delivered to the warmer living space.

The importance of the heat pump as an energy-delivering device is that it delivers more energy (Q_c) than the work (W_c) required to affect the delivery. From the first law of thermodynamics, considering the working fluid as a system and ignoring any superficial heat loss by the connecting tubing, the heat from the condenser (Q_c) is the sum of the heat picked up in the evaporator (Q_e) plus the compressor work (W_c). Thus from the standpoint of energy conservation, the heat pump is superior to electrical-resistance heating, in which the energy delivered (heat) is equal to the energy consumed (electricity).

HEAT PUMP DRYERS

Heat pump sources for grain drying can be classified into two general groups: open- and closed-air systems. In open systems, the air used for drying is taken from ambient and returned to ambient after doing its drying work. In closed systems, the air that is exhausted from the dryer is recirculated through the heat pump and returned to the dryer.

The open-air-loop heat pumps are basically of three types. The first type passes ambient air through the condenser to heat and then through the grain mass to dry. Ambient air is also passed through the evaporator, where energy is extracted. This type is run like a residential heat pump. The second type passes ambient air through the evaporator, where it is cooled and dehumidified. The air then passes through the condenser, where the dehumidified air is heated for drying grain. This type may be useful in warm humid areas.

The third type of open-air-loop heat pump grain dryer first passes ambient air through the condenser, where it is heated for grain drying. The moisture-laden grain dryer exhaust air is then passed through the evaporator for energy recovery, partly sensible heat and partly latent heat as moisture condenses on the evaporator coil. Tests have shown an energy input into the heat pump and fan of 0.46 kWh/kg of water removed.[1] This is possible because of the heat recovery and recycling.

Table 3 gives the energy requirement for drying at various initial moisture contents for such a system. The time required to dry the grain at the recommended airflow rates (Table 4) for the various initial moisture contents was calculated assuming a temperature rise of 5.6°C (10°F). The fan energy requirements were calculated knowing the airflow rates and pressure drops for a grain depth of 4.6 m (15 ft). The heat pump energy requirement was determined based on experimental data.[1]

REFERENCES

1. **Hogan, M. R., Okos, M. R., Williams, E. E., Ayers, D. D., Peart, R. M.,** Low Temperature Heat Pump Grain Dryer Design Performance, and Operational Experience, American Society of Agricultural Engineers, ASAE Paper No. 76-3519, New York, 1976.
2. **McKenzie, B. A., Foster, G. H., Noyes, R. T., and Thompson, R. A.,** Dryeration, AE-72, Cooperative Extension Service, Purdue University, West Lafayette, Ind., 1968.

ENERGY USED FOR TRANSPORTING SUPPLIES TO THE FARM

David Pimentel

Each kilogram of farm supplies (fertilizers, pesticides, machinery, fuel, etc.) is transported an average of 640 km. About 60% of the transporting is by rail and 40% by truck.[3] Based on these assumptions, energy used to transport the supplies to the farm is 257 kcal/kg. The energy inputs for transporting a kilogram 1 km by various methods are shown below.

Table 1
ENERGY USED FOR TRANSPORTING SUPPLIES TO THE FARM[1,2]

Transport	kcal/kg/km
Waterway	0.08
Rail	0.12
Truck	0.83
Air	6.63

REFERENCES

1. **Hirst, E.**, Energy consumption for transportation in the U.S., Oak Ridge National Laboratory, Oak Ridge, Tennessee, 1972.
2. Statistical Abstract of the United States 1976, U.S. Bureau of the Census, U.S. Department of Commerce, 97th ed., U.S. Government Printing Office, Washington, D.C., 1976.
3. Census of Transportation, Vol. 3, Commodity Transportation Survey, Parts 1 and 2, Commodity Groups, U.S. Department of Commerce, 1963.

Energy Inputs and Outputs for Crop Systems—Field Crops

ENERGY INPUTS IN BARLEY PRODUCTION

Robert Bukantis and Nancy Goodman

Barley is one of the most widely cultivated crops in the world today. It is considered by many to be the oldest known cultivated crop. Both barley and wheat were found in an Egyptian archeological site estimated to be 5000—6000 years old.[16] Early people probably used barley for making porridge or bread as well as beer. Clay documents at least 8000 years old have been found with pictures showing the brewing of beer.[14] Until the 16th century, barley flour instead of wheat flour was used to make bread.

Early settlers in the U.S. introduced varieties of barley from England and Europe primarily for brewing beer, although some was used for flour and livestock feed. As the colonists moved westward, barley varieties from the Mediterranean were introduced for livestock feed.

Barley has a wider ecological range than any other cereal grain and nearly every country in the world grows barley. World production in 1976 totalled about 178 million metric t, resulting from 86 million ha planted in barley. In 1976, the U.S.S.R. led the world in production (40%), followed by Canada, China, France, and the U.S.[15] Over the past 25 years, the area planted in barley has increased 81% to about 85 million ha, illustrating the increasing importance of barley as a world grain crop.

Although barley is a hardy plant and can be grown in a wide variety of soils and climates, production is concentrated in the northern latitude. Hot, humid climates are the only areas where barley does not thrive. In the U.S., barley is grown commercially in 36 states and ranks fourth in importance as a grain crop. Since barley is usually less profitable than corn, sorghum, and soybeans, it is grown in areas suih as the northern plains and Pacific coast states, where other grains cannot adapt as well to the climate. North Dakota, California, and Montana were the leading barley-producing states in 1976.

Barley varieties grown in the U.S. are classified as either malting or feed. Livestock and poultry feed production reached a high of 6.3 million in 1970 to 1971 and declined to 4.2 million in 1975. In contrast, the quantity of barley used in the malting industry has been increasing since 1960 and in 1975 was 2.8 million t. Barley malt is used for alcohol and alcoholic beverages, for food, and for export. A small quantity of barley, 14 million bushels in 1975, is used for seed.

Production practices for barley vary depending on whether the crop will be used for feed or malting. Phosphate can improve malting quality, whereas high nitrogen rates can result in barley with too high a protein level for malting. A high protein level is desirable when barley is grown for feed.

In arid production regions, barley is often grown on land that was fallowed the year before. In more humid regions, barley is part of a continuous crop rotation. Energy budgets are included for both practices (Tables 1-9). Irrigated land is often used to grow malting varieties (Tables 7 and 9). Efficiency varies depending on whether or not pumping equipment is needed for irrigation.

Table 1
ENERGY INPUTS PER HECTARE FOR BARLEY FOLLOWING FALLOW IN SOUTH CENTRAL MONTANA

Item	Quantity/ha	kcal/ha	Ref.
	Input		
Labor	5.26 hr	—	3
Machinery	7 kg	126,000	2, 3, 7, 10
Gasoline	35.6 ℓ	360,000	1, 3
Diesel	27.6 ℓ	315,000	1, 3
Seed	102 kg	306,000	3, 4
Nitrogen	30.3 kg	414,000	3, 6
Phosphate	15.7 kg	47,100[a]	3
Herbicide	0.54 kg	48,100	3, 8, 12
Transportation	159 kg[b]	40,900	9
Total		**1,660,000**	
	Output		
Barley yield	2,330 kg	8,120,000	3,11
Protein yield	207 kg		11
kcal output/kcal input		4.89	
kcal output/labor hour		1,540,000	

[a] Assumes triple super phosphate.[6]
[b] Fuels, seed, and machinery.

Table 2
ENERGY INPUTS PER HECTARE FOR BARLEY IN SOUTH CENTRAL MONTANA

Item	Quantity/ha	kcal/ha	Ref.
	Input		
Labor	3.93 hr	—	3
Machinery	7 kg	126,000	2, 3, 7, 10
Gasoline	15.8 ℓ	160,000	1, 3
Diesel	39.5 ℓ	451,000	1, 3
Seed	61.6 kg	185,000	3, 4
Nitrogen	28.0 kg	383,000	3, 6
Phosphate	50.2 kg	151,000[a]	3
Herbicide	0.538 kg	48,000	3, 8, 12
Insecticide	0.775 kg	61,900	3, 8, 12
Transportation	114 kg[b]	29,300	9
Total		**1,600,000**	
	Output		
Barley yield	1,900 kg	6,620,000	3, 11
Protein yield	169 kg		11
kcal output/kcal input		4.14	
kcal output/labor hour		1,680,000	

[a] Assumes triple super phosphate.[6]
[b] Fuels, seed, and machinery.

Table 3
ENERGY INPUTS PER HECTARE FOR BARLEY IN SOUTHEASTERN MINNESOTA

Item	Quantity/ha	kcal/ha	Ref.
	Input		
Labor	6.08 hr	—	3
Machinery	9 kg	162,000	2, 3, 7, 10
Gasoline	31.1 ℓ	314,000	1, 3
Diesel	53.9 ℓ	615,000	1, 3
Seed	107 kg	321,000	3, 4
Nitrogen	26.3 kg	359,000	3, 6
Phosphate	33.1 kg	99,300[a]	3
Potash	14.0 kg	22,400	3, 6
Herbicide	0.557 kg	49,600	3, 8, 12
Transportation	185 kg[b]	47,500	9
Total		**1,990,000**	
	Output		
Barley yield	1,900 kg	6,620,000	3, 11
Protein yield	169 kg		11
kcal output/kcal input		3.33	
kcal output/labor hour		1,090,000	

[a] Assumes triple super phosphate.[6]
[b] Fuels, seed, and machinery.

Table 4
ENERGY INPUTS PER HECTARE FOR BARLEY FOLLOWING CROPPING IN EASTERN NORTH DAKOTA

Item	Quantity/ha	kcal/ha	Ref.
	Input		
Labor	5.12 hr	—	3
Machinery	9 kg	162,000	2, 3, 7, 10
Gasoline	31.6 ℓ	319,000	1, 3
Diesel	45.2 ℓ	516,000	1, 3
Seed	89.7 kg	269,000	3, 4
Nitrogen	57.7 kg	789,000	3, 6
Phosphate	32.7 kg	98,100[a]	3
Potash	9.4 kg	15,100	3, 6
Herbicide	0.308 kg	27,500	3, 8, 12
Insecticide	0.446 kg	35,600	3, 8, 12
Transportation	161 kg[b]	41,400	9
Total		**2,270,000**	
	Output		
Barley yield	1,980 kg	6,900,000	3, 11
Protein yield	176 kg		11
kcal output/kcal input		3.04	
kcal output/labor hour		1,350,000	

[a] Assumes triple super phosphate.[6]
[b] Fuels, seed, and machinery.

Table 5
ENERGY INPUTS PER HECTARE FOR BARLEY FOLLOWING FALLOW IN EASTERN NORTH DAKOTA

Item	Quantity/ha	kcal/ha	Ref.
	Input		
Labor	6.42 hr	—	3
Machinery	10 kg	180,000	2, 3, 7, 10
Gasoline	30.4 ℓ	307,000	1
Diesel	65.8 ℓ	751,000	1
Seed	89.7 kg	269,000	3, 4
Nitrogen	19.4 kg	265,000	3, 6
Phosphate	21.1 kg	63,300[a]	3
Potash	4.71 kg	7,540	3, 6
Herbicide	0.130 kg	11,600	3, 8, 12
Insecticide	0.437 kg	34,900	3, 8, 12
Transportation	178 kg[b]	45,800	9
Total		**1,940,000**	
	Output		
Barley yield	2,000 kg	6,970,000	3, 11
Protein yield	178 kg		11
kcal output/kcal input		3.59	
kcal output/labor hour		1,090,000	

[a] Assumes triple super phosphate.[6]
[b] Fuels, seed, and machinery.

Table 6
ENERGY INPUTS PER HECTARE FOR BARLEY FOLLOWING CROPPING IN CENTRAL IDAHO

Item	Quantity/ha	kcal/ha	Ref.
	Input		
Labor	5.46 hr	—	3
Machinery	8 kg	144,000	2, 3, 7, 10
Gasoline	33.1 ℓ	335,000	1, 3
Diesel	35.3 ℓ	403,000	1, 3
Seed	80.5 kg	242,000	3, 4
3, 6 Nitrogen	53.5 kg	731,000	
Herbicide	0.98 kg	87,000	3, 8, 12
Transportation	143 kg[a]	36,800	9
Total		**1,990,000**	
	Output		
Barley yield	1,200 kg	4,180,000	3, 11
Protein yield	107 kg		11
kcal output/kcal input		2.10	
kcal output/labor hour		766,080	

[a] Fuels, seed, and machinery.

Table 7
ENERGY INPUTS PER HECTARE FOR IRRIGATED BARLEY IN CENTRAL IDAHO

Item	Quantity/ha	kcal/ha	Ref.
	Input		
Labor	15.6 hr	—	3
Machinery	8 kg	144,000	2, 3, 7, 10
Gasoline	33.8 ℓ	342,000	1, 3
Diesel	38.6 ℓ	441,000	1, 3
Seed	145 kg	435,000	3, 4
Nitrogen	82.2 kg	1,120,000	3, 6
Phosphate	30.7 kg	92,100[a]	3
Herbicide	3.6 kg	319,000	3, 8, 12
Transportation	211 kg[b]	54,200	9
Total		**2,950,000**	
	Ouput		
Barley yield	3,620 kg	12,600,000	3, 11
Protein yield	322 kg		11
kcal output/kcal input		4.27	
kcal output/labor hour		808,000	

[a] Assumes triple super phosphate.[6]
[b] Fuels, seed, and machinery.

Table 8
ENERGY INPUTS PER HECTARE FOR BARLEY FOLLOWING CROPPING IN NORTH CENTRAL CALIFORNIA

Item	Quantity/ha	kcal/ha	Ref.
	Input		
Labor	8.53 hr	—	3
Machinery	10 kg	180,000	2, 3, 7, 10
Gasoline	51.5 ℓ	521,000	1, 3
Diesel	49.8 ℓ	568,000	1, 3
Seed	121 kg	363,000	3, 4
Nitrogen	14.6 kg	200,000	3, 6
Phosphate	2.24 kg	6,720[a]	3
Herbicide	0.86 kg	76,700	3, 8, 12
Insecticide	0.18 kg	14,600	3, 8, 12
Transportation	212 kg[b]	54,500	9
Total		**1,980,000**	
	Output		
Barley yield	2,440 kg	8,500,000	3, 11
Protein yield	217 kg		11
kcal output/kcal input		4.29	
kcal output/labor hour		996,000	

[a] Assumes triple super phosphate.[6]
[b] Fuels, seed, and machinery.

Table 9
ENERGY INPUTS PER HECTARE FOR IRRIGATED BARLEY GROWN IN NORTH CENTRAL CALIFORNIA

Item	Quantity/ha	kcal/ha	Ref.
	Input		
Labor	13.4 hr	—	3
Machinery	10 kg	180,000	2, 3, 7, 10
Gasoline	52.6 ℓ	532,000	1, 3
Diesel	68.9 ℓ	786,000	1, 3
Seed	119 kg	357,000	3, 4
Electricity	489 kWh	1,400,000	1, 3
Nitrogen	33.6 kg	459,000	3, 6
Phosphate	7.9 kg	23,600[a]	3
Herbicide	0.63 kg	56,500	3, 8, 12
Insecticide	7.33 kg	585,000	3, 8, 12
Irrigation equipment	$14.00	175,000	3, 5, 13
Transportation	236 kg[b]	60,700	
Total		**4,610,000**	
	Output		
Barley yield	3,680 kg	12,800,000	3, 11
Protein yield	328 kg		11
kcal output/kcal input		2.78	
kcal output/labor hour		955,000	

[a] Assumes triple super phosphate.[6]
[b] Fuels, seed, and machinery.

REFERENCES

1. **Cervinka, V.,** Fuel and energy efficiency, in *Handbook os Energy Utilization in Agriculture,* Pimentel, D., Ed., CRC Press, Boca Raton, Fla., 1980.
2. **Doering, O.,** Accounting for energy in farm machinery and buildings, in *Handbook of Energy Utilization in Agriculture,* Pimentel, D., Ed., CRC Press, Boca Raton, Fla., 1980.
3. **Firm Enterprise Data System,** 1975, *U.S. Department of Agriculture, Economic Research Service, and Department of Agricultural Economics,* Oklahoma State University, Stillwater, 1977.
4. **Heichel, G.,** Assessing the fossil energy costs of propagating agricultural crops, in *Handbook of Energy Utilization in Agriculture,* Pimentel, D., Ed., CRC Press, Boca Raton, Fla., 1980.
5. **Herendeen, R. A. and Bullard, C. W.,** Energy Cost of Goods and Services, 1963 and 1967, Doc. No. 140, Center for Advanced Computation, University of Illinois, Urbana, 1974.
6. **Lockeretz, W.,** Energy inputs for nitrogen, phosphorous, and potash fertilizers, in *Handbook of Energy Utilization in Agriculture,* Pimentel D., Ed., CRC Press, Boca Raton, Fla., 1980.
7. **Gordon, M. I., Ed.,** *National Farm Tractor and Implement Blue Book,* Vol. 36, National Market Reports, Chicago, 1975.
8. **Pimentel, D.,** Energy inputs for the production, formulation, packaging and transport of various pesticides, in *Handbook of Energy Utilization in Agriculture,* Pimentel, D., Ed., CRC Press, Boca Raton, Fla., 1980.

9. **Pimentel, D.,** Energy used for transporting supplies to the farm, in *Handbook of Energy Utilization in Agriculture,* Pimentel, D., Ed., CRC Press, Boca Raton, Fla., 1980.
10. **McDonald, J. L., Ed.,** *Power Farming Technical Annual,* Pacific Publications, Sydney, 1975.
11. Composition of Foods, Agriculture Handbook No. 8, Agricultural Research Service, U.S. Department of Agriculture, Washington, D.C., 1963.
12. Evaluation of Pesticide Supplies and Demand for 1974, 1975, and 1976, Economic Research Service Report No. 300, U.S. Department of Agriculture, Washington, D.C., 1975.
13. Wholesale Prices and Price Indexes Supplement, 1976, for 1975, Bureau of Labor Statistics, U.S. Department of Labor, U.S. Government Printing Office, Washington, D.C., 1976, 81.
14. **Heid, W. G., Jr. and Leath, M. N.,** U.S. Barley Industry, Commodity Economics Division of Economics, Statistics, and Cooperatives Service, U.S. Department of Agriculture. Rep. No. 395, 1978.
15. Agricultural Statistics 1977, U.S. Department of Agriculture, U.S. Government Printing Office, Washington, D.C., 1977.
16. **Weaver, J. C.,** *American Barley Production,* Burgess Publishing, Minneapolis, 1950.

ENERGY INPUTS IN CORN PRODUCTION

David Pimentel and Michael Burgess

Corn is one of the 15 crops that account for 90% of the food protein/calories consumed by humans in the world. Total production in the world is estimated at 335 million t/year. The U.S. produces a major share of this, or 158 million t.

The reason that corn and other cereal grains play a dominant role in the food system of humans occurs for several reasons. Cereals produce relatively large quantities of food nutrients per unit land area and can be cultured under a wide range of environmental conditions, including soil types, moisture levels, and temperatures. Compared to other crops, corn and many other cereal grains are relatively free from pest attack. Furthermore, cereals, with their low moisture (13 to 20%) at harvest, can be more efficiently transported than many other high-moisture crops like the potato which is about 80% water. Corn is nutritious for humans and livestock, contributing not only calories and protein but substantial quantities of iron, thiamine, and riboflavin. Also, corn can be effectively stored for long periods of time with minimal facilities.

Energy inputs in corn production vary with the method of production, i.e., by hand, with assistance of animals, and with full mechanization. Consider first, hand production in Mexico where the swidden or slash/burn technology required only the use of an ax and a hoe. The total input of manpower per hectare was 1144 hr/year (Table 1). The energy input for the manpower that includes food, clothes, heating, and miscellaneous living costs for the farmer was not counted. Hence, the total energy input to produce a hectare of corn in Mexico using only manpower was 52,762 kcal. Based on a yearly corn yield per hectare of 1944 kg, the energy potential of the corn was 6.8 million kcal. Hence, the output/input ratio was about 128.2 (Table 1).

When draft animals were used in Mexico, the number of man-hours necessary to raise a hectare of corn was reduced to about 383 hr (Table 2). About 198 hr of oxen power replaced about 761 hr of manpower (Tables 1 and 2). Assuming that an oxen consumes 20,000 kcal/day, the man/oxen combination required more calories to do the same amount of work accomplished by man alone (Tables 1 and 2).

The output/input ratio for producing corn utilizing the man/oxen combination was about 4.25 cal of corn per input calorie. The total input for production was 770,616 kcal/ha, compared to 52,762 kcal for producing corn by hand. Another reason for the reduction in the output/input ratio was the reduced corn yield, which was less than half (941 kg) the yield from producing corn by slash and burn (1944 kg) (Tables 1 and 2). The 941 kg yield was obtained on bottomland that was cleared and had been planted with corn for some time, so the fertility of the soil was lower than in the swidden corn production. If leaves and other organic matter had been carried onto the land, the corn yield would have been greater, and, of course, the man/oxen input would have also been greater.

Producing corn in the U.S. in 1975 required an average input of about 7.3 million kcal or the equivalent of about 722 ℓ of gasoline equivalents to produce a hectare of corn (Tables 3 to 32). The yield of corn was 5160 kg/ha or 18.0 million kcal of corn energy. This result provides an energy output/input ratio of approximately 2.47.

The manpower input in this type of farming is 11 hr. Man-hours were translated into fossil energy equivalents based on the fact that the per capita utilization of energy in the U.S. is 88 million kcal/year. Eleven hours represents 0.5% of the total annual labor output of man (2080 hr). Hence, 0.5% of 88 million kcal is approximately 465,400 kcal input for labor, if this input were to be included.

Without the labor input, the ratio based on corn output to fossil fuel input is as mentioned, about 2.47 kcal of corn per fossil kcal input. This 2.47 ratio is much less than the 128.2:1 ratio of corn output per energy input in hand-produced corn (Tables 1 and 3). The fossil energy inputs in U.S. corn production are primarily petroleum and natural gas.

Note that the nitrogen fertilizer input represented the largest single input or about 19% of the total energy input in U.S. corn production (Tables 3 to 32). Both the machinery and fuel represented about 50% of the fossil energy input into the system. Taken as a whole, about half of the energy inputs in U.S. corn production were for reducing labor and about half were for increasing corn productivity through the use of such inputs as fertilizers.

Table 1
ENERGY INPUTS AND RETURN IN HAND-POWERED CORN PRODUCTION IN MEXICO

Item	Quantity/ha	kcal/ha
	Input	
Labor	1144 hr	—
Ax and hoe	0.8 kg	16,570
Seeds	10.4 kg	36,192
Total		**52,762**
	Output	
Corn yield (1944 kg)		6,765,120
Output/input ratio		128.2

Table 2
ENERGY INPUTS AND RETURN IN MEXICAN CORN PRODUCTION USING OXEN

Item	Quantity/ha	kcal/ha
	Input	
Labor	383 hr	—
Oxen	198 hr	693,000
Machinery	2 kg	41,424
Seeds	10.4 kg	36,192
Total		**770,616**
	Output	
Corn yield (941 kg)		3,274,680
Output/input ratio		4.25

Table 3
ENERGY INPUTS PER HECTARE FOR IRRIGATED CORN PRODUCTION IN CALIFORNIA

Item	Quantity/ha	kcal/ha	Ref.
	Input		
Labor	31.30 hr	—	3
Machinery	55 kg	990,000	8, 2
Gasoline	65.06 ℓ	657,692	3, 1
Diesel	103.54 ℓ	1,181,806	3, 1
Nitrogen	168.21 kg	2,018,520	3, 5
Phosphorus	35.89 kg	107,670	3, 5
Potassium	24.68 kg	39,488	3, 5
Seeds	22.43 kg	506,750	3, 4
Irrigation	93.98 cm	475,591	3
Insecticides	8.37 kg	727,437	3, 6, 11
Herbicides	4.57 kg	456,589	3, 6, 11
Drying[a]	2,199.02 kg	—	3
Transportation	214.23 kg[b]	55,057	7
Total		**7,216,600**	
	Output		
Total yield	7321.55 kg	25,503,068	3, 10
Protein yield	652.08 kg	2,608,320	10
kcal output/kcal input		3.53	
kcal output/labor hour		814,795	

[a] High-speed to 14% water (−90°C) from 24% wb.
[b] Fuels, seed, and machinery.

Table 4
ENERGY INPUTS PER HECTARE FOR IRRIGATED CORN PRODUCTION IN COLORADO

Item	Quantity/ha	kcal/ha	Ref.
	Input		
Labor	8.47 hr	—	3
Machinery	55 kg	990,000	8, 2
Gasoline	42.85 ℓ	433,171	3, 1
Diesel	59.54 ℓ	679,590	3, 1
LP gas	8.52 ℓ	65,647	3, 1
Nitrogen	125.60 kg	1,507,200	3, 5
Phosphorus	31.96 kg	95,880	3, 5
Potassium	4.72 kg	7,552	3, 5
Seeds	21.54 kg	528,500	3, 4
Irrigation	60.96 cm	823,597	3
Electricity	2,889.90 kWh	8,273,784	3, 1
Insecticides	4.37 kg	379,797	3, 6, 11
Herbicides	1.61 kg	160,855	3, 6, 11
Drying	858.97 kg	—	3
Transportation	164.31 kg[a]	42,228	7
Total		**13,997,801**	
	Output		
Total yield	5721.60 kg	19,929,996	3, 10
Protein yield	509.56 kg	2,038,240	10
kcal output/kcal input		1.42	
kcal output/labor hour		2,353,010	

[a] Fuels, seed, and machinery.

Table 5
ENERGY INPUTS PER HECTARE FOR IRRIGATED CORN PRODUCTION IN TEXAS

Item	Quantity/ha	kcal/ha	Ref.
	Input		
Labor	23.44 hr	—	3
Machinery	55 kg	990,000	8, 2
Gasoline	43.53 ℓ	440,045	3, 1
Diesel	118.05 ℓ	1,347,423	3, 1
Electricity	342.84 kWh	981,551	3, 1
Nitrogen	121.10 kg	1,453,200	3, 5
Phosphorus	44.86 kg	134,580	3, 5
Potassium	4.50 kg	7200	3, 5
Seeds	17.93 kg	448,250	3, 4
Irrigation	71.12 cm	883,833	3
Insecticides	26.36 kg	2,290,948	3, 6, 11
Herbicides	0.49 kg	48,956	3, 6, 11
Drying	7811.99 kg	—	3
Transportation	205.92 kg[a]	52,921	7
Natural gas	2167.06 m^3	25,599,479	3, 1
Total		**34,678,386**	
	Output		
Total yield	7804.61 kg	27,169,864	3, 10
Protein yield	695.06 kg	2,780,240	10
kcal output/kcal input		0.78	
kcal output/labor hour		1,159,124	

[a] Fuels, seed, and machinery.

Table 6
ENERGY INPUTS PER HECTARE FOR IRRIGATED CORN PRODUCTION IN NEBRASKA

Item	Quantity/ha	kcal/ha	Ref.
	Input		
Labor	8.87 hr	—	3
Machinery	55 kg	990,000	8, 2
Gasoline	26.49 ℓ	267,787	3, 1
Natural gas	1.40 m^3	16,538	3, 1
Electricity	303.81 kWh	869,808	3, 1
Nitrogen	192.88 kg	2,314,560	3, 5
Phosphorus	52.71 kg	158,130	3, 5
Potassium	45.97 kg	73,552	3, 5
Seeds	20.18 kg	504,500	3, 4
Irrigation	50.8 cm	472,237	3
Diesel	86.84 ℓ	7,839,592	3, 1
Insecticides	5.06 kg	439,965	3, 6, 11
Herbicides	1.75 kg	174,843	3, 6, 11
Drying	375.66 kg	—	3
Transportation	168.86 kg[a]	49,397	7
Total		**14,170,909**	
	Output		
Total yield	7503.46 kg	26,136,553	3, 10
Protein yield	668.38 kg	2,673,520	10
kcal output/kcal input		1.84	
kcal output/labor hour		2,946,624	

[a] Fuel, seed, and machinery.

Table 7
ENERGY INPUTS PER HECTARE FOR IRRIGATED CORN IN KANSAS

Item	Quantity/ha	kcal/ha	Ref.
	Input		
Labor	10.35 hr	—	3
Machinery	55 kg	990,000	8, 2
Gasoline	29.11 ℓ	294,273	3, 1
Diesel	86.31 ℓ	985,142	3, 1
Electricity	126.96 kWh	363,486	3, 1
Nitrogen	185.03 kg	2,220,360	3, 5
Phosphorus	39.25 kg	117,750	3, 5
Potassium	11.24 kg	17,984	3, 5
Seeds	20.21 kg	505,250	3, 4
Irrigation	53.34 cm	1,023,491	3
Natural gas	1138.09 m^3	13,444,257	3, 1
Insecticides	2.45 kg	212,930	3, 6, 11
Herbicides	2.17 kg	216,805	3, 6, 11
Drying	4791.43 kg		3
Transportation	170.41 kg[a]	43,795	7
Total		**20,435,523**	
	Output		
Total yield	6838.44 kg	23,820,264	3, 10
Protein yield	609.10 kg	2,436,400	10
kcal output/kcal input		1.17	
kcal output/labor hour		2,301,475	

[a] Fuel, seed, and machinery.

Table 8
ENERGY INPUTS PER HECTARE FOR CORN PRODUCTION IN ILLINOIS

Item	Quantity/ha	kcal/ha	Ref.
	Input		
Labor	7.88 hr	—	3
Machinery	55 kg	990,000	8, 2
Gasoline	26.12 ℓ	264,047	3, 1
Diesel	77.23 ℓ	881,503	3, 1
LP gas	44.28 ℓ	341,177	3, 1
Electricity	33.35 kWh	95,481	3, 1
Nitrogen	140.17 kg	1,682,040	3, 5
Phosphorus	72.89 kg	218,670	3, 5
Potassium	84.10 kg	134,560	3, 5
Lime	426.12 kg	134,420	3, 9
Seeds	17.83 kg	445,750	3, 4
Insecticides	1.38 kg	119,936	3, 6, 11
Herbicides	7.78 kg	777,300	3, 6, 11
Drying	4006.69 kg	—	3
Transportation	184.19 kg[a]	47,337	7
Total		**6,132,221**	
	Output		
Total yield	8005.32 kg	27,884,830	3, 10
Protein yield	713.09 kg	2,852,360	10
kcal output/kcal input		4.55	
kcal output/labor hour		3,538,684	

[a] Fuel, seed, and machinery.

Table 9
ENERGY INPUTS PER HECTARE FOR CORN PRODUCTION IN INDIANA

Item	Quantity/ha	kcal/ha	Ref.
	Input		
Labor	8.05 hr	—	3
Machinery	55 kg	990,000	8, 2
Gasoline	26.96 ℓ	272,539	3, 1
Diesel	78.45 ℓ	895,428	3, 1
LP gas	34.54 ℓ	266,131	3, 1
Electricity	31.62 kWh	90,528	3, 1
Nitrogen	120.13 kg	1,441,560	3, 5
Phosphorus	65.04 kg	195,120	3, 5
Potassium	95.32 kg	152,512	3, 5
Lime	354.35 kg	111,780	3, 9
Seeds	23.79 kg	594,750	3, 4
Insecticides	2.47 kg	214,668	3, 6, 11
Herbicides	5.14 kg	513,537	3, 6, 11
Drying	3124.16 kg	—	3
Transportation	186.04 kg[a]	47,820	10
Total		**5,786,373**	
	Output		
Total yield	6242.56 kg	21,744,629	3, 10
Protein yield	556.00 kg	2,224,000	10
kcal output/kcal input		3.76	
kcal output/labor hour		2,701,196	

[a] Fuel, seed, and machinery.

Table 10
ENERGY INPUTS PER HECTARE FOR CORN PRODUCTION IN IOWA

Item	Quantity/ha	kcal/ha	Ref.
	Input		
Labor	9.63 hr	—	3
Machinery	55 kg	990,000	8, 2
Gasoline	96.70 ℓ	977,540	3, 1
Diesel	26.68 ℓ	304,526	3, 1
LP gas	21.53 ℓ	165,889	3, 1
Electricity	27.91 kWh	79,906	3, 1
Nitrogen	106.53 kg	1,278,360	3, 5
Phosphorus	61.68 kg	185,040	3, 5
Potassium	67.28 kg	107,648	3, 5
Lime	632.47 kg	199,513	3, 9
Seeds	18.50 kg	462,500	3, 4
Insecticides	2.15 kg	186,857	3, 6, 11
Herbicides	7.93 kg	792,286	3, 6, 11
Drying	2467.04 kg	—	3
Transportation	181.41 kg[a]	46,622	7
Total		**5,776,687**	
	Output		
Total yield	6160.77 kg	21,459,737	3, 10
Protein yield	548.83 kg	2,195,320	10
kcal output/kcal yield		3.72	
kcal output/labor hour		2,228,425	

[a] Fuel, seed, and machinery.

Table 11
ENERGY INPUTS PER HECTARE FOR CORN PRODUCTION IN MISSOURI

Item	Quantity/ha	kcal/ha	Ref.
	Input		
Labor	9.16 hr	—	3
Machinery	55 kg	990,000	8, 2
Gasoline	49.90 ℓ	504,439	3, 1
Diesel	67.12 ℓ	766,108	3, 1
LP gas	49.15 ℓ	378,701	3, 1
Electricity	30.13 kWh	86,262	3, 1
Nitrogen	140.17 kg	1,682,040	3, 5
Phosphorus	44.86 kg	134,580	3, 5
Potassium	67.28 kg	107,648	3, 5
Lime	650.40 kg	205,169	3, 9
Seeds	15.71 kg	392,750	3, 4
Insecticides	0.41 kg	36,502	3, 6, 11
Herbicides	6.10 kg	609,451	3, 6, 11
Drying	2973.91 kg	—	3
Transportation	194.19 kg[a]	49,907	7
Total		**5,943,557**	
	Output		
Total yield	5941.31 kg	20,695,309	3, 10
Protein yield	529.07 kg	2,116,280	10
kcal output/kcal input		3.48	
kcal output/labor hour		2,259,313	

[a] Fuel, seed, and machinery.

Table 12
ENERGY INPUTS PER HECTARE FOR CORN PRODUCTION IN OHIO

Item	Quantity/ha	kcal/ha	Ref.
	Input		
Labor	7.14 hr	—	3
Machinery	55 kg	990,000	8, 2
Gasoline	55.70 ℓ	563,071	3, 1
Diesel	36.79 ℓ	41,921	3, 1
LP gas	32.77 ℓ	252,493	3, 1
Electricity	32.60 kWh	93,334	3, 1
Nitrogen	112.14 kg	1,345,680	3, 5
Phosphorus	72.89 kg	218,670	3, 5
Potassium	77.39 kg	123,824	3, 5
Lime	67.28 kg	21,223	3, 9
Seeds	14.70 kg	367,500	3, 4
Insecticides	2.00 kg	173,820	3, 6, 11
Herbicides	6.89 kg	688,380	3, 6, 11
Drying	2967.16 kg	—	3
Transportation	162.08 kg[a]	41,655	7
Total		**4,921,571**	
	Output		
Total yield	5928.87 kg	20,651,956	3, 10
Protein yield	528.09 kg	2,112,360	10
kcal output/kcal input		4.20	
kcal output/labor hour		2,892,431	

[a] Fuel, seed, and machinery.

Table 13
ENERGY INPUTS PER HECTARE FOR CORN PRODUCTION IN KANSAS

Item	Quantity/ha	kcal/ha	Ref.
	Input		
Labor	7.34 hr	—	3
Machinery	55 kg	990,000	8, 2
Gasoline	26.96 ℓ	272,539	3, 1
Diesel	54.95 ℓ	629,199	3, 1
LP gas	2.06 ℓ[a]	15,872	3, 1
Electricity	126.96 kWh	363,486	3, 1
Nitrogen	110.46 kg	1,325,520	3, 5
Phosphorus	23.54 kg	70,620	3, 5
Potassium	10.10 kg	16,160	3, 5
Lime	78.50 kg	24,763	3, 9
Seeds	13.68 kg	342,000	3, 4
Insecticides	2.32 kg	201,631	3, 6, 11
Herbicides	1.41 kg	140,873	3, 6, 11
Drying	1107.92 kg	—	3
Transportation	136.82 kg[a]	35,163	7
Total		**4,427,825**	
	Output		
Total yield	2992.63 kg	10,424,186	3, 10
Protein yield	266.51 kg	1,066,040	10
kcal output/kcal input		2.35	
kcal output/labor hour		1,420,189	

[a] Fuel, seed, and machinery.

Table 14
ENERGY INPUTS PER HECTARE FOR CORN PRODUCTION IN NEBRASKA

Item	Quantity/ha	kcal/ha	Ref.
	Input		
Labor	8.42 hr	—	3
Machinery	55 kg	990,000	8, 2
Gasoline	41.09 ℓ	415,379	3, 1
Diesel	0.09 ℓ	1027	3, 1
LP gas	53.92 ℓ	415,454	3, 1
Electricity	303.81 kWh	869,808	3, 1
Nitrogen	84.10 kg	1,009,200	3, 5
Phosphorus	16.82 kg	50,460	3, 5
Potassium	7.50 kg	12,016	3, 5
Lime	11.21 kg	3536	3, 9
Seeds	12.33 kg	308,250	3, 4
Insecticides	2.10 kg	182,511	3, 6, 11
Herbicides	6.72 kg	671,395	3, 6, 11
Drying	165.96 kg	—	3
Transportation	130.04 kg[a]	33,420	7
Total		**4,962,456**	
	Output		
Total yield	3318.76 kg	11,560,211	3, 10
Protein yield	295.66 kg	1,182,640	10
kcal output/kcal input		2.33	
kcal output/labor hour		1,372,947	

[a] Fuel, seed, and machinery.

Table 15
ENERGY INPUTS PER HECTARE FOR CORN PRODUCTION IN NORTH DAKOTA

Item	Quantity/ha	kcal/ha	Ref.
	Input		
Labor	5.20 hr	—	3
Machinery	55 kg	990,000	8, 2
Gasoline	19.09 ℓ	192,981	3, 1
Diesel	54.20 ℓ	618,639	3, 1
LP gas	1.22 ℓ	9400	3, 1
Electricity	16.55 kWh	47,383	3, 1
Nitrogen	130.54 kg	1,566,480	3, 5
Phosphorus	5.25 kg	153,750	3, 5
Potassium	16.60 kg	26,560	3, 5
Seeds	15.71 kg	392,750	3, 4
Insecticides	0.17 kg	14,775	3, 6, 11
Herbicides	4.25 kg	424,618	3, 6, 11
Drying	126.71 kg	—	3
Transportation	131.82 kg[a]	33,878	7
Total		**4,471,214**	
	Output		
Total yield	2541.02 kg	8,851,091	3, 10
Protein yield	226.25 kg	905,000	10
kcal output/kcal input		1.98	
kcal output/labor hour		1,698,866	

[a] Fuel, seed, and machinery.

Table 16
ENERGY INPUTS PER HECTARE FOR CORN PRODUCTION IN SOUTH DAKOTA

Item	Quantity/ha	kcal/ha	Ref.
	Input		
Labor	8.45 hr	—	3
Machinery	55 kg	990,000	8, 2
Gasoline	25.28 ℓ	225,556	3, 1
Diesel	90.52 ℓ	1,033,195	3, 1
LP gas	1.12 ℓ	8630	3, 1
Electricity	50.88 kWh	145,669	3, 1
Nitrogen	49.35 kg	592,200	3, 5
Phosphorus	23.54 kg	70,620	3, 5
Potassium	14.57 kg	23,312	3, 5
Lime	26.92 kg	8492	3, 9
Seeds	12.33 kg	308,250	3, 4
Insecticides	1.61 kg	139,925	3, 6, 11
Herbicides	2.96 kg	295,734	3, 6, 11
Drying	125.71 kg	—	3
Transportation	163.89 kg[a]	42,120	7
Total		**3,913,703**	
	Output		
Total yield	2534.67 kg	8,828,972	3, 10
Protein yield	225.76 kg	903,040	10
kcal output/kcal input		2.26	
kcal output/labor hour		1,044,849	

[a] Fuel, seed, and machinery.

Table 17
ENERGY INPUTS PER HECTARE FOR CORN PRODUCTION IN DELAWARE

Item	Quantity/ha	kcal/ha	Ref.
	Input		
Labor	13.22 hr	—	3
Machinery	55 kg	990,000	8, 2
Gasoline	69.93 ℓ	706,922	3, 1
Diesel	64.97 ℓ	741,568	3, 1
Electricity	12.72 kWh	36,417	3, 1
Nitrogen	116.86 kg	1,402,320	3, 5
Phosphorus	66.94 kg	200,820	3, 5
Potassium	88.13 kg	141,008	3, 5
Lime	381.27 kg	120,272	3, 9
Seeds	16.53 kg	413,250	3, 4
Insecticides	0.35 kg	30,419	3, 6, 11
Herbicides	3.80 kg	379,658	3, 6, 11
Drying	3697.20 kg	—	3
Transportation	179.20 kg[a]	46,054	7
Total		**5,208,708**	
	Output		
Total yield	5677.92 kg	19,777,817	3, 10
Protein yield	505.69 kg	2,022,760	10
kcal output/kcal input		3.80	
kcal output/labor hour		1,496,053	

[a] Fuel, seed, and machinery.

Table 18
ENERGY INPUTS PER HECTARE FOR CORN PRODUCTION IN SOUTHERN MARYLAND

Item	Quantity/ha	kcal/ha	Ref.
	Input		
Labor	13.66 hr	—	3
Machinery	55 kg	990,000	8, 2
Gasoline	78.17 ℓ	790,221	3, 1
Diesel	58.70 ℓ	670,002	3, 1
Electricity	22.97 kWh	65,763	3, 1
Nitrogen	110.46 kg	1,325,520	3, 5
Phosphorus	87.69 kg	263,070	3, 5
Potassium	87.59 kg	140,144	3, 5
Lime	358.84 kg	113,196	3, 9
Seeds	16.25 kg	406,250	3, 4
Insecticides	0.59 kg	51,277	3, 6, 11
Herbicides	3.80 kg	379,658	3, 6, 11
Drying	4326.28 kg	—	3
Transportation	179.77 kg[a]	46,201	7
Total		**5,241,302**	
	Output		
Total yield	5765.55 kg	20,083,058	3, 10
Protein yield	513.51 kg	2,054,040	10
kcal output/kcal input		3.83	
kcal output/labor hour		1,470,209	

[a] Fuel, seed, and machinery.

Table 19
ENERGY INPUTS PER HECTARE FOR CORN PRODUCTION IN CENTRAL NEW JERSEY

Item	Quantity/ha	kcal/ha	Ref.
	Input		
Labor	13.49 hr	—	3
Machinery	55 kg	990,000	8, 2
Gasoline	68.99 ℓ	697,420	3, 1
Diesel	64.41 ℓ	735,175	3, 1
Nitrogen	95.32 kg	1,143,840	3, 5
Phosphorus	66.17 kg	198,510	3, 5
Potassium	63.92 kg	102,272	3, 5
Lime	426.12 kg	134,420	3, 9
Seeds	16.25 kg	406,250	3, 4
Insecticides	0.74 kg	64,313	3, 6, 11
Herbicides	1.14 kg	139,874	3, 6, 11
Drying	3,139.86 kg	—	3
Transportation	177.74 kg[a]	45,679	7
Total		**4,657,753**	
	Output		
Total yield	5207.25 kg	18,138,366	3, 10
Protein yield	463.62 kg	1,854,480	10
kcal output/kcal input		3.89	
kcal output/labor hour		1,344,579	

[a] Fuel, seed, and machinery.

Table 20
ENERGY INPUTS PER HECTARE FOR CORN PRODUCTION IN SOUTHERN NEW YORK

Item	Quantity/ha	kcal/ha	Ref.
	Input		
Labor	13.31 hr	—	
Machinery	55 kg	990,000	8, 2
Gasoline	63.38 ℓ	640,708	3, 1
Diesel	67.59 ℓ	771,472	3, 1
Electricity	11.12 kWh	31,836	3
Nitrogen	72.89 kg	874,680	3, 5
Phosphorus	59.43 kg	178,290	3, 5
Potassium	67.28 kg	107,648	3, 5
Lime	291.56 kg	91,973	3, 9
Seeds	16.25 kg	406,250	3, 4
Insecticides	0.62 kg	53,884	3, 6, 11
Herbicides	1.90 kg	189,829	3, 6, 11
Drying	2724.95 kg	—	3
Transportation	176.24 kg[a]	45,294	7
Total		**4,381,864**	
	Output		
Total yield	5295.14 kg	18,444,492	3, 10
Protein yield	471.52 kg	1,886,080	10
kcal output/kcal input		4.21	
kcal output/labor hour		1,385,762	

[a] Fuel, seed, and machinery.

Table 21
ENERGY INPUTS PER HECTARE FOR CORN PRODUCTION IN SOUTHEASTERN PENNSYLVANIA

Item	Quantity/ha	kcal/ha	Ref.
	Input		
Labor	13.31 hr	—	3
Machinery	55 kg	990,000	8, 2
Gasoline	57.29 ℓ	579,145	3, 1
Diesel	72.18 ℓ	823,863	3, 1
Electricity	24.70 kWh	70,716	3, 1
Nitrogen	88.03 kg	1,056,360	3, 5
Phosphorus	64.15 kg	192,450	3, 5
Potassium	50.68 kg	81,088	3, 5
Lime	381.27 kg	120,272	3, 9
Seeds	16.53 kg	413,250	3, 4
Insecticides	1.04 kg	90,386	3, 6, 11
Herbicides	1.90 kg	189,829	3, 6, 11
Drying	2057.73 kg	—	3
Transportation	175.51 kg[a]	45,106	7
Total		**4,652,465**	
	Output		
Total yield	5138.17 kg	17,897,713	3, 10
Protein yield	457.69 kg	1,830,760	10
kcal output/kcal input		3.85	
kcal output/labor hour		1,344,682	

[a] Fuel, seed, and machinery.

Table 22
ENERGY INPUTS PER HECTARE FOR CORN PRODUCTION IN KENTUCKY

Item	Quantity/ha	kcal/ha	Ref.
	Input		
Labor	9.29 hr	—	3
Machinery	55 kg	990,000	8, 2
Gasoline	58.60 ℓ	592,387	3, 1
Diesel	54.48 ℓ	621,835	3, 1
LP gas	28.18 ℓ	217,127	3, 1
Electricity	24.21 kWh	69,313	3, 1
Nitrogen	106.53 kg	1,278,360	3, 5
Phosphorus	67.28 kg	201,840	3, 5
Potassium	69.53 kg	111,248	3, 5
Lime	67.28 kg	21,223	3, 9
Seeds	13.56 kg	339,000	3, 4
Herbicides	5.21 kg	520,531	3, 6, 11
Drying	1345.66 kg	—	3
Transportation	175.44 kg[a]	45,088	7
Total		**5,007,952**	
	Output		
Total yield	5376.67 kg	18,728,498	3, 10
Protein yield	478.93 kg	1,915,720	10
kcal output/kcal input		3.74	
kcal output/labor hour		2,015,985	

[a] Fuel, seed, and machinery.

Table 23
ENERGY INPUTS PER HECTARE FOR CORN PRODUCTION IN NORTH CAROLINA

Item	Quantity/ha	kcal/ha	Ref.
	Input		
Labor	10.47 hr	—	3
Machinery	55 kg	990,000	8, 2
Gasoline	48.59 ℓ	491,196	3, 1
Diesel	62.63 ℓ	714,859	3, 1
LP gas	55.42 ℓ	427,011	3, 1
Electricity	28.41 kWh	81,338	3, 1
Nitrogen	15.36 kg	184,320	3, 5
Phosphorus	70.64 kg	211,920	3, 5
Potassium	89.71 kg	143,536	3, 5
Lime	897.10 kg	282,990	3, 9
Seeds	15.36 kg	384,000	3, 4
Insecticides	7.51 kg	652,694	3, 6, 11
Herbicides	9.56 kg	955,140	3, 6, 11
Drying	2939.13 kg	—	3
Transportation	192.74 kg[a]	49,534	7
Total		**5,568,538**	
	Output		
Total yield	4736.85 kg	16,499,800	3, 10
Protein yield	421.88 kg	1,687,520	10
kcal output/kcal input		2.96	
kcal output/labor hour		1,575,912	

[a] Fuel, seed, and machinery.

Table 24
ENERGY INPUTS PER HECTARE FOR CORN PRODUCTION IN VIRGINIA

Item	Quantity/ha	kcal/ha	Ref.
	Input		
Labor	10.46 hr	—	3
Machinery	55 kg	990,000	8, 2
Gasoline	68.24 ℓ	689,838	3, 1
Diesel	37.26 ℓ	425,286	3, 1
LP gas	62.90 ℓ	484,645	3, 1
Electricity	21.74 kWh	62,242	3, 1
Nitrogen	89.36 kg	1,072,320	3, 5
Phosphorus	110.68 kg	332,040	3, 5
Potassium	112.14 kg	179,424	3, 5
Lime	986.81 kg	311,289	3, 9
Seeds	17.58 kg	439,500	3, 4
Insecticides	0.57 kg	49,539	3, 6, 11
Herbicides	8.68 kg	867,219	3, 6, 11
Drying	3333.12 kg	—	3
Transportation	192.54 kg[a]	49,483	7
Total		**5,952,825**	
	Output		
Total yield	5376.67 kg	18,728,498	3, 10
Protein yield	478.93 kg	1,915,720	10
kcal output/kcal input		3.15	
kcal output/labor hour		1,790,487	

[a] Fuel, seed, and machinery.

Table 25
ENERGY INPUTS PER HECTARE FOR CORN PRODUCTION IN TENNESSEE

Item	Quantity/ha	kcal/ha	Ref.
	Input		
Labor	11.14 hr	—	3
Machinery	55 kg	990,000	8, 2
Gasoline	51.96 ℓ	525,264	3, 1
Diesel	101.01 ℓ	1,152,928	3, 1
LP gas	20.97 ℓ	161,574	3, 1
Electricity	21.74 kWh	62,242	3, 1
Nitrogen	151.39 kg	1,816,680	3, 5
Phosphorus	77.39 kg	232,170	3, 5
Potassium	77.39 kg	123,824	3, 5
Lime	246.70 kg	77,822	3, 9
Seeds	16.03 kg	400,750	3, 4
Insecticides	0.49 kg	42,586	3, 6, 11
Herbicides	3.26 kg	325,707	3, 6, 11
Drying	1001.39 kg	—	3
Transportation	208.23 kg[a]	53,515	7
Total		**5,965,062**	
	Output		
Total yield	3902.20 kg	13,592,494	3, 10
Protein yield	347.53 kg	1,390,120	10
kcal output/kcal input		2.28	
kcal output/labor hour		1,220,152	

[a] Fuel, seed, and machinery.

Table 26
ENERGY INPUTS PER HECTARE FOR CORN PRODUCTION IN ALABAMA

Item	Quantity/ha	kcal/ha	Ref.
	Input		
Labor	11.93 hr	—	3
Machinery	55 kg	990,000	8, 2
Gasoline	69.37 ℓ	701,261	3, 1
Diesel	68.81 ℓ	785,397	3, 1
LP gas	18.44 ℓ	142,080	3, 1
Electricity	19.02 kWh	54,454	3, 1
Nitrogen	123.35 kg	1,488,200	3, 5
Phosphorus	52.10 kg	156,300	3, 5
Potassium	57.18 kg	91,488	3, 5
Lime	134.56 kg	42,447	3, 9
Seeds	12.32 kg	308,000	3, 4
Insecticides	1.16 kg	100,816	3, 6, 11
Herbicides	1.21 kg	120,891	3, 6, 11
Drying	976.76 kg	—	3
Transportation	188.72 kg[a]	48,501	7
Total		**5,029,835**	
	Output		
Total yield	3902.20 kg	13,592,494	3, 10
Protein yield	347.53 kg	1,390,120	10
kcal output/kcal input		2.70	
kcal output/labor hour		1,139,354	

[a] Fuel, seed, and machinery.

Table 27
ENERGY INPUTS PER HECTARE FOR CORN PRODUCTION IN GEORGIA

Item	Quantity/ha	kcal/ha	Ref.
Input			
Labor	13.71 hr	—	3
Machinery	55 kg	990,000	8, 2
Gasoline	84.25 ℓ	209,795	3, 1
Diesel	67.68 ℓ	851,683	3, 1
LP gas	35.95 ℓ	772,500	3, 1
Electricity	22.23 kWh	63,644	3, 5
Nitrogen	136.81 kg	1,641,720	3, 5
Phosphorus	53.82 kg	161,460	3, 5
Potassium	76.49 kg	122,384	3, 5
Lime	511.35 kg	161,305	3, 9
Seeds	14.35 kg	358,750	3, 4
Insecticides	1.81 kg	157,307	3, 6, 11
Herbicides	2.92 kg	291,737	3, 6, 11
Drying	1906.35 kg	—	3
Transportation	211.28 kg[a]	54,299	7
Total		5,836,584	
Output			
Total yield	3463.04 kg	12,062,752	3, 10
Protein yield	308.50 kg	1,234,000	10
kcal output/kcal input		2.07	
kcal output/labor hour		879,850	

[a] Fuel, seed, and machinery.

Table 28
ENERGY INPUTS PER HECTARE FOR CORN PRODUCTION IN FLORIDA

Item	Quantity/ha	kcal/ha	Ref.
Input			
Labor	14.62 hr	—	3
Machinery	55 kg	990,000	8, 2
Gasoline	83.41 ℓ	843,192	3, 1
Diesel	66.84 ℓ	762,912	3, 1
LP gas	28.36 ℓ	218,514	3, 1
Electricity	18.53 kWh	53,051	3, 1
Nitrogen	114.38 kg	1,372,560	3, 5
Phosphorus	49.34 kg	148,020	3, 5
Potassium	71.77 kg	114,832	3, 5
Lime	291.56 kg	91,973	3, 9
Seeds	11.78 kg	294,500	3, 4
Herbicides	3.70 kg	369,667	3, 6, 11
Drying	1506.01 kg	—	3
Transportation	202.88 kg[a]	52,140	7
Total		5,311,361	
Output			
Total yield	2735.33 kg	9,527,928	3, 10
Protein yield	243.54 kg	974,160	10
kcal output/kcal input		1.79	
kcal output/labor hour		651,705	

[a] Fuel, seed, and machinery.

Table 29
ENERGY INPUTS PER HECTARE FOR CORN PRODUCTION IN SOUTH CAROLINA

Item	Quantity/ha	kcal/ha	Ref.
	Input		
Labor	12.80 hr	—	3
Machinery	55 kg	990,000	8, 2
Gasoline	73.67 ℓ	744,730	3, 1
Diesel	61.04 ℓ	696,711	3, 1
LP gas	21.16 ℓ	163,038	3, 1
Electricity	36.56 kWh	104,671	3, 1
Nitrogen	104.51 kg	1,254,120	3, 5
Phosphorus	82.98 kg	248,940	3, 5
Potassium	94.20 kg	150,720	3, 5
Seeds	14.24 kg	356,000	3, 4
Herbicides	4.55 kg	454,591	3, 6, 11
Drying	1123.62 kg	—	3
Transportation	188.86 kg	48,537	
Total		**5,260,595**	
	Output		
Total yield	4491.99 kg	15,646,896	3, 10
Protein yield	400.14 kg	1,600,000	10
kcal output/kcal input		2.97	
kcal output/labor hour		1,222,414	

[a] Fuel, seed, and machinery.

Table 30
ENERGY INPUTS PER HECTARE FOR CORN PRODUCTION IN MICHIGAN

Item	Quantity/ha	kcal/ha	Ref.
	Input		
Labor	8.23 hr	—	3
Machinery	55 kg	990,000	8, 2
Gasoline	61.50 ℓ	621,704	3, 1
Diesel	58.70 ℓ	670,002	3, 1
LP gas	25.65 ℓ	197,633	3, 1
Electricity	31.37 kWh	89,812	3, 1
Nitrogen	82.53 kg	990,360	3, 5
Phosphorus	81.86 kg	245,580	3, 5
Potassium	111.24 kg	177,984	3, 5
Seeds	15.92 kg	398,000	3, 3
Insecticides	0.06 kg	5215	3, 6, 11
Herbicides	0.26 kg	25,977	3, 6, 11
Drying	2110.44 kg	—	3
Transportation	182.07 kg[a]	46,792	7
Total		**4,459,059**	
	Output		
Total yield	5269.99 kg	18,356,901	3, 10
Protein yield	469.30 kg	1,877,200	10
kcal output/kcal input		4.12	
kcal output/labor hour		2,230,486	

[a] Fuel, seed, and machinery.

Table 31
ENERGY INPUTS PER HECTARE FOR CORN PRODUCTION IN MINNESOTA

Item	Quantity/ha	kcal/ha	Ref.
	Input		
Labor	8.94 hr	—	3
Machinery	55 kg	990,000	8, 2
Gasoline	42.41 ℓ	428,723	3, 1
Diesel	69.74 ℓ	796,012	3, 1
LP gas	17.69 ℓ	136,302	3, 1
Electricity	31.37 kWh	89,812	3, 1
Nitrogen	98.68 kg	1,184,160	3, 5
Phosphorus	61.68 kg	185,040	3, 5
Potassium	60.55 kg	96,880	3, 5
Lime	4.49 kg	1416	3, 9
Seeds	16.82 kg	420,500	3, 4
Insecticides	1.35 kg	117,329	3, 6, 11
Herbicides	4.89 kg	488,560	3, 6, 11
Drying	1778.50 kg	—	3
Transportation	173.34 kg[a]	44,548	7
Total		**4,979,282**	
	Output		
Total yield	4441.95 kg	15,472,599	3, 10
Protein yield	395.69 kg	1,582,760	10
kcal output/kcal input		3.11	
kcal output/labor hour		1,730,716	

[a] Fuel, seed, and machinery.

Table 32
ENERGY INPUTS PER HECTARE FOR CORN PRODUCTION IN WISCONSIN

Item	Quantity/ha	kcal/ha	Ref.
	Input		
Labor	7.36 hr	—	3
Machinery	55 kg	990,000	8, 2
Gasoline	41.66 ℓ	421,141	3, 1
Diesel	56.36 ℓ	643,293	3, 1
LP gas	25.18 ℓ	194,102	3, 1
Electricity	35.82 kWh	102,553	3, 1
Nitrogen	84.10 kg	1,009,200	3, 5
Phosphorus	61.68 kg	185,040	3, 5
Potassium	91.95 kg	147,120	3, 5
Seeds	15.81 kg	395,250	3, 4
Insecticides	3.92 kg	340,687	3, 6, 11
Herbicides	4.61 kg	460,585	3, 6, 11
Drying	2070.07 kg	—	3
Transportation	164.82 kg[a]	42,359	7
Total		**4,931,330**	
	Output		
Total yield	5169.66 kg	18,007,423	3, 10
Protein yield	460.41 kg	1,841,640	10
kcal output/kcal input		3.65	
kcal output/labor hour		2,446,661	

[a] Fuel, seed, and machinery.

REFERENCES

1. **Cervinka, V.,** Fuel and energy efficiency, in *Handbook of Energy Utilization in Agriculture,* Pimentel, D., Ed., CRC Press, Boca Raton, Fla., 1980.
2. **Doering, O., III,** Accounting for energy in farm machinery and buildings, in *Handbook of Energy Utilization in Agriculture,* Pimentel, D., Ed., CRC Press, Boca Raton, Fla., 1980.
3. *Firm Enterprise Data System, 1975, U.S.Department of Agriculture, Economic Research Service, and Department of Agriculture, Oklahoma State University, Norman, Oklahoma, 1977.*
4. **Heichel, G.,** Assessing the fossil energy costs of propagating agricultural crops, in *Handbook of Energy Utilization in Agriculture,* Pimentel, D., Ed., CRC Press, Boca Raton, Fla., 1980.
5. **Lockeretz, W.,** Energy inputs for nitrogen, phosphorus, and potash fertilizers, in *Handbook of Energy Utilization in Agriculture,* Pimentel, D., Ed., CRC Press, Boca Raton, Fla., 1980.
6. **Pimentel, D.,** Energy inputs for the production, formulation, packaging and transport of various pesticides, in *Handboodk of Energy Utilization in Agriculture,* Pimentel, D., Ed., CRC Press, Boca Raton, Fla., 1980.
7. **Pimentel, D.,** Energy used for transporting supplies to the farm, in *Handbook of Energy Utilization in Agriculture,* Pimentel, D., Ed., CRC Press, Boca Raton, Fla., 1980.
8. **Pimentel, D. and Pimentel, M.,** *Food, Energy and Society,* Edward Arnold Ltd., London, in press.
9. **Terhune, E.,** Energy used in the U.S. for agricultural liming materials, in *Handbook of Energy Utilization in Agriculture,* Pimentel, D., Ed., CRC Press, Boca Raton, Fla., 1980.
10. Composition of Foods, Agriculture Handbook No. 8, Agricultural Research Service, U.S. Department of Agriculture, 1963.
11. Evaluation of pesticide supplies and demand for 1974, 1975, and 1976, No. 300, Table 5, Economic Research Service, U.S. Department of Agriculture, 1975.

ENERGY USE IN THE PRODUCTION OF OATS

S. H. Weaver

OATS PRODUCTION

World production of oats has been maintained at the 50.91 million metric ton level since the early 1960s. U.S. production has averaged 20% of world production in recent years (Table 1).

The U.S.S.R. is the only key production area which is increasing its share of the world total. This now is estimated to be 32% of the world oat production.

In many areas of the U.S., climatic conditions are not ideal for maximum yields of oats. As an example, Western Europe has a cool, moist growing season with moderate late-summer temperatures and yields are 30 to 40% greater than those of the U.S.

Oats are grown primarily as a feed grain and are best suited for livestock requiring only moderate quantities of calories, but requiring high palatability and a good protein balance. Typical of these classes of livestock are dairy cattle, lambs, and horses. Oats furnish two additional needs for the livestock industry: a cover crop for establishing small-seeded legumes, such as sweet clover or alfalfa, and a source of absorbent bedding (oats straw).

The U.S. milling industry uses 727,000 metric tons of oats annually. This represents 6 to 8% of current U.S. production. The U.S. production of oats for the years 1975 to 1977 by state is in Table 2.

The leading production area in the U.S. is the north central states — Iowa, Minnesota, North Dakota, South Dakota, and Wisconsin. These five states produce 60 to 65% of the oats harvested for grain each year. The oats grown in this area are spring sown.

In many areas of the southern half of the U.S., oats are grown primarily as a pasture or forage crop. These production areas grow primarily winter oats which are planted in September and October, grazed during December to March, and in some instances allowed to produce grain for harvest in May and June. In other instances, these fields may be planted to other crops after grazing ceases. An example is a state such as Texas where over 600,000 ha are planted to oats, but only 30% are harvested for grain.

VALUE OF THE U.S. OATS CROP

The value of the U.S. oats production is calculated to be about $900 million annually. This figure does not include the additional value of the oats crop in the form of straw, forage (hay silage, green-chop), and pasture. A detailed study of these additional values of the 1975 oats crop was conducted by the Quaker Oats Company. The results of this study are in Table 3.

Assuming the percentage of use to be consistent, the total value of the U.S. oats crop is comprised of approximately 60% grain and 40% pasture, forage, and straw. Obviously, the purpose for which the crop is grown will greatly affect the energy input requirements.

ENERGY REQUIREMENTS

The energy inputs for oats production in westcentral Minnesota, southeast South Dakota, the blacklands area of Texas, and the southeast U.S. are in Tables 4, 5, 6, and 7, respectively. The data are based on the assumption that all of the area of a given farm is planted with oats. In the U.S., oats are rarely, if ever, the only crop a farmer would produce. They are generally grown as an integral part of the overall farming system.

The area required to grow oats as the sole crop on an individual farm was determined by the size of the machinery currently being used, the productive capacity of the soil, and economic factors. The economic factors include prices received for the crop, either as grain or livestock, and energy requirements for a given area.

Tables 8, 9, 10, and 11 detail the machinery budgets for the four areas concerned. The total energy requirements for machinery can be varied to fit a particular cropping system. For example, if a farmer in southeast South Dakota harvests his oats for grain only, then there is no need to include the energy requirements for a baler. Since most of the oats in Texas and the southeast U.S. are used as winter pasture, the energy requirements for combines, grain trailers, and grain trucks can be eliminated. There are no figures shown for irrigation as essentially none are irrigated, and the energy inputs relative to outputs would be far too excessive for profitable oats production.

Table 1
U.S. AND WORLD OATS PRODUCTION

Year	U.S. (Million t)	World (Million t)	U.S. as percent of world
1962	14.7	51.0	29
1963	14.0	50.8	28
1964	12.4	47.9	26
1965	13.4	46.7	29
1966	11.6	48.8	24
1967	11.4	51.2	22
1968	13.6	54.7	25
1969	13.8	55.6	25
1970	13.3	55.1	24
1971	12.8	57.5	22
1972	10.0	50.8	20
1973	9.7	55.0	18
1974	9.0	51.0	18
1975	9.3	46.7	20
1976	7.9	48.7	17
1977	10.8	49.4	22

Table 2
U.S. OATS PRODUCTION 1975 TO 1977 (IN 1000 t)[1]

State	1975	1976	1977
Alabama	15.1	13.4	14.9
Arkansas	33.9	62.2	50.8
California	69.2	71.1	76.9
Colorado	28.6	34.1	20.7
Florida	7.1	8.7	7.8
Georgia	38.2	46.1	39.9
Idaho	36.8	34.9	37.2
Illinois	361.3	325.1	300.7
Indiana	179.4	147.9	115.3
Iowa	1095.1	1197.7	1176.3
Kansas	63.2	121.8	137.0
Kentucky	5.9	5.1	4.7
Louisiana	3.2	6.8	5.5
Maine	22.5	20.3	21.8
Maryland	17.1	17.5	16.9
Michigan	300.4	284.7	271.2
Minnesota	1508.0	1374.0	2346.7
Mississippi	9.9	7.9	8.5
Missouri	48.1	78.9	105.1
Montana	108.5	109.3	81.2
Nebraska	419.2	401.9	563.5
Nevada	2.4	2.1	3.2
New Jersey	4.3	5.9	5.6
New York	295.8	251.2	222.9
North Carolina	59.2	54.5	45.7
North Dakota	814.5	650.2	870.0
Ohio	447.6	413.3	359.3
Oklahoma	46.8	68.5	86.7
Oregon	53.6	55.8	75.4
Pennsylvania	288.2	262.5	269.0
South Carolina	42.6	42.4	36.7
South Dakota	1422.7	617.7	1918.4
Tennessee	8.9	16.3	15.6
Texas	282.8	209.2	348.0
Utah	10.6	9.9	8.0
Virginia	16.3	21.6	21.7
Washington	27.4	28.4	21.8
West Virginia	8.9	9.1	7.1
Wisconsin	1076.6	798.1	1102.7
Wyoming	29.7	35.4	24.8
Total	**9309.6**	**7921.6**	**10844.8**

[a] Annual Crop Summary, Crop Reporting Board, U.S. Department of Agriculture — 1978.

Table 3
VALUE OF 1975 U.S. OATS CROP, INCLUDING GRAIN, STRAW, FORAGE, AND PASTURE

Use of the crop	Value (million $)	Percent of total
Grain	940	60
Straw	269	17
Forage	89	6
Pasture	256	17
Total	**1554**	**100**

Table 4
ENERGY INPUT AND OUTPUT PER HECTARE FOR OATS IN WEST CENTRAL MINNESOTA

Item	Quantity/ha	kcal/ha	Ref.
	Input		
Labor	3.21 hr	—	2
Machinery	7.73 kg	139,140	3
Gasoline	19.62 ℓ	198,339	
Diesel	29.52 ℓ	336,941	
Nitrogen	56.05 kg[a]	823,935	
Phosphorus	25.78 kg[a]	77,340	
Potassium	16.81 kg[a]	26,896	
Seed	107.62 kg[a]	430,480	
Herbicide	0.56 kg[a]	55,950	
Transportation	155.16 kg	39,876	
Total		**2,128,897**	
	Output		
Oats yield	2869.76 kg	10,897,488	
Protein yield	423.29 kg[b]		
kcal output/kcal input		5.12	
kcal output/labor hour		3,394,856	

[a] Fertilizer, seeding, and herbicide rates are those recommended by the Quaker Oats Company to attain maximum yields.
[b] Protein measured "as is" (not dry) from whole grains (hulls included) equals 14.75%.

Table 5
ENERGY INPUT PER HECTARE FOR OATS IN SOUTHEAST SOUTH DAKOTA

Item	Quantity/ha	kcal/ha	Ref.
	Input		
Labor	3.20 hr	—	2
Machinery	5.68 kg	102,240	3
Gasoline	24.76ℓ	257,851	
Diesel	27.94 ℓ	318,907	
Nitrogen	44.84 kg[a]	659,148	
Phosphorus	22.42 kg[a]	67,260	
Potassium	16.81 kg[a]	26,896	
Seed	89.68 kg[a]	358,720	
Herbicide	0.56 kg[a]	55,950	
Transportation	137.68 kg	35,384	
Total	**224.85 kg**	**1,882,356**	
	Output		
Oats yield	2331.68 kg	8,860,384	
Protein yield	355.81 kg[b]		
kcal output/kcal input		4.71	
kcal output/labor hour		2,768,870	

[a] Fertilizer, seeding, and herbicide rates are those recommended by the Quaker Oats Company to attain maximum yields.
[b] Protein measured "as is" (not dry) from whole grain (hulls included) is 15.25%.

ENERGY INPUT PER HECTARE FOR OATS IN BLACKLANDS AREA OF TEXAS

Item	Quantity/ha	kcal/ha	Ref.
	Input		
Labor	5.18 hr	—	2
Machinery	12.23 kg	220,140	4
Gasoline	26.16 ℓ	272,430	
Diesel	35.32 ℓ	403,142	
Nitrogen	67.26 kg	988,722	5
Phosphorus	67.26 kg	201,780	5
Potassium	33.63 kg	53,808	5
Seed	89.68 kg	358,720	5
Insecticide	0.56 kg	55,950	5
Transportation	151.55 kg	38,948	
Total	**323.53 kg**	**2,593,640**	
	Output		
Oats yield	2511.04 kg	9,541,952	
Protein yield	361.59 kg[a]		
kcal output/kcal input		3.68	
kcal output/labor hour		1,184,076	

[a] Protein measured "as is" (not dry) from whole grain is 14.4%.

ENERGY INPUT PER HECTARE FOR OATS IN SOUTHEAST U.S.

Item	Quantity/ha	kcal/ha	Ref.
	Input		
Labor	5.68 hr	—	2
Machinery	31.60 kg	568,800	6
Gasoline	28.50 ℓ	288,107	
Diesel	35.50 ℓ	405,197	
Nitrogen	67.26 kg[a]	988,722	
Phosphorus	44.84 kg[a]	134,520	
Potassium	44.84 kg[a]	71,744	
Lime	1121.00 kg[a]	353,619	
Seed	89.68 kg[a]	358,720	
Herbicide	0.56 kg[a]	55,950	
Transportation	172.83 kg	44,417	
Total	**1454.59 kg**	**3,269,796**	
	Output		
Oats yield	2152.32 kg	8,178,816	
Protein yield	309.93 kg		
kcal output/kcal input		2.50	
kcal output/labor hour		1,439,932	

[a] Dates recommended by North Carolina State University Extension Service and Coker's Pedigreed Seed Co. for maximum production.
[b] Protein measured "as is" (not dry) from whole grain equals 14.4%.

Table 8
TOTAL MACHINERY SUMMARY FOR OATS PRODUCTION IN WESTCENTRAL MINNESOTA

Implement	Total energy (kcal)	Life (years)	kg/ha/ year	Annual energy depreciation (kcal)
Tractors	266 122 144	15	2.67	17 741 476
Drill	24 928 076	15	0.47	1 661 872
Combine	115 691 948	12	1.64	9 640 996
Swather	23 935 822	15	0.28	1 595 721
Baler	21 787 567	15	0.21	1 461 548
Plow	22 129 940	10	0.34	2 212 994
Disc	49 565 209	10	0.63	4 956 521
Harrow	14 611 621	15	0.12	0 974 108
Chisel plow	41 461 052	10	0.53	4 146 105
Sprayer	28 345 355	15	0.25	1 889 690
Fertilizer spreader	17 257 755	15	0.15	1 150 517
Wagon (1)	23 914 447	20	0.16	1 195 722
Truck (2-ton)	42 269 856	20	0.28	2 113 493
			7.73	

Total annual machinery energy = 50 740 763 kcal/year

Total annual machinery energy per hectare, 405 ha = 125,286 kcal/ha/year

Table 9
TOTAL MACHINERY SUMMARY FOR OATS PRODUCTION IN SOUTHEAST SOUTH DAKOTA

Implement	Total energy (kcal)	Life years	kg/ha/ year	Annual energy depreciation (kcal)
Tractors	266 122 144	15	2.06	17 741 476
Drill	51 973 217	15	.17	3 464 881
Combine	115 691 948	12	1.27	9 640 996
Swather	31 879 986	15	.16	2 125 332
Baler	21 923 215	15	.16	1 461 548
Plow	26 538 139	10	.22	2 653 814
Disc	49 565 209	10	.49	4 956 521
Drag harrow	14 611 221	15	.09	0 974 108
Chisel plow	41 461 052	10	.41	4 146 105
Sprayer	28 345 355	15	.20	1 889 690
Fertilizer spreader	17 257 755	15	.11	1 150 517
Wagon	23 914 447	20	.12	1 195 722
Truck (2-ton)	42 269 856	20	.22	2 113 493
			5.68	

Total annual machinery energy = 53 514 203 kcal/year

Total annual machinery energy per hectare, 5.26 ha = 101,738 kcal/ha/year

Table 10
TOTAL MACHINERY SUMMARY FOR OATS PRODUCTION IN BLACKLANDS AREA OF TEXAS

Implement	Total energy (kcal)	Life years	kg/ha/ year	Annual energy depreciation (kcal)
Tractors	167 815 462	12	4.22	13 984 622
Chisel plow	47 832 729	10	1.23	4 783 273
Offset disc	39 401 326	10	1.01	3 940 133
Tandem disc	32 856 401	10	0.84	3 285 640
Drill	24 928 076	15	0.45	1 661 872
Fertilizer spreader	17 509 594	15	0.30	1 167 306
Combine	81 996 487	10	2.81	8 199 649
Truck (2-ton)	42 269 856	20	0.56	2 113 493
Grain trailer (1)	34 389 968	20	0.47	1 719 498
Truck (3/4-ton)	24 109 121	20	0.34	1 205 450
			12.23	

Total annual machinery energy = 42 060 936 kcal/year

Total annual machinery energy per hectare, 202 ha = 208,222 kcal/ha/year

Table 11
TOTAL MACHINERY SUMMARY FOR OATS PRODUCTION IN THE SOUTHEAST U.S.

Implement	Total energy (kcal)	Life (years)	kg/ha/ year	Annual energy depreciation (kcal)
Tractors	117 282 630	12	7.36	9 773 553
Disc	35 031 626	10	2.24	3 503 163
Drill	18 315 192	10	1.23	1 831 519
Herbicide Applicator	8 279 082	10	5.60	0 827 908
Top dress N Applicator	10 391 347	10	6.71	1 039 135
Combine	72 163 397	10	6.16	7 216 340
Truck (1½ ton)	33 867 158	20	1.12	1 693 358
Truck (1½ ton)	33 684 000	20	1.18	1 684 200
			31.60	

Total annual machinery energy = 27 569 176 kcal/year

Total annual machinery energy per hectare, 81 ha = 340,360 kcal/ha/year

REFERENCES

1. Annual Crop Summary, Crop Reporting Board, Economics, Statistics and Cooperatives Service, U.S. Department of Agriculture, 1978.
2. Firm Enterprise Data System, 1975, Economic Research Service, U.S. Department of Agriculture, and Department of Agricultural Economics, Oklahoma State University, Stillwater, 1977.
3. **Fisher, D. C.,** Field Representative, The Quaker Oats Co., personal communication, 1978.
4. **Le Pouri, W.,** Agricultural Engineering Department, Texas A&M University, College Station, and McDaniel, M.E. Department of Soil and Crop Sciences, Texas A&M University, College Station, personal communications.
5. **McDaniel, M. E.,** Department of Soil and Crop Sciences, Texas A&M, University, College Station, personal communication.
6. **Murphy, C. F.,** Department of Crop Science, North Carolina State University, Raleigh, and Harrison, H. F., Coker's Pedigreed Seed Co., Hartsville, S. C., personal communication.

ENERGY USE IN RICE PRODUCTION

J. N. Rutger* and W. R. Grant**

Rice is grown on about 1 million hectares per year in the U.S. Average paddy rice yields (1974—76) at 5.1 MT/ha were more than double the world average yield of 2.4 MT/ha.[1] Farm value of the U.S. crop (1974—76) averaged $ 1 billion per year. Over half of the crop is exported. Farm area in rice in the U.S. is less than 1% of the world total, and the production less than 2%. However, the United States is often the world's largest exporter of rice, because most of the world rice crop is consumed within the producing countries. After seed uses are satisfied, almost all of the rice crop is used for human food or alcoholic beverages.

Rice production in the U.S. is concentrated in five states, Arkansas (31% of the total in 1974—76),[1] California (22%), Louisiana (20%), Mississippi (5%), and Texas (21%). Highest average yields (6.3 MT/ha) were obtained in California, which has high light intensity and little cloud cover during the growing season, and is further characterized by relative freedom from traditional rice diseases and insect pests.

All U.S. rice is grown in lowland conditions, with 10—15 cm of irrigation water standing in the fields during most of the growing season. Irrigation energy accounts for about 20—40% of total energy inputs in U.S. rice production (Tables 1—5). Irrigation is supplemented by rainfall during the growing season in all areas except California. Even though the greatest amount of irrigation water per hectare is used in California, irrigation energy costs in California are lowest, since at least three-fourths of the irrigation is from surface water collected behind dams and distributed largely by gravity or with relatively low lifts. California's low irrigation energy costs and high grain yields result in the highest output/input ratios (1.76) among the U.S. rice areas (Tables 1—5).

U.S. rice production is highly mechanized, with only 20—30 man-hours of labor per hectare, compared to 800 or more man-hours per hectare in developing countries (Tables 1—7). Energy return per hour of human labor ranges from about 400,000—800,000 kcal/hour in the U.S. (Tables 1—5), compared to only 12,000—15,000 kcal/hour in the Philippines (Tables 6,7). Fossil energy inputs are much lower in developing countries because rice is grown in conditions ranging from no mechanization, irrigation, or fertilization to conditions as intensive as those in the U.S. In the Philippines, for example, 13% of the rice crop is grown in upland conditions with no irrigation, 45% is rainfed (without irrigation but in fields that are bunded to impound rainfall), and 42% is irrigated.[2] Fossil fuel inputs in upland rice are minimal: little mechanization is used in land preparation; fertilizer is not generally used; weed control is commonly by hand with about 300 man-hours/ha per weeding; and harvesting is by hand.[3] Although upland rice in the Philippines averages only 0.9 MT/ha,[2] energy output/fossil fuel input ratios must be very large, easily as high as the 17.3 ratio estimated by Heichel.[4] However, as yields are raised through irrigation and more intensive cultural practices, output/input ratios decline to about 3.4 in Laguna Province in the Philippines (Tables 6 and 7).

* Research Geneticist, Science and Education Administration, U.S. Department of Agriculture, Department of Agronomy and Range Science, University of California, Davis.

** Agricultural Economist, CED, U.S. Department of Agriculture, Department of Agricultural Economics, Texas A & M University, College Station.

Table 1
ENERGY INPUTS PER HECTARE OF RICE, SACRAMENTO VALLEY, CALIFORNIA, 1977

Item	Quantity/ha[a]	kcal/ha
Input		
Labor	23.6 hr	—
Machinery	37.7 kg[b]	742,460
Gasoline	55.2 ℓ[c]	558,017
Diesel	225.4 ℓ	2,572,716
Electricity	29.7 kwh	85,031
Nitrogen (units N)	132.3 kg	1,944,810
Phosphate (units P_2O_5)	56.0 kg	168,000
Zinc	9.8 kg[d]	49,000
Furadan	0.5 kg[e]	42,650
Parathion	0.085 kg[f]	7,387
Molinate	3.4 kg[g]	294,440
MCPA	0.6 kg[h]	59,946
Copper sulphate	11.2 kg[i]	56,000
Seed	180.5 kg	722,000
Irrigation	250.0 cm[j]	2,138,886
Drying	6,969.0 kg[k]	1,393,800
Transportation	451.3 kg	115,984
Total		**10,951,127**
Output		
Rice yield	6,513.0 kg[l]	19,226,376
Protein yield	373.8 kg[m]	
kcal output/ kcal input		1.76[n]
kcal output/hour of labor		814,677

[a] All input quantities except machinery and irrigation are from FEDS, 1977.[5] FEDS inputs are 104% of actual in order to account for approximately 4% reseeding. Energy values per unit are from the Advisory Board, except as noted.

[b] Kg/ha from FEDS, 1977. Kcal/kg for machinery calculated by Dr. Otto Doering of the Advisory Board. Slightly different machines, with different energy values, were used in each state. Thus, the kcal/kg of machinery were as follows: California 19,694; Arkansas 19,444; Louisiana 16,619; Mississippi 20,358; Texas 22,401.

[c] Includes fuel for custom hauling of production and custom aerial service.

[d] Assumes that 56 kg/ha (50 lb/a) of $ZnSO_4$ are applied on half the area with Zn being 35% $ZnSO_4$ (56 kg/ha × 0.50 × 0.35 = 9.8 kg/ha). Assumes energy value of 5000 kcal/kg for zinc.

[e] Assumes 12.1 kg/ha (10 lb/a) of 10% technical material on 40% of area (12.1 kg/ha × 0.10 × 0.40 = 0.5 kg/ha). Assumes Advisory Board value of 85,300 kcal/kg for an insecticide in dust form.

[f] Assumes Advisory Board value of 86,910 kcal/kg for an insecticide in miscible oil.

[g] Assumes Advisory Board value of 86,600 kcal/kg for herbicide in granular form.

[h] Assumes 0.84 kg/ha (¾ lb/a) of technical material on ⅔ of area. Assumes Advisory Board value of 99,910 kcal/kg for herbicide in miscible oil.

[i] Assumes 28.0 kg/ha (25 lb/a) on 40% of area. Assumes energy value of 5000 kcal/kg for copper sulphate.

[j] Estimated for Sacramento River Valley. Knutson et al. calculated that it took 95.0 × 10⁶ kwh to pump an average of 250 cm of water (8.2 acre-feet/acre) to 127,162 ha (314,090 acres) of rice (95.0 × 10⁶ kwh × 2,863 kcal/kwh ÷ 127,162 ha = 2,138,886 kcal/ha).

[k] Assumes moisture content of all the crop is lowered from 20% down to 12%.

[l] Brown rice is 82% of rough rice. Energy value of 3,600 kcal/kg is for brown rice. Hence energy value of rough rice = 0.82 × 3600 = 2952 kcal/kg.

[m] One obvious way to increase this ratio is to use higher yielding cultivars. Hence Brandon et al. (1978) showed that the combination of using short stature cultivars with additional nitrogen will result in 15% higher grain yields. Thus a short stature cultivar at 151 kg/ha (135 lb/a) yields 15% more than a tall cultivar at the same N rate. After adding the energy cost of the additional 19 kg N/ha (17 lb N/a), and the energy required to dry an additional 15% of grain, the energy output/input ratio becomes 1.94.

[n] Assumes brown rice protein content is 7%.

Table 2
ENERGY INPUTS PER HECTARE OF RICE, GRAND PRAIRIE, ARKANSAS, 1977

Item	Quantity/ha[a]	kcal/ha
Input		
Labor	29.5 hr	—
Machinery	35.3 kg	686,364
Gasoline	86.1 ℓ	870,385
Diesel	205.4 ℓ	2,344,436
Electricity	29.7 kwh	85,031
Nitrogen	134.5 kg	1,977,150
Potash (units K_2O)	33.6 kg	53,760
Zinc	3.9 kg	19,500
Propanil	4.5 kg	449,595
Molinate	3.4 kg	294,440
2,4,5-T	1.1 kg	109,901
Insecticide	1.1 kg	95,601
Seed	156.9 kg	627,600
Irrigation	61.0 cm[b]	3,803,139
Drying	5,074.0 kg	1,014,800
Transportation	431.4 kg	110,870
Total		**12,542,572**
Output		
Rice yield	4,742.0 kg[c]	13,998,384
Protein yield	272.2 kg	
kcal output/kcal input		1.12
kcal output/hour of labor		474,521

[a] Input explanations similar to those for California, except for irrigation.

[b] 0.61 m/ha (2 ft/a) delivered to the field (FEDS, 1977). Assumes that 0.76 m/ha (2.5 ft/a) were lifted 38.1 m (125 ft), with 0.15 m/ha (0.5 ft/a) loss (20% of total pumped) in conveyance and application. Energy use calculated by formula adapted from Knutson et al. (1977): kcal/ha = 131,342 × L × A, where L = lift in meters, and A = amount of water pumped in m/ha.

[c] 1977 rice yields in Arkansas were about 8% lower than the 1974—76 average.

Table 3
ENERGY INPUTS PER HECTARE OF RICE, SOUTHWEST LOUISIANA, 1977

Item	Quantity/ha[a]	kcal/ha
Input		
Labor	24.8 hr	—
Machinery	29.4 kg	488,596
Gasoline	89.0 ℓ	899,701
Diesel	130.2 ℓ	1,486,103
Electricity	29.7 kwh	85,031
Nitrogen	76.2 kg	1,120,140
Phosphate	53.8 kg	161,400
Potash	53.8 kg	86,080
Propanil	4.5 kg	449,595
Molinate	5.0 kg	433,000
Furadan	1.0 kg	93,830
Seed	151.3 kg	605,253
Irrigation	91.0 cm[b]	4,566,761
Drying	4,402.0 kg	880,400
Transportation	358.1 kg	92,032
Total		**11,447,922**
Output		
Rice yield	4,114.0 kg	12,144,528
Protein yield	236.1 kg	
kcal output/kcal input		1.06
kcal output/hour of labor		489,699

[a] Input explanations similar to those for California, except for irrigation.

[b] 0.91 m/ha (3 ft/a) delivered to the field (FEDS, 1977). Assumes that 1.14 m/ha (3.75 ft/a) were lifted 30.5 m (100 ft), with 0.23 m/ha (0.75 ft/a) loss (20% of total pumped) in conveyance and application. Energy use calculated as in Table 2.

Table 4
ENERGY INPUTS PER HECTARE OF RICE, MISSISSIPPI RIVER DELTA, 1977

Item	Quantity/ha[a]	kcal/ha
Input		
Labor	31.1 hr	—
Machinery	37.1 kg	755,306
Gasoline	74.5 ℓ	753,120
Diesel	206.7 ℓ	2,359,274
Electricity	29.7 kwh	85,031
Nitrogen	134.5 kg	1,977,150
2,4,5-T	1.1 kg	109,901
Propanil	4.5 kg	449,595
Molinate	3.4 kg	294,440
Seeds	181.6 kg	726,400
Irrigation	107.0 cm[b]	4,259,160
Drying	4,798.0 kg	959,600
Transportation	450.3 kg	115,727
Total		**12,844,704**
Output		
Rice yield	4,484.0 kg	13,236,768
Protein yield	257.4 kg	
kcal output/kcal input		1.03
kcal output/hour of labor		425,620

[a] Input explanations similar to those for California, except for irrigation.

[b] 1.07 m/ha (3.5 ft/a) delivered to the field (FEDS, 1977). Assumes that 1.34 m/ha (4.375 ft/a) were lifted 24.2 m (80 ft), with 0.27 m/ha (0.875 ft/a) loss (20% of total pumped) in conveyance and application. Energy use calculated as in Table 2.

Table 5
ENERGY INPUTS PER HECTARE OF RICE, TEXAS GULF COAST, 1977

Item	Quantity/ha[a]	kcal/ha
Input		
Labor	24.4 hr	—
Machinery	37.2 kg	833,304
Gasoline	67.3 ℓ	680,336
Diesel	181.5 ℓ	2,071,641
Electricity	29.7 kwh	85,031
Nitrogen	182.7 kg	2,685,690
Phosphate	58.3 kg	174,900
Potash	31.4 kg	50,240
Propanil	6.2 kg	619,442
Molinate	3.3 kg	285,780
Furadan	1.9 kg	162,070
Insecticide	1.1 kg	95,601
Seed	134.5 kg	538,000
Irrigation	122.0 cm[b]	3,653,409
Drying	5,602.0 kg	1,120,400
Transportation	376.5 kg	96,761
Total		**13,152,605**
Output		
Rice yield	5,235.0 kg	15,453,720
Protein yield	300.5 kg	
kcal output/kcal input		1.17
kcal output/hour of labor		633,349

[a] Input explanations similar to those for California, except for irrigation.

[b] Assumes 1.22 m/ha (4 ft/a) delivered to the field (FEDS, 1977). Assumes that 1.52 m/ha (5 ft/a) were lifted 18.3 m (60 ft), with 0.30 m/ha (1 ft/a) loss (20% of total pumped) in conveyance and application. Energy use calculated as in Table 2.

Table 6
ENERGY INPUT PER HECTARE OF RICE, WET SEASON, PHILIPPINES, 1972—73

Item	Quantity/ha[a]	kcal/ha
Input		
Labor	814.4 hr	—
Machinery	4.5 kg[b]	81,000
Gasoline	131.3 ℓ[c]	1,327,312
Nitrogen	33.0 kg	485,100
Herbicide	0.7 kg	69,937
Insecticide	3.2 kg	255,664
Seed	88.0 kg	352,000
Irrigation	15.0 cm[d]	227,090
Total		**2,798,103**
Output		
Rice yield	3,232.0 kg	9,540,864
Protein yield	185.5 kg	
kcal output/kcal input		3.41
kcal output/hour of labor		11,715

[a] Input data developed from Pecadizo et al., 1973.[9]
[b] Assumes 18,000 kcal/kg.
[c] A total of 69.4 hours of small tractor use per hectare.
[d] Well irrigation was used on about 25% of the farms. Assume 0.152 m/ha (0.61 m/ha × 25% of area) delivered to the field. Assumes that 0.190 m/ha were lifted 9.1 m (30 ft), with 0.038 m/ha loss (20% of total pumped) in conveyance and application. Energy use calculated as in Table 2.

Table 7
ENERGY INPUT PER HECTARE OF RICE, DRY SEASON, PHILIPPINES, 1972—73

Item	Quantity/ha[a]	kcal/ha
Input		
Labor	814.4 hr	—
Machinery	4.5 kg[b]	81,000
Gasoline	131.3 ℓ[c]	1,327,312
Nitrogen	88.0 kg	1,293,600
Herbicide	0.3 kg	26,073
Insecticide	0.7 kg	69,937
Seed	103.9 kg	415,600
Irrigation	30.0 cm[d]	454,180
Total		**3,667,702**
Output		
Rice yield	4,175.0 kg	12,324,600
Protein yield	239.6 kg	
kcal output/kcal input		3.36
kcal output/hour of labor		15,133

[a] Input data developed from Romon et al., 1973.[10]
[b] Assumes 18,000 kcal/kg.
[c] A total of 69.4 hours of small tractor use per hectare.
[d] Well irrigation was used on about 25% of the farms. Assume 0.305 m/ha (1.22 m/ha × 25% of area) delivered to the field. Assumes that 0.380 m/ha were lifted 9.1 m (30 ft), with 0.076 m/ha loss (20% of total pumped) in conveyance and application. Energy use calculated as in Table 2.

REFERENCES

1. Agricultural Statistics, U.S. Department of Agriculture, Washington, D.C., 1977.
2. **Herdt, R. W. and Wickham, T. H.,** Exploring the gap between potential and actual rice yield in the Philippines, *Food Res. Inst. Stud. (Stanford),* 14(2), 163, 1975.
3. International Rice Research Institute, *Major Research in Upland Rice,* Los Banos, Philippines, 1975, chap. 1.
4. **Heichel, G. H.,** Comparative efficiency of energy use in crop production, *Conn. Agric. Exp. Stn New Haven Bull.,* 739, 1973.
5. Farm Enterprise Data System 1975, U.S. Department of Agriculture, Economic Research Service, and Department of Agricultural Economics, Oklahoma State University, Stillwater, 1977.
6. **Knutson, G. D., Curley, R. G., Roberts, E. B., Hagan, R. M., and Cervinka, V.,** Pumping energy requirements for irrigation in California, *Special Publication 3215, Division of Agricultural Science,* University of California, 1977.
7. **Sorenson, J. W., Jr.,** Energy to Perform Agricultural Operations: Harvesting and Processing, Proc. Texas Chapter, American Society of Agronomy, 1974.
8. **Brandon, D. M., Rutger, J. N., Mueller, K. F., Williams, J., and Wick, C. M.,** Differential Grain/Straw Response of Tall and Short Stature Rice Cultivars to Nitrogen, Proc. 17th Rice Tech. Work Group, College Station, Tex., 1978.
9. **Pecadizo, L. M., Romon, E. R., Fortuna, N. M., and Abarientos, E. P.,** Cost of Producing Palay in Laguna, Series No. 3, University of Phillipines, Los Banos, 1973.
10. **Romon, E. R., Fortuna, N. M., and Abarientos, E. P.,** Cost of Producing Palay in Laguna, Series No. 4, University of Philippines, Los Banos, 1973.

ENERGY REQUIREMENTS IN RYE PRODUCTION

D. L. Reeves

The primary use of rye in the north central plains of the U.S. is for grain production, whereas in many areas rye is often used as a winter cover crop or for pasture. The four states leading in rye grain production are South Dakota, North Dakota, Nebraska, and Minnesota, producing 26, 16, 10, and 9%, respectively, of the total U.S. production (Table 1). These four states account for over 60% of the rye grain production. The average yield in these states is about 18.2 q/ha. A considerable acreage of rye is planted in other states, as indicated by the fact that South Dakota, North Dakota, Nebraska, and Minnesota account for only 21% of the planted acreage. The southern states plant considerable rye, but most is not cut for grain. A good example is Georgia, which ranks fifth in rye grain production and produces just under 6% of the grain. Farmers in Georgia plant more rye than any north central state, but under 25% of the planted acreage is cut for grain. This compares to about 90% of the seeded acreage cut for grain in the Dakotas and 30% for a national average.

The following discussion pertains to the South Dakota, North Dakota, Nebraska, and Minnesota region unless otherwise noted. The calculations made for the estimates of energy involved in rye grain are for this region, since this is the primary area of rye grain production in the U.S. Inputs for rye used for other than grain production are presented in a different section. Energy inputs are calculated for the two systems of production common in the north central plains. Most rye is grown on land that produced a crop the previous year. In areas of limited rainfall this continuous cropping practice usually results in lower yields. There are also some common management practices that reduce the efficiency of rye production in terms of energy inputs. Many farmers will put their rye on their poorest ground, since rye has a reputation of doing better than other grains under these conditions. This ground may be poorly drained, sandy, low in fertility, or have other problems. In addition, less fertilizer is applied to rye than other crops since rye has a reputation of producing better under low fertility. Producers are correct, because rye does better than other small grain crops under low fertility; however, rye does have the ability to respond to good fertility. Since many people treat rye as a secondary crop, the input/output ratio for rye is larger than it need be. As long as rye is treated as a "second rate" grain crop this will not change.

The two sets of calculations are included because they require different inputs and also differ in return per unit of input. For both cropping sequences, the inputs of seed, machinery, and time are essentially the same for a given land area regardless of grain yield. Table 2 shows typical inputs when rye follows another grain crop. In this system rye is usually fertilized lightly at planting. Stands will be erratic at times due to limited fall moisture; therefore, about half of the crop will be sprayed to control weeds.

Inputs required for rye grown on summer fallow are summarized in Table 3. When grown on summer fallow, rye is usually not fertilized, although some producers will apply some fertilizer if moisture prospects are good. Spraying for weed control is not usually needed when rye is planted on fallow, because it is an excellent competitor with weeds.

Table 1
RYE GRAIN PRODUCTION IN SELECTED STATES IN THE U.S. (AVERAGES FOR 1969—1975)

	Hectares planted	Percentage cut for grain	Average yield (q/ha)	Grain production (t)
South Dakota	113,117	88	19.4	194,119
North Dakota	75,897	91	17.6	122,831
Nebraska	86,322	57	15.1	74,046
Minnesota	46,170	90	16.9	71,114
Georgia	154,653	24	11.9	43,756
U.S.	1,530,495	31	15.7	758,223

Table 2
INPUTS PER HECTARE FOR RYE FOLLOWING ANOTHER CROP IN THE NORTH CENTRAL PLAINS

Item	Quantity/ha	kcal/ha[a]
Input		
Labor	3 hr	—
Machinery	13.0 kg[b]	234,000
Gasoline	3 ℓ[c]	30,327
Diesel	37.5 ℓ	428,025
Seed	78.4 kg	261,856
Transportation	131 kg	33,667
Herbicide	0.28 kg	17,576
Nitrogen	33.6 kg	493,920
Phosphorus	33.6 kg	30,000
Total		**1,529,371**
Output		
Rye yield	1,258 kg	4,201,720
Protein yield	152 kg	
kcal output/kcal input		2.75

[a] Data on kcal per unit input supplied by Advisory Board.

[b] Assumes 34.5 t of equipment were used on 203 ha and machine life was 10 to 15 years, depending on machine involved.

[c] Fuel estimates from Allen, H. R., Economics Pamphlet 153, South Dakota State University, 1976.

Table 3
INPUTS PER HECTARE FOR RYE GROWN ON SUMMER FALLOW IN THE NORTH CENTRAL PLAINS

Item	Quantity/ha	kcal/ha[a]
	Input	
Labor	3.7 hr	—
Machinery	15.3 kg[b]	275,400
Diesel	20.8 l[c]	237,411
Seeds	78.4 kg	261,856
Transportation	188.3 kg	48,393
Total		**823,060**
	Output	
Rye yield	2509 kg	8,380,060
Protein yield	304 kg	
kcal output/kcal input		10.18

[a] Data on kcal per input supplied by Advisory Board.

[b] Assumes 26.7 t of equipment used on 284 ha with half being in summer fallow each year. Calculations for the combine assume it will cut an equal acreage each year for neighbors.

[c] Fuel estimates from Allen, H. R., Economics Pamphlet 153, South Dakota State University, 1976.

ENERGY INPUTS IN SORGHUM PRODUCTION

Robert Bukantis

Sorghum, a member of the grass family (Gramineae), is one of the most important grain crops grown in the world today. It is thought that sorghum had its beginnings as a crop plant with the Cushite people, around 3,000 to 4,000 years before the time of Christ. Eventually, sorghum spread throughout those parts of Africa, India, Asia, and Europe suitable for its cultivation. Sorghum first reached the Western Hemisphere with the slave trade.

In the U.S., sorghum is used mainly for livestock feed. Worldwide, sorghum is important in the human diet, with over 300 million people[15] dependent on it, primarily in Africa, India, China, and the Middle East.

Sorghum, being better adapted to hotter climates and areas of limited or variable rainfall, is grown primarily in areas unsuitable for cultivation of corn and rice.

World production of grain sorghum in 1977 was 54 million tonnes. The developing countries produced more grain sorghum (32 million t) than the developed countries (22 million t). While producing 45% more grain sorghum than the developed countries, the developing countries invest six times as much acreage in grain sorghum production. The U.S. leads the world in grain sorghum produced (20 million t), while India leads in sorghum cropland (16 million ha).[4]

In the U.S., sorghum is used primarily as a food for livestock, and secondarily as an export commodity. One fourth of the grain sorghum produced is used on the farms where it is grown, the rest being sold. Exports accounted for another one fourth of the grain sorghum produced in the U.S. from 1962 to 1975. Japan imported the largest quantity (2.25 million t in 1975). India, Israel, Netherlands, and Mexico are other prominent importers of U.S. grain sorghum.[19]

Texas, Kansas, and Nebraska (7.44 million t, 4.31 million t, 3.04 million t in 1976, respectively) together produce about 80% of the grain sorghum in the U.S.[19] Accordingly, areas from these three states have been selected for grain sorghum energy budgets. Budgets from Sudan and Nigeria have been included also in order to contrast labor intensive and energy intensive grain sorghum production.

It is interesting to note that in the U.S, irrigation brings about a significant increase in grain sorghum yield while lowering the energy efficiency of production. In comparing the budgets for Nigeria and the Sudan with the U.S., it becomes apparent that the total yield of the manual labor system is relatively poor, but the energy efficiency is excellent compared to the low labor fuel based U.S. systems.

Table 1
ENERGY INPUTS PER HECTARE FOR NONIRRIGATED GRAIN SORGHUM IN WESTERN KANSAS

Item	Quantity/ha	kcal/ha	Ref.
	Input		
Labor	4.7 hr	—	5
Gasoline	20.3 ℓ	205,000	1,5
Diesel	42.9 ℓ	490,000	1,5
L.P. gas	2.6 ℓ	20,000	1,5
Nitrogen	37.8 kg	453,000[a]	5
Phosphate	3.4 kg	10,000[b]	5
Potash	0.9 kg	1,000[b]	5
Herbicide	1.0 kg	89,000	5,12, 18
Insecticide	0.6 kg	49,000	5,12, 18
Seed	3.4 kg	47,000	5,6
Machinery	7.0 kg	126,000	2,3,5, 9,10
Transportation	63.6 kg[c]	16,000	13
Total		**1,506,000**	
	Output		
Grain sorghum yield	1,840 kg	6,035,000	5,17
Protein yield	202 kg		17
kcal output/kcal input		4.01	
kcal output/labor hour		1,284,000	

[a] Assumes anhydrous ammonia.[8]
[b] Assumes triple super phosphate.[8]
[c] Fuels, seed, and machinery.

Table 2
ENERGY INPUTS PER HECTARE FOR IRRIGATED GRAIN SORGHUM IN WESTERN KANSAS

Item	Quantity/ha	kcal/ha	Ref.
	Input		
Labor	15.9 hr	—	5
Gasoline	34.7 ℓ	351,000	1,5
Diesel	64.9 ℓ	741,000	1,5
L.P. gas	15.8 ℓ	122,000	1,5
Natural gas	902.8 m^3	10,644,000	1,5
Nitrogen	111.4 kg	1,337,000[a]	5
Phosphate	2.9 kg	9,000[b]	5
Herbicide	1.4 kg	121,000	5,12, 18
Insecticide	2.0 kg	160,000	5,12, 18
Seed	6.7 kg	94,000	5,6
Machinery	9.0 kg	162,000	2,3,5, 9,10
Irrigation equipment	57.55 dollar	691,000	7,20, 5
Transportation	126.3[b]	32,000	13
Total		**14,464,000**	
	Output		
Grain sorghum yield	4,170 kg	13,678,000	5,17
Protein yield	459 kg		17
kcal output/kcal input		0.95	
kcal output/labor hour		860,000	

[a] Assumes anhydrous ammonia.[8]
[b] Assumes triple super phosphate.[8]
[c] Fuels, seed and machinery.

Table 3
ENERGY INPUTS PER HECTARE FOR NONIRRIGATED GRAIN SORGHUM IN EASTERN NEBRASKA

Item	Quantity/ha	kcal/ha	Ref.
	Input		
Labor	7.4 hr	—	5
Gasoline	46.3 ℓ	468,000	1,5
Diesel	42.8 ℓ	488,000	1,5
Nitrogen	74.3 kg	891,000[a]	5
Phosphate	9.2 kg	28,000[b]	5
Potash	1.2 kg	2,000[b]	5
Lime	49.3 kg	16,000	5,16
Herbicide	1.9 kg	172,000	5,12, 18
Insecticide	2.0 kg	162,000	5,12, 18
Seed	7.1 kg	99,000	5,6
Machinery	11.0 kg	198,000	3,5
Transportation	89.3 kg[c]	23,000	13
Total		2,547,000	
	Output		
Grain sorghum yield	3,471 kg	11,385,000	5,17
Protein yield	382 kg		17
kcal output/kcal input		4.47	
kcal output/labor hour		1,539,000	

[a] Assumes anhydrous ammonia.[8]
[b] Assumes triple super phosphate.[8]
[c] Fuels, seed, and machinery.

Table 4
ENERGY INPUTS PER HECTARE FOR IRRIGATED GRAIN SORGHUM IN EASTERN NEBRASKA

Item	Quantity/ha	kcal/ha	Input
Labor	13.2 hr	—	5
Gasoline	49.2 ℓ	497,000	1,5
Diesel	39.7 ℓ	453,000	1,5
Natural gas	2.1 m^3	25,000	1,5
Electricity	565.8 kwh	1,620,000	1,5
Nitrogen	97.0 kg	1,163,000[a]	5
Phosphate	10.8 kg	32,000[b]	5
Potash	2.7 kg	4,000[b]	5
Herbicide	2.1 kg	187,000	5,12, 18
Insecticide	1.9 kg	153,000	5,12, 18
Seed	5.7 kg	80,000	5,6
Machinery	10.0 kg	180,000	2,3,5, 9,10
Irrigation equipment	30.02 dollar	376,000	5,7, 20
Transportation	86.4 kg[c]	22,000	13
Total		**4,792,000**	
	Output		
Grain sorghum yield	4,994 kg	16,367,000	5,17
Protein yield	549 kg		17
kcal output/kcal input		3.42	
kcal output/labor hours		1,240,000	

[a] Assumes anhydrous ammonia.[8]
[b] Assumes triple super phosphate.[8]
[c] Fuels, seed, and machinery.

Table 5
ENERGY INPUTS PER HECTARE FOR NONIRRIGATED GRAIN SORGHUM IN THE SOUTHERN HIGH PLAINS OF TEXAS

Item	Quantity/ha	kcal/ha	Ref.
Input			
Labor	9.1 hr	—	5
Gasoline	52.7 ℓ	532,000	1,5
Diesel	63.0 ℓ	719,000	1,5
Nitrogen	8.1 kg	97,000[a]	5
Phosphate	5.2 kg	16,000[b]	5
Herbicide	0.1 kg	8,000	5,12, 18
Insecticide	0.1 kg	10,000	5,12, 18
Seed	6.1 kg	85,000	5,6
Machinery	11.0 kg	198,000	2,3,5, 9,10
Transportation	110.3 kg[c]	28,000	13
Total		**1,693,000**	
Output			
Grain sorghum yield	1,655 kg	5,392,000	5,17
Protein yield	181 kg		17
kcal output/kcal input		3.18	
kcal output/labor hour		593,000	

[a] Assumes anhydrous ammonia.[8]
[b] Assumes triple super phosphate.[8]
[c] Fuels, seed, and machinery.

Table 6
ENERGY INPUTS PER HECTARE FOR IRRIGATED GRAIN SORGHUM IN THE SOUTHERN HIGH PLAINS OF TEXAS

Item	Quantity/ha	kcal/ha	Ref.
Input			
Labor	18.2 hr	—	5
Gasoline	27.9 ℓ	282,000	1,5
Diesel	107.3 ℓ	1,225,000	1,5
Electricity	2,212 kwh	6,333,000	1,5
Nitrogen	134.5 kg	1,614,000[a]	5
Phosphate	35.9 kg	108,000[b]	5
Potash	3.4 kg	5,000[b]	5
Herbicide	0.2 kg	18,000	5,12, 18
Insecticide	3.1 kg	244,000	5,12, 18
Seed	9.0 kg	126,000	5,6
Machinery	12.0 kg	216,000	2,3,5, 9,10
Irrigation equipment	71.17 dollar	890,000	5,7, 20
Transportation	132.7 kg[c]	34,000	13
Total		**11,095,000**	
Output			
Grain sorghum yield	5,272 kg	17,292,000	5,17
Protein yield	580 kg		17
kcal output/kcal input		1.56	
kcal output/labor hours		950,000	

[a] Assumes anhydrous ammonia.[8]
[b] Assumes triple super phosphate.[8]
[c] Fuels, seed, and machinery.

Table 7

ENERGY INPUTS PER HECTARE FOR MANUAL LABOR PRODUCED GRAIN SORGHUM IN SUDAN

Item	Quantity/ha	kcal/ha	Ref.
	Input		
Labor	240 hrs	—	
Hoe	0.8 kg	16,500	11
Seed	19.0 kg	62,700	11
Total		**79,200**	
	Output		
Grain sorghum yield	900 kg	2,952,000	11,17
Protein yield	99 kg		17
kcal output/kcal input		37.27	
kcal output/labor hour		12,300	

Table 8

ENERGY INPUTS PER HECTARE FOR GRAIN SORGHUM IN NIGERIA USING DRAFT ANIMALS

Item	Quantity/ha	kcal/ha	Ref.
	Input		
Labor	116 hrs	—	11
Equipment	2.0 kg	41,000	11
Oxen (pair)	365 hrs ea	2,555,000	11
Seeds	19.0 kg	63,000	11
Total		**2,659,000**	
	Output		
Grain sorghum yield	749 kg	2,457,000	11,17
Protein yield	82 kg		17
kcal output/kcal input		0.92	
kcal output/labor hours		21,000	

REFERENCES

1. **Cervinka, V.,** Fuel and energy efficiency, in *Handbook of Energy Utilization in Agriculture,* Pimentel, D., Ed., CRC Press, Boca Raton, Fla., 1980.
2. **Culpin, C.,** *Farm Machinery,* 9th ed., Crosby Lockwood Staples, London, 1976.
3. **Doering, O.,** Accounting for energy in farm machinery and buildings, in *Handbook of Energy Utilization in Agriculture,* Pimentel, D., Ed., CRC Press, Boca Raton, Fla., 1979.
4. *Production Yearbook 1976,* Vol. 30, Food and Agriculture Organization, Rome, 1977.
5. Firm Data Enterprise Data System, 1975, U.S. Department of Agriculture, Economic Research Service, and Department of Agricultural Economics, Oklahoma State University, Stillwater, 1977.
6. **Heichel, G.,** Assessing the fossil energy costs of propagating agricultural crops, in *Handbook of Energy Utilization in Agriculture,* Pimentel, D., Ed., CRC Press, Boca Raton, Fla., 1980.
7. **Herendeen, R. A. and Bullard, C. W.,** Energy Cost of Goods and Services, 1963, and 1967, Doc. No. 140, Center for Advanced Computation, University of Illinois, Urbana, Ill., 1974.
8. **Lockeretz, W.,** Energy inputs for nitrogen, phosphorous, and potash fertilizers, in *Handbook of Energy Utilization in Agriculture,* Pimentel, D., Ed., CRC Press, Boca Raton, Fla., 1980.
9. *National Farm Tractor and Implement Blue Book,* Vol. 36, Gordon, M. I., Ed., National Market Reports, Inc, Chicago, 1975.
10. **Oehlschlaeger, R. E. and Whittlesey, N. K.,** Operating costs for tillage implements on eastern Washington grain farms, *Wash. Agric. Exp. Stn. Cir.,* 554, 1972.
11. **Pimentel, D.,** Energy use in world food production, Environmental Biology Report, 74-1, Cornell University, Ithaca, N.Y., 1974.
12. **Pimentel, D.,** Energy inputs for the production, formulation, packaging and transport of various pesticides, in *Handbook of Energy Utilization in Agriculture,* Pimentel, D., Ed., CRC Press, Boca Raton, Fla., 1980.
13. **Pimentel, D.,** Energy used for transporting supplies to the farm, in *Handbook of Energy Utilization in Agriculture,* Pimentel, D., Ed., CRC Press, Boca Raton, Fla., 1980.
14. **McDonald, J. L., Ed.,** *Power Farming, Power Farming Technical Annual,* Pacific Publishers, Sydney, 1975.
15. **Schmeck, H. M.,** Research finds rich sorghums to bolster diet of world's poor, *New York Times,* p. 1, September 29, 1973.
16. **Terhune, E.,** Energy used in the U.S. for agricultural liming materials, in *Handbook of Energy Utilization in Agriculture,* Pimentel, D., Ed., CRC Press, Boca Raton, Fla., 1980.
17. Composition of Foods, Agriculture Handbook No. 8, Agricultural Research Service, U.S. Department of Agriculture, U.S. Government Printing Office, Washington, D.C., 1963.
18. Evaluation of Pesticide Supplies and Demand for 1974, 1975, and 1976, Economic Research Service Report No. 300, U.S. Department of Agriculture, 1975.
19. Agricultural Statistics 1977, U.S. Department of Agriculture, U.S. Government Printing Office, Washington, D.C., 1977.
20. Wholesale Prices and Price Indexes Supplement, 1976, for 1975, Bureau of Labor Statistics, U.S. Department of Labor, U.S. Government Printing Office, Washington, D.C., 1976, 81.

INTRODUCTION TO ENERGY USE IN WHEAT PRODUCTION

L. W. Briggle*

Wheat produced in the U.S. is extremely valuable. The number of bushels produced, categorized by market class, in the past 3 years is as follows:[1]

	Winter (1000 Bushels)			Spring (1000 Bushels)			
Year	Hard red	Soft red	White	Hard red	Durum	White	Total
1975	1,058,063	326,208	256,125	326,594	123,362	32,107	2,122,459
1976	975,840	336,555	247,528	411,127	134,914	36,398	2,142,362
1977	993,072	341,334	192,307	397,479	79,964	21,637	2,025,793

Hard red winter wheat is grown primarily in the southern and central plains states (Texas, Oklahoma, Kansas, and Nebraska). Significant acreages occur also in southern South Dakota, most of Montana, eastern Colorado, parts of Illinois and Missouri, and in smaller areas of Idaho, Utah, Oregon, and Washington. Hard red winter wheat grown in California, Arizona, and southern Texas is actually hard red spring varieties managed as winter wheat (planted in the fall and harvested in spring or early summer).

Soft red winter wheat is grown primarily in those states east of the Mississippi River. Leading in production are Ohio, Indiana, Illinois, Pennsylvania, and North Carolina. Important acreages west of the Mississippi are in Missouri and north central Texas.

Most white wheat produced in the U.S. is soft winter wheat, and most of this class is grown in the Pacific northwest (Washington, Oregon, and Idaho), but a significant acreage is grown also in the eastern states of Michigan and New York. There is a limited amount of hard white winter and soft white spring wheat; both types are produced principally in Washington, Oregon, and Idaho.

Hard red spring wheat is the predominant class in the northern plains states (North Dakota, South Dakota, western Minnesota, and eastern Montana).

About 80% of U.S. durum wheat is grown in North Dakota. Lesser acreages are in South Dakota, Montana, western Minnesota, northern California, and Arizona.

The major U.S. wheat growing area is in the plains states, where virtually every year the wheat plants are subjected to stress conditions at some stage of growth. The highest nonirrigated production areas (yield of grain in bushels per acre) are the Palouse in eastern Washington and the western edge of Idaho, and in the eastern states of Illinois, Indiana, Michigan, and Ohio.

Approximately 5% of the U.S. wheat acreage is irrigated.[5] Most of this is in the high plains (panhandle) of Texas and contiguous areas in New Mexico and Oklahoma, plus southeastern Colorado and southwestern Kansas. Most wheat grown in Arizona and nearly one fourth of the California acreage is irrigated. Lesser wheat acreages in intermountain valleys of other western states are irrigated.

More energy is required for producing wheat under irrigation than under any other management system. Heavier rates of fertilizer are used, more tillage operations are required if one includes land leveling, ditching, field corrugation, and similar operations relating to water application, and of course the energy required to pump water is probably the largest expenditure.

* Staff Scientist, Science and Education Administration, Agricultural Research, U.S. Department of Agriculture, Beltsville, Md., 20705.

Under rainfed management systems in the eastern states, in the Palouse region of the Pacific northwest, and in the higher rainfall area of the plains states heavier rates of fertilizer (particularly nitrogen) are required than in the central and western parts of the plains states where little or no fertilizer is used in some parts. Generally weeds are less of a problem where there is lower rainfall, so fewer herbicides are required.

Considerable wheat acreage in those states below an imaginary line drawn from southern Illinois to Maryland is double-cropped, primarily with soybeans. Considering only the wheat part of the double cropping system, less energy is required than if wheat was the only crop during that season. In most cases wheat is drilled into the soybean stubble with little or no fertilizer applied. Yields are lower, however, than if wheat was seeded in tilled ground and usual rates of fertilizer were applied.

Limited-till or no-till management systems are used not only as described for double-cropping, but also in limited rainfall areas to combat wind and water erosion. Plant refuse left on the soil surface (or at least only partially covered) does help hold soil in place.

Much of the wheat in the plains states and western states, where the average rainfall is under about 15 in. annually, is grown under a system that is commonly called dry-land farming. Alternate crop and fallow (no crop for 1 year) is a common sequence.[4] Sometimes a second wheat (or barley) crop in 3 years is included. This system is used for both spring wheat and winter wheat. Generally, the only tillage during the idle season (after a heavy initial tillage following the previous crop) is light tillage operations required to control weeds during the summer.

Wheat is an energy frugal crop, compared to most other food crops. Wheat grain can be produced with the energy equivalent of less than two barrels of oil per acre per year.[3] Corn requires just under 4 barrels; potatoes nearly 10 barrels.

The energy inputs for wheat production in various regions of the U.S. are those from FEDS,[2] calculated by Dr. John Krummel and Mr. Sterling Chick of Cornell University. Tables 1 through 10 give a range of energy inputs for wheat production in different regions of the U.S. Note that the inputs range from about 2 to 18 million kcal and that yields range from 900 to nearly 5000 kg/ha.

Table 1
ENERGY INPUTS PER HECTARE FOR WINTER WHEAT PRODUCTION IN KANSAS

Item	Quantity/ha	kcal/ha	Ref.
	Input		
Labor	8.47 hr	—	2
Machinery	17.54 kg	315,720	7
Gasoline	34.82 ℓ	351,995	2,6
Diesel	49.52 ℓ	565,221	2,6
Electricity	14.33 kWh	41,027	2,6
Nitrogen	89.71 kg	1,076,520	2,10
Phosphorus	1.12 kg	3,360	2,10
Seeds	72.55 kg	217,650	2,8
Irrigation	35.6 cm	1,136,864	2,9, 15
Natural Gas	928.94 m^3	10,973,568	2,6
Insecticides	0.24 kg	20,858	2,11
Herbicides	0.22 kg	21,980	2,11
Drying	—	—	
Transporation	147.86 kg[a]	38,000	12
Total		14,762,763	
	Output		
Wheat yield	2,600.05 kg	8,540, 699	2,14
Protein yield	320.11 kg	1,280,440	14
kcal output/kcal input		0.58	
kcal output/labor hour		1,008,347	

[a] Fuels, seed, and machinery.

Table 2
ENERGY INPUTS PER HECTARE FOR WINTER WHEAT PRODUCTION IN TEXAS

Item	Quantity/ha	kcal/ha	Ref.
	Input		
Labor	14.65 hr	—	2
Machinery	22.48 kg	404,640	7
Gasoline	30.24 ℓ	305,696	2,6
Diesel	70.49 ℓ	804,573	2,6
Electricity	94.85 kWh	271,556	2,6
Nitrogen	78.50 kg	942,000	2,10
Phosphorus	6.73 kg	20,190	2,10
Potassium	2.24 kg	3,584	2,10
Seeds	77.38 kg	232,140	2,8
Irrigation	40.64 cm	869,796	2,9, 15
Natural gas	1,205.25 m^3	14,237,618	2,6
Insecticides	1.55 kg	134,711	2,11
Herbicides	0.34 kg	33,969	2,11
Transportation	182.46 kg[a]	46,892	12
Total		18,307,365	
	Output		
Wheat yield	2,378.37 kg	7,812,519	2,14
Protein yield	292.70 kg	1,170,800	14
kcal output/kcal input		0.43	
kcal output/labor hour		533,278	

[a] Fuels, seed, and machinery.

Table 3
ENERGY INPUTS PER HECTARE FOR WHEAT PRODUCTION FOLLOWING FALLOW IN NEBRASKA

Item	Quantity/ha	kcal/ha	Ref.
	Input		
Labor	8.40 hr	—	2
Machinery	21.24 kg	382,320	7
Gasoline	50.46 ℓ	510,100	2,6
Diesel	46.34 ℓ	528,925	2,6
Electricity	14.82 kWh	42,430	2,6
Nitrogen	13.46 kg	161,520	2,10
Phosphorus	29.16 kg	87,480	2,10
Seeds	94.87 kg	284,610	2,8
Transportation	193.34 kg[a]	49,688	12
Total		**2,047,073**	
	Output		
Wheat yield	2,324.6 kg	7,676,197	2,14
Protein yield	286.0 kg	1,144,104	14
kcal output/kcal input		3.75	
kcal output/labor hour		913,833	

[a] Fuels, seed, and machinery.

Table 4
ENERGY INPUTS PER HECTARE FOR WHEAT FOLLOWING CROP IN NEBRASKA

Item	Quantity/ha	kcal/ha	Ref.
	Input		
Labor	6.03 hr	—	2
Machinery	20.01 kg	360,180	7
Gasoline	37.07 ℓ	374,741	2,6
Diesel	45.50 ℓ	519,337	2,6
Electricity	14.82 kWh	42,430	2,6
Nitrogen	40.37 kg	484,440	2,10
Phosphorus	8.07 kg	24,210	2,10
Potassium	4.60 kg	7,360	2,10
Lime	40.37 kg	12,735	2,13
Seeds	74.57 kg	223,710	2,8
Herbicides	0.19 kg	18,983	2,11
Transportation	161.06 kg[a]	41,392	12
Total		**2,109,518**	
	Output		
Wheat yield	1,881.15 kg	6,179,252	2,14
Protein yield	231.44 kg	925,760	14
kcal output/kcal input		2.93	
kcal output/labor hour		1,024,752	

[a] Fuels, seed, and machinery.

Table 5
ENERGY INPUTS PER HECTARE FOR DRYLAND WHEAT IN NEW MEXICO

Item	Quantity/ha	kcal/ha	Ref.
	Input		
Labor	7.48 hr	—	2
Machinery	19.02 kg	342,360	7
Gasoline	72.18 ℓ	729,668	2,6
Diesel	19.28 ℓ	220,062	2,6
Electricity	178.58 kWh	511,275	2,6
Seeds	44.86 kg	134,580	2,8
Transportation	134.40 kg[a]	34,541	12
Total		**1,972,486**	
	Output		
Wheat yield	994.43 kg	3,266,534	2,14
Protein yield	122.51 kg	490,040	14
kcal output/kcal input		1.66	
kcal output/labor hour		436,702	

[a] Fuels, seed, and machinery.

Table 6
ENERGY INPUTS PER HECTARE FOR WINTER WHEAT IN OHIO

Item	Quantity/ha	kcal/ha	Ref.
	Input		
Labor	5.93 hr	—	2
Machinery	25.94 kg	466,920	7
Gasoline	52.42 ℓ	529,914	2,6
Diesel	20.88 ℓ	238,324	2,6
Electricity	12.60 kWh	36,074	2,6
Nitrogen	56.07 kg	672,840	2,10
Phosphorus	61.68 kg	185,040	2,10
Potassium	61.68 kg	98,688	2,10
Lime	112.14 kg	35,375	2,13
Seeds	168.21 kg	504,630	2,8
Transportation	251.21 kg[a]	64,561	12
Total		**2,832,366**	
	Output		
Wheat yield	2,895.71 kg	9,511,903	2,14
Protein yield	356.42 kg	1,425,680	14
kcal output/kcal input		3.36	
kcal output/labor hour		1,604,031	

[a] Fuels, seed, and machinery.

Table 7
ENERGY INPUTS PER HECTARE FOR OTHER SPRING WHEAT FOLLOWING CROPS IN NORTH DAKOTA

Item	Quantity/ha	kcal/ha	Ref.
	Input		
Labor	4.47 hr	—	2
Machinery	19.02 kg	342,360	7
Gasoline	26.59 ℓ	268,798	2,6
Diesel	46.34 ℓ	528,925	2,6
Electricity	13.34 kWh	38,192	2,6
Nitrogen	67.28 kg	807,360	2,10
Phosphorus	25.79 kg	77,370	2,10
Potassium	6.95 kg	11,120	2,10
Seeds	104.29 kg	312,870	2,8
Insecticides	0.30 kg	26,073	2,11
Herbicides	1.74 kg	173,843	2,11
Transportation	182.64 kg[a]	46,938	12
Total		**2,633,849**	
	Output		
Wheat yield	1,760.11 kg	5,816,911	2,14
Protein yield	246.51 kg	986,040	14
kcal output/kcal input		2.21	
kcal output/labor hour		1,301,322	

[a] Fuels, seed, and machinery.

Table 8
ENERGY INPUTS PER HECTARE FOR OTHER SPRING WHEAT FOLLOWING FALLOW IN NORTH DAKOTA

Item	Quantity/ha	kcal/ha	Ref.
	Input		
Labor	6.03 hr	—	2
Machinery	22.97 kg	413,460	7
Gasoline	27.90 ℓ	282,041	2,6
Diesel	73.49 ℓ	838,815	2,6
Electricity	13.34 kWh	38,192	2,6
Nitrogen	28.04 kg	336,480	6,10
Phosphorus	25.79 kg	77,370	2,10
Potassium	25.79 kg	41,264	2,10
Seeds	106.53 kg	319,590	2,8
Insecticides	0.30 kg	26,073	2,11
Herbicides	0.71 kg	70,936	2,11
Transportation	212.89 kg[a]	54,713	12
Total		**2,498,934**	
	Output		
Wheat yield	2,022.32 kg	6,683,470	2,14
Protein yield	283.31 kg	1,133,240	14
kcal output/kcal input		2.67	
kcal output/labor hour		1,108,370	

[a] Fuels, seed, and machinery.

Table 9
ENERGY INPUTS PER HECTARE FOR HARD RED SPRING WHEAT IN IDAHO

Item	Quantity/ha	kcal/ha	Ref.
	Input		
Labor	24.13 hr	—	2,7
Machinery	16.06 kg	289,080	7
Gasoline	33.23 ℓ	335,922	2,6
Diesel	34.08 ℓ	388,989	2,6
Electricity	167.71 kWh	480,154	2,6
Nitrogen	63.25 kg	759,000	2,10
Phosphorus	19.18 kg	57,540	2,10
Seeds	114.38 kg	343,140	2,8
Irrigation	106.68 cm	—[a]	2
Insecticides	0.81 kg	70,397	2,11
Herbicides	5.78 kg	577,480	2,11
Transportation	184.33 kg[b]	47,373	2,12
Total		**3,349,075**	
	Output		
Wheat yield	4,702.88 kg	15,542,327	2,14
Protein yield	658.75 kg	2,635,000	14
kcal output/kcal input		4.64	
kcal output/labor hour		644,108	

[a] Water purchased from irrigation district— gravity distribution.
[b] Fuels, seed, and machinery.

Table 10
ENERGY INPUTS PER HECTARE FOR WHITE WINTER WHEAT FOLLOWING CROP IN WASHINGTON

Item	Quantity/ha	kcal/ha	Ref.
	Input		
Labor	4.10 hr	—	2
Machinery	24.21 kg	435,780	7
Gasoline	13.11 ℓ	132,529	2,6
Diesel	54.30 ℓ	619,780	2,6
Electricity	144.74 kWh	414,391	2,6
Nitrogen	56.07 kg	672,840	2,10
Phosphorus	11.21 kg	33,630	2,10
Seed	78.50 kg	235,500	2,8
Herbicides	4.45 kg	444,600	2,11
Transportation	158.70 kg[a]	40,786	12
Total		**3,029,836**	
	Output		
Wheat yield	3,332.27 kg	11,179,528	2,14
Protein yield	313.20 kg	1,252,800	14
kcal output/kcal input		3.69	
kcal output/labor hour		2,726,714	

[a] Fuels, seed, and machinery.

REFERENCES

1. Crop Reporting Board, Economics, Statistics, and Cooperatives Service, Crop Production, 1977 Annual Summary, Acreage, Yield, Production. U.S. Department of Agriculture, CrPr 2-1 (78), 1978.
2. Firm Enterprise Data System, 1975, U.S. Department of Agriculture, Economic Research Service, and Department of Agricultural Economics, Oklahoma State University, Stillwater, 1977.
3. **Heichel, G. H.,** Managing energy use in food production, in *Conference Proceedings, Second Midwestern Conference on Food and Social Policy,* Morningside College, Sioux City, Iowa, in press, 1978.
4. **Reitz, L. P.,** Wheat in the United States, *Agric. Inf. Bull.,* 386, Agricultural Research Service, U.S. Department of Agriculture, 1976.
5. U.S. Department of Commerce, Census of Agriculture, 1974.
6. **Cervinka, V.,** Fuel and energy efficiency, in *Handbook of Energy Utilization in Agriculture,* Pimentel, D., Ed., CRC Press, Boca Raton, Fla., 1980.
7. **Doering, O.,** Accounting for energy in farm machinery and buildings, in *Handbook of Energy Utilization in Agriculture,* Pimentel, D., Ed., CRC Press, Boca Raton, Fla., 1980.
8. **Heichel, G.,** Assessing the fossil energy costs of propagating agricultural crops in *Handbook of Energy Utilization in Agriculture,* Pimentel, D., Ed., CRC Press, Boca Raton, Fla., 1980.
9. **Herendeen, R. A. and Bullard, C. W.,** Energy Cost of Goods and Services, 1963, and 1967, Document No. 140, Center for Advanced Computation, University of Illinois, Urbana, 1974.
10. **Lockeretz, W.,** Energy inputs for nitrogen, phosphorous, and potash fertilizers, in *Handbook of Energy Utilization in Agriculture,* Pimentel, D. Ed., CRC Press, Boca Raton, Fla., 1980.
11. **Pimentel, D.,** Energy inputs for the production, formulation, packaging and transport of various pesticides, in *Handbook of Energy Utilization in Agriculture,* Pimentel, D., Ed., CRC Press, Boca Raton, Fla., 1980.
12. **Pimentel, D.,** Energy used for transporting supplies to the farm, in *Handbook of Energy Utilization in Agriculture,* Pimentel, D., Ed., CRC Press, Boca Raton, Fla., 1980.
13. **Terhune, E.,** Energy used in the U.S. for agricultural liming materials, in *Handbook of Energy Utilization in Agriculture,* Pimentel, D., Ed., CRC Press, Boca Raton, Fla., 1980.
14. Composition of Foods, Agricultural Research Service, Agriculture Handbook No. 8, U.S. Department of Agriculture, Washington, D.C., 1963.
15. Wholesale Prices and Price Indexes Supplement, 1976, for 1975, Bureau of Labor Statistics, U.S. Department of Labor, Washington, D.C., 1976.

ENERGY USED IN PRODUCING SOYBEANS

W. O. Scott and John Krummel

Soybeans are the most important source of protein supplement for livestock in the U.S. They are also the most important source of edible oil. About 98% of the soybean meal consumed domestically is used in producing livestock. Only 2% is consumed as food. Conversely, 93% of the soybean oil used domestically is consumed as salad oil or in other edible products. Only 7% is used in non-feed products.

The average of 25,746,916 hectares of soybean harvested annually during 1977—1979 ranks second in area among all crops harvested for food or feed in the U.S. Only corn occupies more land each year.[1]

The average annual production during the period 1976—1978 was 44,379,127 metric tons, with a value of over $10.3 billion. About 46% of the 1976—1978 production was exported, with an annual average of over $4.25 billion.[2]

The Economics, Statistics and Cooperative Service of the U.S. Department of Agriculture reports production in 30 states. In contrast to corn, wheat and hay, all production is located east of the Rocky Mountains. Ten Corn Belt states (66%) and five southcentral states (25%) produce over 90% of the crop.

Soybeans are adapted to a wide range of soil and climatic conditions, but they thrive especially well on the relatively deep soils and under the climatic conditions typical of the central Corn Belt states of Ohio, Indiana, Illinois, and Iowa. These four states historically have the highest yields per hectare and produce about 50% of the crop each year.

The optimum temperature for photosynthesis in the leaves of the Lee variety is reported to be 35°C.[3] However, the ideal under non-irrigated field conditions may be more in the range of 24 to 30°C. Soybeans are sensitive to moisture stress throughout the growing season. However, the maximum effect of moisture stress on yield occurs during the last week of pod development and during the bean-filling stage.[3] In the central Corn Belt, this stage of growth is normally reached in late July or early August. Runge and Odell[4] reportd 2.5 cm of precipitation above normal at Urbana in August increases yield by 11 to 22 kilos per hectare.

With the exception of fertilizer, the energy input in producing soybeans is comparable to that in producing corn. Soybeans respond to the same cultural practices as corn, and, for the most part, the machinery used to produce the crop is the same as employed in producing corn or cotton. Soybeans are a relatively new crop in the U.S. Corn and cotton were the most important crops in the regions where soybeans eventually became established. Therefore, the production of soybeans was adapted to the same machinery used to produce corn and cotton. For the most part this is still true. However, in recent years there has been renewed interest in the use of the cereal grain drill for planting soybeans. The drill was a relatively common planting tool early in its history when the crop was grown primarily for forage or green manure. As the proportion of the crop used for processing increased, so did the use of row crop planters.

The fertility requirements for producing soybeans are similar to those of corn with the exception of nitrogen. A sizable amount of nitrogen is removed in the harvested soybean crop. Research at the University of Illinois indicates that symbiotic fixation of nitrogen will vary with the amount of residual nitrogen in the soil. If the residual is high, very little nitrogen will be fixed; if low, the symbiotically-fixed nitrogen may almost equal the amount removed in the harvested grain. Current research has not demonstrated a consistent response to nitrogen used on soybeans when the pH is favorable for nitrogen fixation and the soil is well supplied with other essential elements.

Liming medium to strongly acid soils is the first step in increasing soybean yields. Most state extension services suggest that the pH be kept between 6.0 and 6.5. This is the same pH needed for the production of corn.

Maintenance requirements for phosphorus and potassium will vary with the size of the crop. In Illinois, about 14 kg P_2O_5 and 22 kg K_2O per metric ton of soybeans harvested per hectare are recommended as maintenance.[5]

The inputs for producing soybeans in different regions of the nation vary some as well as the yields (Tables 1-4). A significant advantage in soybean production compared with other non-leguminous crops is that little or no nitrogen fertilization is used. This is one of the important reasons the output/input ratios range from 1.8 to 4.5. Information on machinery inputs for soybean production for Illinois are presented in Table 5.

Table 1
ENERGY INPUTS PER HECTARE FOR SOYBEANS IN GEORGIA

Item	Quantity/ha	kcal/ha	Ref.
	Input		
Labor	11.86 hr	—	
Machinery	11.69 kg[a]	210,420	
Gasoline	68 ℓ	687,412	
Diesel	62 ℓ	707,668	
Nitrogen	27 kg	324,000	
Potassium	67 kg	107,200	
Phosphate	51 kg	153,000	
Lime	314 kg	99,051	
Herbicide	2.94 kg	293,735	
Insecticide	4.92 kg	427,597	
Seed	73 kg	584,000	
Transportation	188 kg[b]	48,316	
Total		**3,642,399**	
	Output		
Soybean yield	1,664 kg	6,691,942	6,10
Protein yield	566 kg		10
kcal output/kcal input		1.84	
kcal output/labor hour		307,116	

[a] See Table 5.

[b] Fuels, seed and machinery.

Table 2
ENERGY INPUTS PER HECTARE FOR SOYBEANS IN ILLINOIS

Item	Quantity/ha	kcal/ha	Ref.
	Input		
Labor	7.71 hr	—	
Machinery	11.69 kg	210,420	
Gasoline	50 ℓ	505,450	
Diesel	51 ℓ	582,114	
Phosphate	2.25 kg	6,750	
Potassium	3.37 kg	5,392	
Lime	63 kg	19,873	
Herbicide	2.99 kg	298,731	
Insecticide	0.01 kg	869	
Seed	84 kg	672,000	
Transportation	177 kg[b]	45,489	
Total		**2,347,088**	
	Output		
Soybean yield	2,600 kg	10,456,160	6,10
Protein yield	885 kg		10
kcal output/kcal input		4.45	
kcal output/labor hour		1,356,182	

[a] See Table 5.
[b] Fuels, seed, and machinery.

Table 3
ENERGY INPUTS PER HECTARE FOR SOYBEANS IN OHIO

Item	Quantity/ha	kcal/ha	Ref.
	Input		
Labor	7.31 hr	—	
Machinery	11.69 kg[a]	210,420	
Gasoline	66 ℓ	667,194	
Diesel	25 ℓ	285,350	
Nitrogen	5.28 kg	63,360	
Potassium	17 kg	27,200	
Phosphate	18 kg	54,000	
Lime	180 kg	56,781	
Herbicide	2.86 kg	285,743	
Seed	73 kg	584,000	
Transportation	155 kg[b]	39,835	
Total		**2,273,883**	
	Output		
Soybean yield	1,994 kg	8,019,070	6,10
Protein yield	679 kg		10
kcal output/kcal input		3.53	
kcal output/labor hour		1,097,000	

[a] See Table 5.
[b] Fuels, seed, and machinery.

Table 4
ENERGY INPUTS PER HECTARE FOR IRRIGATED SOYBEANS IN NEBRASKA

Item	Quantity/ha	kcal/ha	Ref.
Input			
Labor	11.98 hr	—	
Machinery	11.69 kg[a]	210,420	
Gasoline	24 ℓ	242,616	
Diesel	54 ℓ	616,356	
Irrigation			
Water	1 ha-ft	—	
Electricity	566 kwh	1,620,458	
Nitrogen	5.61 kg	67,320	
Phosphate	4.49 kg	13,470	
Herbicide	3.08 kg	307,733	
Seed	73.2 kg	585,600	
Transportation	149 kg[b]	38,293	
Total		**3,702,266**	
Output			
Soybean yield	2,210 kg[a]	8,887,736	6,10
Protein yield	722 kg		10
kcal output/kcal input		2.40	
kcal output/labor hour		741,881	

[a] See Table 5.
[b] Fuels, seed, and machinery.

Table 5
MACHINERY INPUTS FOR SOYBEAN PRODUCTION IN ILLINOIS

Type	Weight (kg)	Hr/ha/yr	Lifetime (hr)	Depreciation (kg/ha/yr)	Ref.
Tractor—60 hp	4700	0.27	8084	0.16	7
Tractor—70 hp	4400	2.25	8081	1.23	7
Tractor—80 hp	5300	0.02	8625	0.01	7
Tractor—100 hp	7100	1.75	8082	1.54	7
Pickup—½ ton	1900	1.48	3859	0.73	7
Truck—2-ton	3700	0.86	4010	0.79	7
SP combine	7000	0.87	1241	4.91	8
Shredder—4R	500	0.05	1057	0.02	9
MB plow—5/16 in.	800[a]	0.74	1223	0.48	
Tandem disk	2100	0.44	1223	0.76	11
Chisel plow—9	800	0.27	1223	0.18	7
Harrow—5-sec	200	0.02	1223	0.01	7
Dry fert. sprd.	300	0.01	700	0.01	9
Sprayer	200	0.35	700	0.01	11
Planter w/fert—6R	1000	0.42	734	0.57	8
Rotary hoe—6R	400	0.20	1223	0.07	11
Row cultivator—6R	400	0.65	1223	0.21	11
Total				**11.69**	

[a] Estimate.

REFERENCES

1. Crop Production, 1978 Annual Summary, Crop Reporting Board, U.S. Department of Agriculture, 1979.
2. Fats and Oils Situation, Economics, Statistics, and Cooperative Services, U.S. Department of Agriculture, 1977 and 1978.
3. **Mederski, H. J., Jeffers, D. L., and Peters, D. B.,** *Soybeans: Improvement, Production and Uses,* American Society of Agronomy, Madison, Wis., 1973, Chap. 8.
4. **Runge, E. C. A. and Odell, R. T.,** The relation between precipitation and temperature and the yield of soybeans on the Agronomy South Farm, Urbana, Ill., *Agron. J.*, 52, 245, 1960.
5. Illinois Agronomy Handbook—1979 and 1980. University of Illinois College of Agriculture Cooperative Extension Service, Circular 1165, University of Illinois, 1979.
6. Firm Enterprise Data System, Commodity Economics Division, U.S. Department of Agriculture, Economic Research Service, and Oklahoma State University, Stillwater, 1975.
7. McDonald, Ed., 1975.
8. National Market Reports, Inc., 1975.
9. National Market Reports, Inc., 1974.
10. Composition of Foods, Agriculture Handbook No. 8, U.S. Department of Agriculture, 1963.

ENERGY INPUTS IN DRY BEAN PRODUCTION

M. W. Adams

The U.S. ranks fifth in dry bean production in the world, but has the world's highest per hectare yields, due to the technology invested in their production. Within the U.S., however, production areas and average yields vary considerably. Production is confined primarily to some 13 to 15 states, and, in practice, to particular regions within those states. In Michigan, the leading state in bean production, most of the crop is produced on the lake-plain soils of the east-central portion of the state. The growers are among the most advanced producers in the state, with large investments in land, machinery, and commitment to use of high-input technology excepting irrigation.

Colorado production is found in the north-eastern irrigated area and also in the nonirrigated south-western portion of the state. In Idaho, production is centered, under irrigated conditions, in the Snake River basin. In most producing states technological inputs are high. In the western U.S., irrigation is a must for efficient production. In the midwestern and eastern states, dependent largely upon seasonal rainfall, yield per hectare tends to be more variable, and usually lower than in the west. In the U.S. beans are always grown in monoculture, as a summer crop.

The U.S. produces some 14 to 16 commercial classes of dry beans, with usually several varieties of each class.[2] The navy beans (small ovoid white seed) and the pinto (larger flattened seed with tan mottling) together represent about 60% of U.S. production. The diversity in seed size, shape, and seed coat color patterns is extensive, and is related to consumer preferences and uses. Consumer preference is to a great extent influenced by ethnic origin, many Latin Americans, for example, preferring small black beans, or medium-small red beans, for special dishes to meet special tastes.

Brazil, India, and mainland China are the biggest producers of dry beans in the world, followed by Mexico and the U.S. Following their introduction into Portuguese colonies of Africa in the middle 15th century, beans have come to be grown quite generally in the east-central highlands, as well as in South Africa. All Central and South American countries grow and consume beans. In these countries, beans are produced under two distinct systems: one, on large commercial farms where technological inputs are similar to those prevailing in North America and Europe; and the other, on small farms where the rule is hand planting, tilling, harvesting, and threshing. A large percentage of the bean hectarage in this second category is planted in association with another crop species, often maize. Yields tend to be lower than under monoculture, and often more variable than those obtained under high level input agriculture. Even so, there are advantages to mixed cropping, which will probably assure the continuance of these systems into the future.

In summary, dry beans represent a palatable and nutritious food. They are grown and consumed in most countries of the world. Technology applied to their production ranges from the highest level to the most primitive, correlated well with the general economic and developed status of the countries. Production efficiency as measured by input/output ratios based on kcal may not range as widely as suggested by the differential in technology invested, since relatively good yields of beans are often obtained where hand labor on small parcels of land takes the place of the machinery and chemicals used by the more developed producers (Table 1). In Michigan the average yield per hectare is 1176 kg. Based on the energy inputs the return per kcal input is 1.31.

Table 1
ENERGY INPUTS PER HECTARE FOR DRY BEAN PRODUCTION IN MICHIGAN (NONIRRIGATED)

Item	Quantity/ha	kcal/ha
Input		
Labor	18.77 hr	—
Machinery	40 kg	720,000
Gasoline	85.38 ℓ	863,106
Diesel	—	—
Nitrogen	45 kg	643,500
Phosphorus	56 kg	168,000
Potassium	56 kg	89,600
Lime	61.65 kg	952
Seeds	44.83 kg	179,320
Insecticides	0.56 kg	34,423
Herbicides	3.92 kg	391,647
Transportation	148.87 kg	38,258
Seed treatment chemicals	0.08 kg	3,036
Total		**3,131,842**
Output		
Dry bean yield	1,176 kg	4,092,480
Protein yield	284.6 kg	
kcal output/kcal input		1.31
kcal output/labor hour		218,033

Table 2
ENERGY INPUT AND OUTPUT PER HECTARE FOR THE COMMON FIELD BEAN IN CALIFORNIA (IRRIGATED)[1]

Item	Quantity/ha	kcal/ha
Input		
Labor	19 hr	—
Machinery	40 kg	720,000
Gasoline	62.08 ℓ	627,567
Diesel	230.21 ℓ	2,627,617
Nitrogen	87.7 kg	1,254,110
Phosphorus	20.8 kg	62,400
Irrigation	—	692,500
Insecticide	1.98 kg	172,082
Herbicide	3.37 kg	336,697
Seed treatment	0.81 kg	50,784
Transportation	364.34 kg	93,635
Seed	82.1 kg	328,400
Total		**6,965,792**
Output		
Dry bean yield (average of pink, white, and red beans)	2146.96 kg	7,331,868
Protein yield	480.92 kg	
kcal output/kcal input		1.05
kcal output/labor hour		385,888

From Bluestein, C., Williams, W., and Lyons, J., Department of Agronomy and Range Science Series XX, University of California, Davis, 1975. Wih permission.

Table 3
FOOD LEGUME PRODUCTION IN 100-POUND BAGS (000 OMITTED) OF THE PRINCIPAL COMMERCIAL CLASSES GROWN IN THE U.S., 1965—1974 INCLUSIVE[3]

Class	1974	1973	1972	1971	1970	1969	1968	1967	1966	1965	10-Year Mean	Coefficient Var. (%)	% of Total U.S. Production
Navy (pea)	6,943	4,882	6,450	5,022	5,163	7,224	5,615	4,787	7,290	5,480	5,886.6	16.92	33.36
Great northern	2,128	1,776	1,515	1,517	1,427	1,707	1,383	1,500	1,949	1,432	1,633.4	15.28	9.26
Small white	620	360	302	309	275	488	441	423	617	504	433.9	28. 73	2.46
Flat small white[a]	45	61	95	69	67	—	—	—	—	—	—	—	—
Black turtle[a]	246	135	144	—	—	—	—	—	—	—	—	—	—
Pinto	4,928	4,622	5,613	4,843	5,301	4,421	4,658	4,039	4,761	4,523	4,770.9	9.31	27.04
Red kidney	1,469	1,145	816	1,123	1,302	1,548	1,124	1,158	1,633	1,362	1,268.0	19.53	7.19
Pink	1,030	804	624	724	678	501	682	488	450	410	639.1	29.54	3.62
Small red	446	318	371	371	585	465	354	266	636	397	420.9	27.47	2.39
Cranberry	184	205	257	112	155	165	184	137	149	132	168.0	24.88	0.95
Large lima	670	533	471	398	558	770	814	774	597	755	634.0	22.67	3.59
Baby lima	574	378	317	400	478	430	589	280	340	211	399.7	30.53	2.27
California black-eye	1,092	766	801	413	712	513	781	565	851	668	716.2	26.80	4.06
Garbanzo	83	98	60	85	68	101	58	88	92	87	82.0	18.37	0.46
Other	347	306	282	531	527	541	706	672	687	496	509.5	30.54	2.89
Total	**20,805**	**16,389**	**18,118**	**15,917**	**17.296**	**18,894**	**17,389**	**15,177**	**19,962**	**16,457**	**17,640.4**	**10.24**	**100%**

[a] Production in prior years included under "other".

REFERENCES

1. **Bluestein, C., Williams, W., and Lyons, J.,** Energy efficiency of protein production in the common field bean, *Phaseolus vulgaris,* Department of Agronomy and Range Science Series XX, University of California, Davis, 1975.
2. Agricultural Statistics 1977, U.S. Department of Agriculture, U.S. Government Printing Office, Washington, D.C., 1978.
3. Crop Reporting Board, Statistical Research Service, U.S. Department of Agriculture, Washington, D.C. 1965—1974.

ENERGY INPUTS IN SNAP BEAN PRODUCTION

Roger Sandsted

Snap beans are of the same genus and species as the dry bean *Phaseolus vulgaris* but are grown for their immature pods instead of dry seed. They are marketed fresh or processed by canning and freezing. Snap beans are a popular vegetable in the U.S. Table 1 shows production by states for fresh market and processing. Snap beans also are grown in many other countries of the world, but production figures for this bean alone are difficult to obtain.

Production practices for snap beans are very similar to those for dry beans until they are ready for harvest. The bush type varieties are picked by machine. Those beans destined for fresh market are hauled by truck to grading and sorting sheds from where they are generally sent in crates or boxes to wholesale or retail markets in refrigerated trucks. Those destined for processing are delivered by truck to the processing plants for canning or freezing. The energy inputs for snap bean production in New York State total 4.6 million kcal (Table 2). The largest inputs are diesel fuel, machinery, pesticides, and seeds. Snap bean yields are about 5,000 kg/ha, providing 1.6 million kcal.

Table 1
SNAP BEAN PRODUCTION IN THE UNITED STATES, 1977

State	Processing (tonnes)	Fresh market (tonnes)	Total (tonnes)
Alabama	5,715	1,633	7,348
Arkansas	7,666	—	7,666
California	14,243	14,651	28,894
Colorado	5,488	—	5,488
Florida	—	51,211	51,211
Georgia	—	5,942	5,942
Illinois	18,280	—	18,280
Maryland	11,703	2,404	14,107
Michigan	37,467	4,173	41,640
New Jersey	—	10,206	10,206
New York	87,180	12,066	99,246
North Carolina	1,678	9,525	11,203
Ohio	408	1,451	1,859
Oklahoma	3,266	—	3,266
Oregon	127,187	—	127,187
Pennsylvania	9,389	1,678	11,067
South Carolina	—	4,445	4,445
Tennessee	23,678	3,992	27,670
Virginia	4,944	6,169	11,113
Washington	4,128	—	4,128
Wisconsin	168,237	—	168,237
Hawaii	—	544	544
Other states	82,508	—	82,508
Total	**613,165**	**130,090**	**743,255**

From Crop Reporting Board, Statistical Research Service, U.S. Department of Agriculture, December, 1977.

Table 2
ENERGY INPUTS FOR SNAP BEAN PRODUCTION IN NEW YORK STATE

Item	Quantity/ha	kcal/ha
Input		
Labor	12 hrs	—
Machinery	50 kg	900,000
Diesel	122 ℓ	1,392,508
Nitrogen	27 kg	386,100
Phosphorus	82 kg	246,000
Potassium	55 kg	88,000
Lime	5 kg	9,465
Seeds	84 kg	672,000
Fungicides	1.7 kg	110,347
Herbicides	7.3 kg	729,343
Electricity	10 kwh	28,630
Transportation	238 kg	61,089
Total		**4,623,482**
Outputs		
Bean yield	4,995 kg	1,595,319
Protein yield	95 kg	
kcal out/kcal input		0.345
kcal output/labor hour		132,943

ENERGY INPUTS IN PEA PRODUCTION

David Pimentel

Peas are utilized fresh, frozen, canned, and dry in cooking in the U.S. There is hardly a more popular vegetable consumed by the American public. An estimated 2,150,000 tons of dry peas are produced annually.

The only data available to date on the energy inputs in pea production is for dry pea production (Table 1). The total inputs are 3,212,928 kcal/ha and the yield is 1612.2 kg/ha. The output/input ratio is calculated to be 1.7.

Table 1
ENERGY INPUTS PER HECTARE FOR DRY PEA PRODUCTION IN WASHINGTON

Item	Quantity/ha	kcal/ha
	Input	
Labor	5.3 hr	—
Machinery	48.5 kg[a]	1,487,488
Gasoline	13.5 ℓ	136,269
Diesel	74.6 ℓ	851,599
Electricity	45.0 kWh	128,692
Seed	196.4 kg	164,687
Insecticide	1.5 kg	128,627
Herbicide	2.3 kg	233,789
Transportation	318.4 kg	81,777
Total		**3,212,928**
	Output	
Dry pea yield	1,612.2 kg	5,475,020
Protein yield	388.8 kg	1,555,280
kcal output/kcal input		1.7
kcal output/labor hour		1,033,022

[a] Calculated from Firm Enterprise Data System data.

SAFFLOWER

P. F. Knowles and Robert Bukantis

Safflower is grown on a commercial scale for the oil contained in its seeds. One of the ancient crop plants of the world, safflower was originally cultivated for use as a dye for food and clothing and only later was recognized as a food source. Safflower, *Carthamus tinctorius* L., is known only in cultivated plantings or as a recent escape from cultivation. The origins of safflower have been largely obscured, due to both its extensive area of cultivation and its long term domestication.[20]

Safflower is a relatively minor oilseed crop and is relatively new to American agriculture. Early varieties of safflower contained relatively little oil and high fiber seeds and thus were uncompetitive with other oilseed crops. However, advances in the last two decades have resulted in high yields per hectare along with increased oil content of the seed. The present usage of safflower oil stems principally from its high unsaturated fatty acid content, mainly linoleic acid (approximately 80% of oil). This imparts desirable characteristics to the oil for use in industrial coatings. A major proportion of safflower oil is used as edible oil. The oil is of interest in reducing the chances of heart disease in humans through substitution of unsaturated for saturated fats in the diet.

Cultural and equipment requirements for producing safflower are about the same as those for cereal crops. Safflower does well under relatively dry conditions similar to the conditions required by some of the cereals. Safflower has a deeper root system than wheat and other grains. The roots will often penetrate to depths of 3 to 4 m, whereas cereals do not reach much beyond 2 m. In deep soils well supplied with water, safflower will produce yields of up to 3400 kg/ha. Safflower requires more water than wheat and barley because safflower develops later than the cereals when the temperatures are high and transpiration is high. Safflower does poorly on shallow soils.

Excess moisture results in disease problems when growing safflower. High humidity will encourage rust or blight development on aboveground safflower parts. Irrigation must be carefully controlled because waterlogging or standing water will kill safflower. Excess soil moisture encourages root rot. Because of the high soil moisture requirements of safflower, and its low tolerance for wet soil, it is usually grown on deep soils. Safflower may be grown after an irrigated crop; as a preirrigated crop; as a rainfed crop; or as a dry-farmed crop. When irrigated, subirrigation is the preferred method. Safflower grown on raised beds is less susceptible to root rot. Safflower yields on irrigated land will vary from 1700 to 4500 kg/ha, depending upon management practices, soil, and incidence of disease.

World safflower seed production for 1976 was 702,000 t, while area harvested was 1,061,000 ha. India devoted the most land to safflower cultivation (645,000 ha; 209,000 t). Mexico produced the most safflower seed (240,000 t on 170,000 ha). The U.S. harvested 95,000 ha, producing 180,000 t of safflower seed, an average of 1895 kg/ha. World average production was 661 kg/ha. Average production for developing countries was 543 kg/ha, and for the developed countries, 1319 kg/ha.[3]

Safflower plays a minor role in world trade. Of world safflower oil production of 210,000 t in 1975, only 15% was exported. This amounted to about one half of 1% of world trade in edible vegetable oils for that year.[15]

California is a major safflower producing state in the U.S. Accordingly, energy budgets are included for production of irrigated safflower in Fresno County, Cal. and for dry-land safflower production in Glenn County, Cal. (Tables 1 and 2). Both budg-

Table 1
ENERGY INPUTS FOR IRRIGATED SAFFLOWER PRODUCTION IN FRESNO COUNTY, CAL.

Item	Quantity/ha	kcal/ha	Ref.
Input			
Labor	8.6 hr	—	14
Machinery	16 kg	288,000	2,4,9,12
Fuel	180 ℓ[a]	2,100,000	1,4,12,15
Electricity	2,700 kWh	7,700,000	1,4,13
Seed	34 kg	170,000[b]	14
Nitrogen	110 kg	1,500,000	6,14
Insecticide	4.3 kg	340,000	4,10,14,18,19
Irrigation equipment	$44	480,000	4,5,14,19
Transportation	200 kg[c]	51,000	11
Total		**13,000,000**	
Output			
Safflower seed yield	2,800 kg	17,000,000	17
Protein yield	530 kg		17
kcal output/kcal input		1.3	
kcal output/labor hours		2,000,000	

[a] Assumes diesel.
[b] Estimated.
[c] Includes seed, machinery, and fuel.

Table 2
ENERGY INPUTS FOR DRYLAND SAFFLOWER GROWN IN GLENN COUNTY, CAL.

Item	Quantity/ha	kcal/ha	Ref.
Input			
Labor	4.0 hr	—	12
Machinery	16 kg	288,000	2,4,9,12
Fuel	180 ℓ[a]	2,100,000	1,4,12,15
Seed	28 kg	140,000[b]	12
Nitrogen	120 kg	1,600,000	6,12
Phosphate	71 kg	210,000[c]	6,12
Transportation	190 kg[d]	49,000	11
Total		**4,400,00**	
Output			
Safflower seed yield	1,700 kg	10,000,000	17
Protein yeld	320 kg		17
kcal output/kcal input		2.3	
kcal output/labor hour		2,500,000	

[a] Assumes diesel.
[b] Estimated.
[c] Assumes triple sugar phosphate.
[d] Includes seed, machinery, and fuel.

ets reflect average yields for their respective areas and management systems. It is apparent from the budgets that irrigated safflower yield was 2800 kg/ha, whereas the yield was only 1700 kg/ha under dry-land conditions. The energy output/input ratios, however, were only 1.3:1 for irrigated safflower and 2.3:1 for dry-land safflower. These differences in ratios are due primarily to the increased cost of irrigation in Fresno County.

REFERENCES

1. **Cervinka, V.,** Fuel and energy efficiency, in *Handbook of Energy Utilization in Agriculture,* Pimentel, D., Ed., CRC Press, Boca Raton, Fla., 1980.
2. **Doering, O.,** Accounting for energy in farm machinery and buildings, in *Handbook of Energy Utilization in Agriculture,* Pimentel, D., Ed., CRC Press, Boca Raton, Fla., 1980.
3. *Production Yearbook 1976,* Vol. 30, Food and Agriculture Organization, Rome, 1977.
4. Firm Enterprise Data System. 1975, U.S. Department of Agriculture, Economic Research Service, and Department of Agricultural Economics, Oklahoma State University, Stillwater, 1977.
5. **Herendeen, R. A. and Bullard, C. W.,** Energy cost of Goods and Services, 1963 and 1967, Document No. 140, Center for Advanced Computation, University of Illinois, Urbana, 1974.
6. **Lockeretz, W.,** Energy inputs for nitrogen, phosphorous and potash fertilizers, in *Handbook of Energy Utilization in Agriculture,* Pimentel, D., Ed., CRC Press, Boca Raton, Fla., 1980.
7. **Gordon, M. F., Ed.,** National Farm Tractor and Implement Blue Book, Vol. 35, National Market Reports, Chicago, 1974.
8. **Gordon, M. F., Ed.,** National Farm Tractor and Implement Blue Book, Vol. 36, National Market Reports, Chicago, 1975.
9. **McDonald, J. L., Ed.,** Power Farming Technical Annual, Pacific Publications, Sydney, 1975.
10. **Pimentel, D.,** Energy inputs for the production, formulation, packaging and transport of various pesticides, in *Handbook of Energy Utilization in Agriculture,* Pimentel, D., Ed., CRC Press, Boca Raton, Fla., 1980.
11. **Pimentel, D.,** Energy used for transporting supplies to the farm, in *Handbook of Energy Utilization in Agriculture,* Pimentel, D., Ed., CRC Press, Boca Raton, Fla., 1980.
12. **Sailsbery, R. L. and Benson, L. C.,** Safflower production costs, Agriculture Extension Service, University of California, Berkeley, 1977.
13. **Sheesley, R., Stromberg, L., and Yeary, E. A.,** What Shall I Plant in 1978?, Mimeograph, University of California, Berkeley, 1977.
14. **St. Andre, J.,** Safflower, Cost Analysis Worksheet — 1977, Mimeograph, University of California, Berkeley, 1977.
15. Agricultural Statistics 1977, U.S. Department of Agriculture, U.S. Government Printing Office, Washington, D.C., 1977.
16. Agricultural Prices, Crop Reporting Board, Statistical Reporting Service, Washington, D.C., 1977.
17. Composition of Foods, Agriculture Handbook No. 8, U.S. Department of Agriculture, 1963.
18. Wholesale Prices and Price Indexes, Supplement 1976 Data for 1975, U.S. Bureau of Labor Statistics, Department of Labor,Washington, D.C., 1976.
19. Wholesale Prices and Price Indexes, Supplement 1977 Data for 1976, U.S. Bureau of Labor Statistics, Department of Labor, Washington, D.C., 1977.
20. **Weiss, E. A.,** *Castor, Sesame, and Safflower,* Barnes and Noble, New York, 1971.

ENERGY INPUT AND OUTPUT FOR SUGARCANE IN LOUISIANA

Ray Ricaud

The sugar consumption in the U.S. was 9,872,150 tonnes or 43.0 kilograms per capita in 1976. About 60% of this consumption was produced in the U.S. and Puerto Rico, and the balance was imported as raw sugar from foreign producers.

There are only four states, three on the mainland and Hawaii, which produce and process sugar from sugarcane. Florida, Louisiana, and Texas produced 827,000, 645,000 and 100,000 t of sugar, respectively, in 1976. Hawaii and Puerto Rico produced 954,961 and 283,756 t, respectively. Nineteen states produced 3,298,182 tonnes of sugar from sugar beets in 1976.

Sugarcane is one of the major crops produced in Louisiana. There are 130,000 hectares planted on 1,213 farms producing an average yield of approximately 62.5 t of cane per hectare. Most of the farm sizes are between 80 and 600 hectares with a few below and above that range. The sugarcane crop was valued at about $86 million to the growers and an additional $134 million at the processing, refining, and distribution levels in 1976.[1,5]

Sugarcane in Louisiana is grown on relatively productive soils along the lower Mississippi River, Bayou Lafouche, and Bayou Teche. Seedbeds are prepared in rows six feet wide during the summer prior to planting time. Sugarcane is propagated vegetatively by planting stalks of seed cane in rows during the early fall. The buds at each node of the stalks sprout and tiller in the spring to produce a crop after remaining mostly dormant during the winter months. The crop is fertilized and cultivated several times in the spring. Herbicides are applied in the fall and spring to control annual weeds and johnsongrass. Insecticides are applied during the growing season to control cane borers.[2-4]

Sugarcane is harvested during October, November, and December and usually prior to a killing freeze. Normally, three crops are harvested from one planting; therefore, plant cane, first and second ratoon crops are harvested each year. At harvest time, the cane is cut and piled on heap rows with a mechanical harvester, burned to remove leaves, loaded in carts, and transported to raw sugar mills for processing.

The data on energy input and output for sugarcane sugar in Louisiana are presented in Table 1. The quantity of each input is the best estimate available of the actual use of inputs by growers. Sugarcane requires large inputs in labor for planting and in heavy equipment for cultivating and harvesting. Most of the crop is planted with hand labor. Diesel fuel is the largest single input because of the numerous operations required with large equipment.[6]

The output in sugar yield is a state average over several years. There are many growers producing more and less than the average with similar inputs. The yield per hectare is usually smaller on large than on medium size farms. Louisiana produces more hectares of sugarcane, but less sugar per hectare, than the other producing states. It has a relatively low ratio of energy output to energy input. This is due mainly to the short growing season in Louisiana as compared to the other states.

Table 1
ENERGY INPUT AND OUTPUT PER HECTARE PER YEAR FOR SUGARCANE IN LOUISIANA

Item	Quantity/ha	Units	kcal/unit[a]	kcal/ha	Ref.
		Input			
Labor	30[a]	hr	—	—	
Machinery	72[b]	kg	18,000	1,296,000	
Gasoline	54	ℓ	10,109	545,886	6
Diesel	284	ℓ	11,414	3,241,576	6
N (ammonia)	158[c]	kg	12,000	1,896,000	
P (triple)	97[c]	kg	3,000	291,000	
K (muriate)	149[c]	kg	1,600	238,400	
Lime	1120[c]	kg	315	352,800	
Seed	215[d]	kg	3,730	801,950	
Insecticide	2.5[e]	kg	86,910	217,275	
Herbicide	6.2[e]	kg	99,910	619,442	
Transportation	568.9[f]	kg	257	146,207	
Total				**9,646,536**	
		Output			
Sugar yield	4687[g]	kg	3,730	17,482,510	
kcal input/kg sugar[h]				2,185	
kcal output/hr labor[h]				582,750	
Protein yield	0				
kcal output/kcal input[h]				1.81	

Sources:

[a] Labor — 2.5, 3.0, 7.5, 8.5, and 8.5 hr/ha for land preparation, planting, cultivation, harvesting, and supervision, respectively.

[b] Machinery — 57,727 kg used for 80 hectares and depreciated over 10 years.

[c] 158 kg N as anhydrous ammonia, 45 kg P_2O_5 as 46% triple superphosphate, 90 kg K_2O as 60% muriate, and 1120 kg ground limestone.

[d] Seed — 8.6 tons for 3 crop years or 2.8 tons per year times 75 kg sugar per ton of cane.

[e] 2.5 kg Guthion for borers, 0.67, 1.68, 1.12, and 2.70 kg of Sinbar, Fenac, Silvex, and Asulox, respectively, for weed control.

[f] Transportation — sum of machinery, fuel and seed inputs.

[g] Average yield — 62.5 tons cane per hectare times 75 kg sugar per ton.

[h] kcal input per unit input supplied by Advisory Board.

REFERENCES

1. **Loupe, D. T.,** Giant Step 11 Report—Sugarcane, Louisiana State University Cooperative Extension Service, Baton Rouge, 1976.
2. Fertilizer Recommendations for Louisiana, Louisiana State University Cooperative Extension Service, Baton Rouge, 1977.
3. Weed Control Recommendations for Louisiana, Louisiana State University Cooperative Extension Service, Baton Rouge, 1977.
4. Insect Control Guide, Louisiana State University Cooperative Extension Service, Baton Rouge, 1977.
5. Louisiana Sugar, Louisiana State University Cooperative Extension Service, Baton Rouge, 1975.
6. **Nolen, B. and Cochran, B. J.,** Energy Consumption in Louisiana Agriculture, Agricultural Engineering Department Report, Louisiana State University, Baton Rouge, 1976.

ENERGY REQUIREMENTS FOR SUGAR BEET PRODUCTION AND PROCESSING

W. J. Chancellor, P. K. Avlani, V. Cervinka, and J. T. Alexander

Sugar beets are grown in temperate climates.* They are planted in the spring and harvested in the fall or, in some cases, in the following spring. Occasionally fall tillage is used in preparation for spring planting. Traditional practice involved overplanting of seed followed by hand thinning and weeding. Recent developments such as monogerm seed, precision planters, electronic thinners, selective herbicides, etc., have caused hand work to be eliminated to a great extent. In some areas, sugar beets are grown in irrigated culture. Where furrow irrigation is practiced, sugar beets are grown on raised beds — bed spacing is typically 50 to 75 cm and irrigation water is supplied in the between-bed furrows. Row culture is generally practiced wherever sugar beets are grown.

During harvest or immediately prior to harvest the green leafy tops and a small bit of beet crown are removed from the beets. The fresh weight of these tops is from 70 to 100% of that of the remaining beet roots. These tops may be collected and fed to livestock or ensiled. In some cases, cattle are allowed to eat the tops in the field after harvest. Sometimes no use is made of the tops and they decompose in the field. The beet roots are lifted from the ground, cleaned of excess soil, and transported to the processing plant. The sugar content of the fresh beet root parts typically ranges from 12 to 20% by weight.

The processing plant usually produces refined white sugar (though some European plants produce raw sugar which is given final refining in some other location). In addition, molasses is produced as a byproduct, as is a wet beet pulp. Usually the pulp is subsequently dried in the processing plant to approximately 10% moisture and sold as livestock feed. The energy required for processing constitutes 60 to 80% of total energy inputs for sugar production. Of this processing energy, 20 to 30% is used for drying mechanically dewatered beet pulp. Usually the molasses is combined with the dried pulp to enhance its palatability and feeding value. Typically, the weight of the dried pulp amounts to from 5 to 7% of the fresh weight of the beet roots processed. If the molasses is not mixed with the pulp, its wet weight is approximately 2 to 6% of that of the fresh weight of beet roots, its sugar content is about 50% by weight.

In the process of purifying the sugar, lime and carbon (usually in the form of coke), are used. The weight of lime used ranges from 3 to 7% of the weight of fresh beet roots, and the weight of coke used is on the order of 10% of the weight of lime used.

The four production and processing systems illustrated here represent:

1. British practice in the period 1968 to 1973 for which data are drawn from those developed by Leach.[1] Irrigation is not involved.
2. California practice using irrigation, for the period 1972 to 1976. Data are drawn from Avlani et al.[2] in which total primary energy values are determined using a system based on input-output analysis of the U.S. economy.[3,4] Data reflect a production-weighted average of energy account figures for five California regions.
3. A historic nonirrigated system based on use of animal power and extensive human labor as reported for northern Germany in the year 1950 (approximate). Data on

* In 1975, 609,716 ha of sugar beets were harvested in the U.S. with an average yield of 43,780 kg per ha. California using irrigated production had an average yield of 58,380 kg per ha.

field practice are from Kreher[5] while those for processing are from Silin[6] and Verein Der Zuckerindustrie.[7] In this case, an attempt is made to determine the land requirements to provide feed for the work animals used, and food for the human labor employed.

4. Nonirrigated practice in western Minnesota in 1976 to 1977. Data on field operations are taken from Hvinden and Johnson.[8]

In each of these four cases, input energy requirements have been adjusted to reflect not only the fuel energy involved, but also the energy expended in the general economy to get the fuel to the point of use. This means that energy rejected in electrical generation, as well as that used for petroleum refining, is included in the figures shown. Assessments are also made for the energy used in the general economy to provide the nonenergy materials consumed in producing and processing sugar beets. Such materials are, farm tractors and machinery, irrigation equipment, fertilizers, lime, transport equipment, processing plant structure, and processing equipment.

No energy assignment is made for the life-style support of the human labor used, since this, for the most part, constitutes the final consuming sector of the economy. It is for consumption by this sector that all production expenditures of energy are made. To make an energy charge for the life-style support of this labor would thus amount to double accounting. Furthermore, the energy supplied by human labor is not tabulated because it is an insignificant part of the human contribution to the production process. In Case 3, values are indicated for food and feed energy charges that might be made, although it is illustrated how these could be converted to the form of a land requirement.

The four cases illustrated are summarized in the following tables.

Table 1
SUMMARY OF SUGAR BEET PRODUCTION ENERGY AND PROCESSING VALUES

Location	Inputs (millions of kcal/ha)	Refined sugar (kg/ha)	Energy Inputs per kg sugar[a] (kcal)	Outputs (millions of kcal/ha)	Energy Ratio (out/in)
Britain	29.7	5,016	5,924	33.8	1.14
California	72.9	7,081	10,281	51.3	0.70
Germany	32.4	4,300	7,714	33.9	1.05
Minnesota	42.4	3,564	11,878	24.1	0.57

[a] Food energy content of refined sugar is 3943 kcal/kg.[1]

Table 2
CASE 1: ENERGY REQUIREMENTS FOR SUGAR BEET PRODUCTION AND PROCESSING IN GREAT BRITAIN

Item	Per ha	kcal/ha[a]	Comments
		Input	
Labor	50 man-hr	—	Leach[1] cites a British Ministry of Agriculture bulletin giving 50 man-hr per ha as tractor driving time.
Machinery	27.38 kg	1,139,419	Machinery weight consumed is obtained by applying to the energy quantity from Leach,[1] an assumed value (in 1967) of $2.20 per kg and a value (in 1967) of 19,033 kcal/$ (75,528 Btu/$) from Bullard et al.,[3] for farm machinery. This amounts to 41,873 kcal per kg of farm machinery.
Gasoline	60 ℓ	595,101	The volume of fuel used for transport and harvest (assumed as gasoline) is obtained by applying the value of 10,109 kcal/ℓ to the energy quantity from Leach.[1]
Diesel	52 ℓ	586,089	The volume of fuel used for field work (assumed as diesel) is obtained by applying the value of 11,414 kcal/ℓ to the energy quantity from Leach.[1]
Electricity	0	—	Although electricity is used in processing, this is usually generated in excess by the processing plant itself. Processing fuel inputs are adjusted to discount fuel used for generating electricity sold out from the plant.[1]
Nitrogen	160 kg	3,057,216	
Phosphorous	50 kg (P_2O_5)	167,191	
Potassium	150 kg (K_2O)	322,440	

Table 2 (continued)
CASE 1: ENERGY REQUIREMENTS FOR SUGAR BEET PRODUCTION AND PROCESSING IN GREAT BRITAIN

Item	Per ha	kcal/ha[a]	Comments
Kainit (17% K_2O)	280 kg	102,703	
NaCl	70 kg	23, 884	
Lime (for fields)	0		
Seeds	4.5 kg	256,796	An assumed 4.5 kg/ha (4 lb per acre) is used — energy values computed as 34,394 kcal/ℓ (144 MJ/ℓ) and seed costs of £ 7.50/ha.
Irrigation	0		
Insecticides, Herbicides	10.9 kg	260,341	
Transportation	115.88 kg	29,781	Transportation includes the sum of the fuel, machinery, and seed weights — the sum of the weights is multiplied by 257 kcal/kg transported.
Processing			
Direct — includes coke for processing		19,094,702	
Indirect		4,078,480	
Total		**29,715,144**	
		Output	
Sugar beets	35,500 kg		
Refined sugar	5,016 kg	19,778,088	The sugar content of the 35,500 kg of beets harvested from one hectare was 6000 kg, but only 83.6% of this was extracted as refined sugar.[1]
Fresh beet tops	35,500 kg (20% utilization)	3,821,520	35,500 kg of fresh tops per ha at 539.8 kcal/kg (2.26 MJ/kg) were available, but these received only 20% usage as livestock feed. A 13% animal conversion ratio was used.[1]

(As human food energy after animal conversion)		(501,574)	
Protein in total tops (digestible)		711 kg	Dry matter content of tops is 16.7%. Crude protein content is 16% and digestible (by cattle) protein content is 12% on a dry matter basis.[9] However, only 20% of the tops were utilized as feed.[1]
Beet pulp	2,059 kg	5,948,451 (digestible)	The ratio of dry beet pulp to fresh beet weight was assumed to be 1 to 17.2 (5.8%). Gross energy in beet pulp (91% dry matter) is 3837 kcal/kg, but that digestible by cattle is 2889 kcal/kg.[9] Crude protein content is 9.1%, digestible (by cattle) protein content is 4.1%.[9]
Protein in pulp (digestible)	84 kg		
Molasses	1,404 kg	4,241,484 (digestible)	For each kg of sugar contained in the beets, 0.117 kg was contained in molasses.[1] Since molasses is approximately 50% sugar, the molasses produced was 1404 kg with a 6000 kg sugar yield per hectare. Digestible (by cattle) energy content of molasses is 3021 kcal/kg and digestible protein content is 3.8% as fed.[9]
Protein in molasses (digestible)	53 kg		
Total		33,789,543	
Energy ratio kcal output/ kcal input		1.14	

[a] Energy values are taken directly from Leach[1] and regrouped in the format shown.

Table 3
CASE 2: ENERGY REQUIREMENTS FOR PRODUCTION AND PROCESSING OF SUGAR BEETS IN CALIFORNIA

Item	Per ha	kcal/ha	Comments
		Input	
Labor	37.25 man-hrs	—	
Machinery			
Depreciation	23.9 kg	775,904	Depreciation of farm field machinery only, at $47.27 (1963 dollars) per hectare annually.[10] Actual dollar values were deflated to 1963 dollars so that coefficients developed by Herendeen[4] for 1963 could be applied.
Repairs		478,729	Repairs of farm field machinery only, at $56.49 (1963 dollars) per hectare annually.[10]
Vehicles; depreciation, repairs, sales		458,686	Energy consumed in manufacture, repair, and sales of farm vehicles at 75% of that in fuel consumed as computed for automobiles by Hirst.[11]
Gasoline (vehicles only)	68 ℓ	689,912	In 1972, California farm vehicles used 445,866,020 ℓ (117,798,156 gal) of gasoline and 39,542,675 ℓ (10,447,206 gal) of diesel fuel for farm-related work.[12] 1976 gross farm income in California was 9095 million dollars and income from 126,356 ha (312,100 acres) of sugar beets was $176 million. Using these figures, farm vehicle fuel use was allocated to sugar beets on a per-acre basis.
Diesel (6 ℓ for vehicles)	209 ℓ	2,385,526	Diesel fuel reported here covers that for on-farm operations and farm vehicles. Diesel fuel energy use for sugar beet transport and for processing are included under those headings.

Electricity	(2,481 kWh)		The electrical energy used here includes both direct and indirect use associated with on-farm production exclusive of transport and processing. No energy values are shown since they are included under various headings such as irrigation, nitrogen, etc. Electricity used in processing is generally generated within the processing plant itself.
Nitrogen	190 kg	3,933,046	
Phosphorous	31 kg (P_2O_5)	68,817	
Potassium	0		
Lime (for processing)	3,580 kg	1,129,311	The lime specified here is that used in the refining process in the amount of 1 kg for every 15 kg of beets. A value of 315.45 kcal per kg of lime was used. Coke is also used in the refining process at the rate of 0.1 kg per 15 kg of beets. A heating value of 6930 kcal/kg (12,500 Btu per lb) was used for coke.[13] An additional 7.8% was added to cover coke production inputs.
Coke (for processing)	358 kg	2,674,453	
Seeds	4.5 kg	58,293	Seeding rates vary from 2.24 kg to 7.28 kg per ha (2 lb to 6½ lb per acre).[10] A value of 4.48 kg per ha (4 lb per acre) was chosen. Energy content of that seed was based on the energy required for its production at a seed yield of 3368 kg per ha (3000 lb per acre) of general-run seed or 1684 kg per ha (1500 lb per acre) of highly selected seed. This results in 12,954 kcal per kg of highly selected seed. This value includes energy contained in the special fertilizer application for seed production amounting to 447 kg of total N per ha.

Table 3 (continued)
CASE 2: ENERGY REQUIREMENTS FOR PRODUCTION AND PROCESSING OF SUGAR BEETS IN CALIFORNIA

Item	Per ha	kcal/ha[a]	Comments
Irrigation			
Pumping energy	114 cm	7,222,431	Energy values presented here represent direct energy inputs to pumping obtained from production-weighted averages of values from five main production regions in California.[2]
Depreciation of irrigation equipment	19.75 kg	393,401	Indirect energy costs for depreciation of irrigation equipment represent 1976 costs of \$37.47 per hectare, or \$16.49 1963 dollars per hectare.
Insecticides, Herbicides	1.91 kg active ingredient	67,452	
Processing			
Direct		47,673,514	Direct energy requirements for processing include all inputs to the energy supply systems. Current California practice draws 48% of this energy from residual fuel oil and the balance from natural gas. Of all the energy inputs required for processing, those for pulp drying constitute 20.4% of the total.
Indirect		1,948,161	Indirect energy for beet processing was assumed to be 17.6% of the total processing energy requirements, in accordance with reports cited by Leach.[1] From this value was deducted the production input supply energy for coke (7.2% of coke total), lime, and direct processing energy (14.5% of direct processing total).
Transportation	236.15 kg	60,691	Transportation includes the sum of fuel, machinery, and seed weights — the sum is multiplied by 257 kcal/kg transported.

Transportation (haulage of beets)		2,844,648	Energy values include both direct and indirect energy inputs to the transport system. Although 60% of the beets go to the processor by truck, 88% of the kilogram-kilometers involved in beet transport occur on railroads.
Total		**72,862,945**	
		Output	
Sugar beets	53,644 kg		
Refined sugar	7,081 kg	27,951,700	Total sugar was 15% of sugar beet weight, average extraction was 88%, and energy values were obtained using 3943 kcal/kg.[1]
Beet pulp	3,832 kg	11,070,648	Beet pulp (100% dry matter) was 6.5% of sugar beet weight and the digestable (by cattle) energy content of (91% dry matter) pulp was 2889 kcal/kg.[9] Digestable protein is 4.1% (crude protein 9.1%). Molasses is 5.5% of sugar beet weight. Digestable energy in molasses is 3021 kcal/kg.[9] Digestable protein is 3.8% (crude protein 6.7%).[9]
Protein in pulp (digestible)	157 kg		
Molasses	2,950 kg	8,911,950	
Protein in molasses (digestible)	112 kg		
Beet tops	8,047 kg (dry matter)		Of fields in California 25% (approximate) have livestock pastured on beet tops after harvest. Livestock consumed 60% of the dry matter on the field (approximate). For each kg of beets produced there are also produced 0.15 kg of beet top dry matter. Caloric value (digestable by cattle) of dry matter is assumed to be 2822 kcal per kg. Crude protein content of top dry matter is 16%, while digestable (by cattle) protein content is 12%.[9]

Table 3 (continued)
CASE 2: ENERGY REQUIREMENTS FOR PRODUCTION AND PROCESSING OF SUGAR BEETS IN CALIFORNIA

Item	Per ha	kcal/ha[a]	Comments
	1,207 kg d.m. consumed	3,406,126	
Protein in total tops (digestable)	966 kg		
Total		51,340,424	
Energy ratio kcal output/ kcal input		0.70	

[a] Energy values are based on information developed by Avlani et al.[2] using an input-output method[4] for determining both direct and indirect energy inputs and categorizing each according to that as refined petroleum, electricity and natural gas. Values given here are those of total primary energy as defined by Herendeen.[4] Values represent an average for all California production.

Table 4
CASE 3: ENERGY REQUIREMENTS FOR SUGAR BEET PRODUCTION AND PROCESSING IN NORTHERN GERMANY IN 1950 USING A TEAM OF TWO HORSES FOR PRODUCTION AND TRANSPORT[a]

Item	1 ha Sugar Beets: Per ha	kcal/ha	0.41 ha to feed horses[b]	0.22 ha to feed workers[c]	Comments
Input					
Labor					
Direct	574 man-hr	(419,020)	25.5 man-hr	29.0 man-hr	Food energy for workers is 730 kcal per man-hr based on 1500 working hours per year and 3000 kcal per person per day. These energy figures are not included in the total energy figure.
Indirect	299 man-hr	(291,270)	17.7 man-hr	20.1 man-hr	Kreher[5] indicates that in-field work constitutes only 59% of total human labor on farms 19 ha and less in size. The remaining 41% is spent in barn, house, and poultry work. Energy values are not included in the total energy figure.
Horse-hr	311 horse-hr	5,417,761	20 horse-hr	22.2 horse-hr	Horse-hours are determined as the product of horses and hours. A team of two horses working for 3 hr does 6 horse-hr of work. Feed consumed by horses is assumed to be 47,727 gross kcal (assumed as hay of which approximately only 50% is digestible) per horse per day. Working hours per horse is assumed to be 1000 annually because of reduced field activity in winter.
Machinery	19.5 kg	633,060	1.7 kg	1.2 kg	
Manure	20,000 kg	3,085,952	4,800 kg	3,500 kg	Manure applied is reported by Kreher[5] to be typically 20,000 kg per crop hectare (no application is shown for pasture although this takes place naturally). Values shown are for that applied to cropland only. Manure assumed to be 30% dry matter, and of that dry matter 2.1% is N, 0.5% P, and 1.8% K.[14] Twenty tonnes per hectare is the equivalent of 126 kg N, 69 kg P_2O_5, and 130 kg K_2O per ha. The energy equivalent of these nutrients is 3,085,952 kcal. The energy available from combustion of the dry matter contained in 20,000 kg of manure (30% dry matter) would be 10,851,624 kcal.[14] The total manure produced by the livestock indicated for the 0.41 ha and 0.22 ha is only 4364 kg. Normally sugar beet production areas would be in a much smaller proportion to livestock raised than is the case for the example presented here.
Lime (for processing)	900 kg	430,200			Lime required was 3% of harvested beet weight. Much of the lime used was recaptured and used again.[6] Energy required was 478 kcal/kg.[1]
Coke (for processing)	90 kg	672,349			Coke requirement by weight is 10% that of lime.[6] Coke heat content is 6930 kcal/kg (12,500 Btu/lb).[13] An additional 7.8% is included to represent production inputs.
Seed	32 kg	98,088			Overseeding by 15 to 20 times was the European practice in the early 1950s.[6] Energy shown is 1/100 of the energy to produce 1 ha of beets since seed yield is approximately 3200 kg/ha, while seeding rate is 32 kg/ha.

Table 4 (continued)
CASE 3: ENERGY REQUIREMENTS FOR SUGAR BEET PRODUCTION AND PROCESSING IN NORTHERN GERMANY IN 1950 USING A TEAM OF TWO HORSES FOR PRODUCTION AND TRANSPORT[a]

Item	1 ha Sugar Beets		0.41 ha to feed horses[b]	0.22 ha to feed workers[c]	Comments
	Per ha	kcal/ha			
Transportation (included above)					All haulage of materials (including beets to the factory — 2 hr one way) is assumed done by horse and wagon, and is included in figures for horse power, man power, and for equipment depreciation. Per hectare values for factory haulage are 54 man-hr, 84 horse-hr, and 5.3 kg.
Coal (for processing)	2,571 kg	20,015,384			Data were taken from Verein Der Zuckerindustrie.[7] The processing plant produces its own electrical and mechanical power. A plant to process the beet to raw sugar requires 295.48 kcal per kg beet, to process the beet to refined white sugar requires 407.53 kcal per kg. Drying of pulp requires an additional 211.40 kcal per kg beet. Coal (hard) is the main fuel used. Of all the energy inputs required for processing those for pulp drying constitute 29.5% of the total.
Indirect processing energy		2,046,242			Indirect energy for processing was assumed to be 17.6% of the total processing energy requirements in accordance with Leach.[1] Deducted from this amount is 7.2% of the total coal and coke energy, as this represents indirect production supply input energy already accounted for in the energy values for these fuels. Lime supply energy is also deducted.
Total		**32,399,036**			
			Output		
Sugar beets	30,000 kg				
Refined sugar	4,200 kg	16,560,600			Traditional sugar yield is reported to have been 14% of beet weight, and the weight of fresh beet tops 74% of beet weight, according to Verein Der Zukerindustrie.[7] Refined sugar energy content was obtained using 3943 kcal/kg.[1]
Fresh beet tops	22,258 kg	8,699,472[p] (digestible)			
Protein in total tops (digestible)	370kg				The dry matter content of fresh tops was 13.85%.[7] A value of 2822 kcal digestible energy (cattle) per kg dry matter was used to estimate energy content.[9] Dry matter has a crude protein content of 16% while digestible (by cattle) protein content is 12%.[9]
Dried beet pulp	1,740 kg	5,026,860 (digestible)			Dried beet pulp yield (90% dry matter) was 5.8% of beet weight.[7] Gross energy content was 3837 kcal/kg and digestible (by cattle) energy content was 2889 kcal/kg.[9] Crude protein content is 9.1% while digestible (by cattle) protein content is 4.1%.[9]

Protein in pulp (digestible)	71 kg			
Molasses	1,200 kg	3,625,200 (digestible)		Molasses weight (contains approximately 50% sugar, 20% water, 10% ash, and 13% nonsugar, nitrogen-free organic material) is 4% of sugar beet weight.[7] Digestible (by cattle) energy in molasses is 3021 kcal/kg. Crude protein is 6.7% while digestible (by cattle) protein is 3.8%.[9]
Protein in molasses (digestible)	46 kg			
Total		33,912,132		
Energy ratio kcal output/kcal input		1.05		

[a] Data based on values given by Kreher[5] for a 25-ha farm powered by two horses. The two horses are able (at 1000 hr each of field work annually) to power the production of their own feed (on 2.5 ha) and the production and transport of 6.04 ha of sugar beets. Practical scheduling requires that the 6.04 ha be actually distributed in small portions on a number of farms growing an assortment of crops and having extensive livestock production.

[b] The 0.41 ha shown is that required to feed the two horses (16.5% of their annual feed requirement) for production of their own feed on the 0.41 ha, and for the production of 1 ha of sugar beets through the use of 16.5% of their annual work output. This 0.41 ha consists of 0.165 ha hay (yield = 6900 kg/ha), 0.165 ha meadow pasture (yield = 6900 kg hay equivalent per ha), and 0.08 ha small grain (yield = 3800 kg/ha).

[c] The 0.22 ha shown is that required to feed the workers that operate: (a) the 0.22 ha, (b) the 0.41 ha for horses feed production, and (c) the 1 ha of sugar beets. The 0.22 ha includes land needed for producing the feed for the horses which power operations on the 0.22 ha. Food produced is that for workers only (not their families) computed at 3000 kcal per day and 1500 farm-working hours per man-year. The 0.22 ha is composed of 0.044 ha hay (yield = 6900 kg/ha), 0.044 ha meadow pasture (yield = 6900 kg hay equivalent/ha), 0.015 ha fodder beets (yield = 65,000 kg/ha), 0.022 ha potatoes (yield = 30,000 kg/ha), and 0.095 ha of small grain (yield = 3800 kg/ha). Of the small grain, 0.022 ha is used for food grains and the balance for feed. The livestock fed are 0.029 horse (plus 0.33 horse on the 0.41 ha), 0.088 cows (giving 3636 kg of milk per year per cow), and 0.293 pigs (yielding 63.6 kg of dressed pork per pig). Consumable food energy is assumed to be 80% of that produced due to losses, etc. (edible part of potatoes is 81% of tuber weight).

Table 5
MACHINERY REQUIREMENTS FOR SUGAR BEET FARMING WITH A TEAM OF OF TWO HORSES IN NORTHERN GERMANY IN 1950 (CASE 3)

			Hours used annually		
Machine	Assumed Mass-kg	Assumed Life-hr	1 ha Sugar beets	0.41 ha Horse feed	0.22 ha Human food
Plow (1 bottom)	100	2500	19.9		0.8
Plow (2 bottom)	300	2500		0.8	0.9
Wagon	500	4000	98.0	4.8	3.7
Drag	100	2500	1.5		0.3
Harrow	300	2500	7.9	0.7	0.5
Digger	300	2500	6.6		0.2
Roller	1000	2500	3.0		
Drill	500	1500	4.9	0.3	0.3
Cultivator	300	2500	5.5		0.2
Mower	400	1500		0.7	0.2
Rake	300	2500		1.0	0.2
Tedder	400	1500		0.5	0.1
Binder	1000	1500		0.5	0.3
Beet digger	300	2500	9.9		0.1
Sprayer	300	1500			0.1

Table 6
CASE 4: ENERGY REQUIREMENTS FOR SUGAR BEET PRODUCTION AND PROCESSING IN WESTERN MINNESOTA

Item	Per ha	kcal/ha	Comments
		Input	
Labor[a]	31.9 man-hrs		
Machinery[a]	12.82 kg	536,808	Based on as assumed value (in 1967) of \$2.20 per kg and a value (in 1967) of 19,033 kcal/\$ (75,528 Btu/\$). This amounts to 41,873 kcal/kg of farm machinery.[3] (See Table 7.)
Gasoline[a]	121.8 ℓ	1,231,276	Of this amount of gasoline, 121.5 ℓ was used for vehicles. This use includes haulage from field to processor.
Diesel[a]	150.9 ℓ	1,722,282	
Electricity (for processing)	982 kWh	2,810,540	
Nitrogen	168 kg	3,477,641	
Phosphorus	50 kg (P_2O_5)	183,500	Computed using 3670 kcal/kg P_2O_5 which represents an averaged of values form several references.
Lime (for processing)	18.3 kg	571,911	Computed using 315.45 kcal/kg.
Coke (for processing)	217 kg	1,621,107	Computed using 6430 kcal/kg (12,500 Btu/1b) coke.[13] An additional 7.8% was included to represent production energy inputs. Coke requirement is 0.007 kg/ kg of beets.

Table 6 (continued)
CASE 4: ENERGY REQUIREMENTS FOR SUGAR BEET PRODUCTION AND PROCESSING IN WESTERN MINNESOTA

Item	Per ha	kcal/ha	Comments
Seed	2.24 kg	28,997	Seed is produced in southeastern U.S. Energy equivalent of highly selected seed used is 12,945 kcal/kg.
Herbicides	4.18 kg active ingredient	253,857	Computed using a mean value of 60,700 kcal/kg active ingredient for 14 herbicides.
Insecticides	1.96 kg active ingredient	86,701	Computed using a mean value of 44,128 kcal/kg active ingredient for five insecticides.
Lignite (for processing)	5887 kg	25,366,219	Computed using a heat content of North Dakota lignite of 3997 kcal/kg (7210 Btu/lb), from di Lorenzi.[13] An additonal 7.8% was added to cover production inputs for this lignite.
Coal-Hard (for processing)	310 kg	2,413,350	
Transportation (haulage of beets—indirect)		745,431	Computed as 75% of the energy content of fuels used by highway transport vehicles (hauling truck and pickup truck included) as is suggested by Hirst[11] for automobiles, to cover energy use represented by depreciation, sales, repairs, highways, etc.
Transportation	219.59 kg	56,435	The transportation figure includes the sum of weights of machinery, fuel (for production only), and seeds multiplied by 257 kcal/kg transported.
Depreciation of processing plant		788,697	Based on \$0.0032 per kg (\$2.91 per short ton) of beets (1977 to 1978 average), \$2 (in 1977) = \$1 (in 1967) and 15,903 kcal/\$ (63,106 Btu/\$) (in 1967) from Bullard et al.[3] for an average on new construction and special industry machinery sectors.
Repairs and maintenance of processing plant		494,630	Based on \$0.0025 per kg (\$2.27 per short ton) of beets (1977 to 1978 average), \$2 (in 1977) = \$1 (in 1967) and 12,785 kcal/\$ (50,735 Btu/\$) (in 1967) from Bullard et al.[3] for an average of maintenance and repair construction and auto repairing sectors.
Total		**42,389,382**	
		Output	
Sugar beets	30,987 kg		
Refined sugar	3,564 kg	14,051,024	Refined sugar output is 0.115 kg/ kg of beets. Energy computed at 3943 kcal/kg refined sugar.[1]
Beet tops	6,182 kg (dry matter)		The amount shown for dry matter of tops is for reference only, since generally, none of the tops are used for livestock feed.

Table 6
CASE 4: ENERGY REQUIREMENTS FOR SUGAR BEET PRODUCTION AND PROCESSING IN WESTERN MINNESOTA

Item	Per ha	kcal/ha	Comments
Molasses	1,549 kg	4,679,529 (digestible)	The weight of molasses produced is 5% of the weight of sugar beets. Energy content (digestible by cattle) of molasses as fed is 3021 kcal/kg[9] and digestible (by cattle) protein is 3.8% of the weight of molasses as fed.[9]
Protein in molasses (digestible)	59 kg		
Dried beet pulp	1,859 kg	5,371,286 (digestible)	The weight of dried beet pulp (91% dry matter) produced is 6% of weight of sugar beets.
Energy (digestable by cattle)			Content of the pulp as fed is 2889 kcal/kg,[9] and digestable (by cattle) protein is 4.1% of the weight of pulp as fed.[9]
Protein in beet pulp (digestible)	76 kg		
Total		**24,101,839**	
Energy Ratio kcal Output/kcal Input		0.57	

[a] Data on machinery, labor, and fuel from Hvinden and Johnson.[8]

Table 7
MACHINERY USE FOR SUGAR BEET PRODUCTION IN WESTERN MINNESOTA (CASE 4)

Machine	Mass-kg	Life-hr	Hr/ha-year[8]	Kg/ha-year
Plow	1,818	2,500	.37	.27
Disc harrow	2,727	2,500	.52	.56
Multiweeder	2,614	2,500	.35	.36
Drag	864	2,500	.12	.04
Planter	2,636	,1200	.37	.81
Cultivator	2,273	2,500	1.36	1.23
Beet thinner	1,545	2,500	.20	.12
Rotary hoe	2,273	2,500	.22	.20
Rotobeater	1,364	2,000	.74	.50
Harvester	5,136	2,500	.81	1.67
Sprayer	1,477	2,500	.04	.02
Pickup truck	1,818	2,500	.62	.45
Truck	3,636	4,000	3.95	3.59
275 Hp tractor	12,500	12,000	1.23	1.28
120 Hp tractor	6,202	12,000	1.14	.58
100 Hp tractor	5,087	12,000	2.69	1.14
Total				**12.82**

Note: Figures given are for the approximate mass (weight/gravitational acceleration) of the machine used.

REFERENCES

1. **Leach, G.,** *Energy and Food Production,* International Institute for Environment and Development, 1525 New Hampshire Ave., N.W., Washington, D.C., 1975.
2. **Avlani, P. K., Singh, R. P., and Chancellor, W. J.,** Energy Consumption in Sugar Beet Production and Processing in California, presented at the First International Congess on Engineering and Food, Boston, August, 1976.
3. **Bullard, C., Hannon, B., and Herendeen, R.,** Energy Flow Through the United States Economy — wall chart, University of Illinois Press, Urbana, 1975.
4. **Herendeen, R. A.,** The Energy Cost of Goods and Services, ORNL-NSF-EP-58, Oak Ridge National Laboratory, Oak Ridge, Tenn., 1973.
5. **Kreher, G.,** Leistungszahlen fur Arbeitsvoranschlage und Der Arbeitsvoranschlag im Bauernhof, "Studiengesellschaft fur landwirtschaftliche Arbeitswirtschaft e. V.," Stutgart, 1955.
6. **Silin, P. M.,** Technology of Beet-Sugar Production and Refining (Tekhnologiya sveklosakharnogo i rafinadnogo proizvodstva), PPI Pishchpromizdat, Moskova, 1958, translated from Russian by Israel Program for Scientific Translation, Jerusalem, 1964.
7. **Verein Der Zuckerindustrie,** *Technologie Des Zuckers,* M. and H. Schaper, Hannover, 1955.
8. **Hvinden, S. C. and Johnson, R. G.,** *Preliminary 1977 Sugar Beet Production Costs,* Department of Agricultural Economics, North Dakota State University, Fargo, 1978, 93.
9. National Academy of Sciences United States-Canadian Tables of Feed Consumption, Publication 1684, (2nd revision) Washington, D.C., 1969.
10. **Reed, A. D.,** *Sugar Beet Production Costs in California,* Leaflet No. 2877, Division of Agricultural Sciences, University of California, Berkeley, March, 1976.
11. **Hirst, E.,** How Much Overall Energy Does the Automobile Require?, *Automot. Eng.,* 80(7), 36, 1972.
12. **Cervinka, V., Chancellor, W. J., Coffelt, R. J., Curley, R. G., and Dobie, J. B.,** Energy Requirements for Agriculture in California, California Department of Food and Agriculture, and Agricultural Engineering Department, University of California, Davis, January, 1974.
13. **de Lorenzi, O., Ed.,** *Combustion Engineering,* 1st ed., Combustion Engineering-Superheater, Inc., New York, 1951.
14. **Azevedo, J. and Stout, P. R.,** *Farm Animal Manures,* Manual 44, Division of Agricultural Sciences, University of California, Berkeley, August, 1974.

ALFALFA

G. H. Heichel and N. P. Martin*

INTRODUCTION

Alfalfa (*Medicago sativa* L.), a perennial legume introduced into the southwestern U.S. in 1840 and into Minnesota in 1858, was the first domesticated perennial forage crop of this nation. Its principal uses are as hay, silage, and pasture for ruminant animals and as a source of semiprocessed protein for nonruminants like hogs and poultry. Currently, alfalfa sprouts sometimes garnish salads and sandwiches, but alfalfa has substantial potential as a primary source of purified plant protein for human diets.

The alfalfas introduced into the southwest lacked winterhardiness, but during the period of their introduction and spread, varieties became classified as "hardy", "medium", and "nonhardy".[1] For example, those introduced into Minnesota have given rise to the more winterhardy types adapted throughout the Great Plains, north central, and northeastern states. Alfalfa is grown in monoculture, in rotations with grain crops, and in mixtures with various species of forage grasses.

Approximately 10.9 million ha of alfalfa for hay and 19 million ha of alfalfa in mixture with other forage species were grown in the U.S. in 1976.[4] The value of alfalfa is difficult to estimate because only 21% of the 1976 crop entered commerce; the remainder was used directly for animal feed. Using a recent market price of $66/metric ton, the 64 million tons of alfalfa and alfalfa-based hay produced in 1976 would be valued at $4.2 billion before being marketed through animals. To sustain this acreage, more than $80 million of alfalfa seed was produced in 1976,[4] chiefly in California, Idaho, and Washington. Energy analyses of seed production are discussed elsewhere in this volume.

The leading north central and Great Plains states in 1976 alfalfa hectarage were Wisconsin (1.22 million), South Dakota (0.93 million), Minnesota (0.89 million), Iowa (0.71 million), Nebraska (0.69 million), and North Dakota (0.66 million). The average alfalfa yield in these six states was 4.5 tons/ha,[4] although individual farmers in some of these states often attain 18 tons/ha. In the eastern states where rainfall is more plentiful, yields in New York (0.40 million ha), Pennsylvania (0.33 million ha), and Ohio (0.22 million ha) average 6.3 tons/ha. The irrigated nonhardy alfalfa in California (0.45 million ha) and Arizona (0.09 million ha) averaged 14.5 tons/ha in 1976.[4] Lack of adaptation precludes large-scale alfalfa production in the deep southeastern U.S.

Alfalfa is sown, managed, and harvested with a variety of cultural techniques that greatly influence the energetics of its production. Alfalfa may be established in the spring by seeding it with a marketable small grain companion crop that also suppresses weeds or by direct seeding with herbicides. In both arid and humid regions of the U.S., alfalfa may also be direct seeded in the spring or summer without a herbicide or companion crop. In the energy audits that follow, the fossil energy for seed production, processing, and distribution is charged to the establishment year of the crop rather than being prorated over establishment and several production years.

Cultural practices required to manage the post-seeding alfalfa stands include fertil-

* Cooperative investigation of Science and Education Administration — Agricultural Research, U.S. Department of Agriculture, and the Minnesota Agricultural Experiment Station (Scientific Journal Series Paper No. 10432).

ization, disease and insect control, and cutting frequency. Disease-resistant alfalfa varieties have reduced the need for fungicides and chemicals to control diseases, but over 1.0 million kg of insecticides are applied annually in the U.S. to alfalfa.[2] This is only 8% of the U.S. cropland, but insecticide usage modifies the energetics of alfalfa production. Insect and disease problems occur more readily in areas of high rainfall and humidity or under irrigated conditions.

During the 2 to 4 post-seeding years of full production, the crop may be harvested two to four times per season if a hardy variety is used, or seven to nine times annually for irrigated nonhardy varieties in the southwestern U.S. Nationwide, 2.0 million ha or 22% of the alfalfa grown in 1969 were irrigated, the largest hectarage of any crop.[3] After harvest, the crop may be preserved in the form of bales of different sizes, chopped dry hay, chopped silage, dehydrated pellets, or compressed cubes. The harvesting and preservation methods strongly modify the energetics of crop culture.

Because alfalfa is a legume capable of deriving 50 to 85% of its total nitrogen requirements from symbiotic fixation, little nitrogen fertilizer is used on the crop. Thus, the subsequent energy analyses will show little energy attributed to nitrogen, except when alfalfa is seeded with a marketable small grain, in contrast with annual grains, vegetables, and fruits. Harvested alfalfa contains 5 to 7 kg P_2O_5 and 20 to 27 kg k_20 per ton of hay. Thus, the energy budgets reflect P_2O_5 and K_2O applications to maintain the soil nutrient levels. Alfalfa requires substantial amounts of calcium for optimum growth and symbiotic N_2-fixation capacity, a need usually met by limestone additions to maintain soil water pH at 6.5 to 7.0. Because the benefit of limestone is seldom realized when it is applied in the establishment year, the fossil energy for limestone is prorated over all of the full production years of the alfalfa crop.

The energy analyses in the following tables are intended to exemplify the diversity of alfalfa management techniques and harvesting methods in the north central U.S. They clearly show that the energetic efficiency of the cropping system is dependent upon the environment of growth, the methods of culture and management, and the form of the harvested product. Alfalfa hay production in southeastern Minnesota appears to be one of the most energy efficient cropping systems in the U.S.

Table 1
AN ENERGY AUDIT OF ALFALFA ESTABLISHMENT WITH AN OAT COMPANION CROP IN SOUTHEASTERN MINNESOTA (ESTABLISHMENT YEAR, OAT GRAIN HARVESTED; RAINFED, TEMPERATE CLIMATE)

Item	Quantity/ha	kcal/ha
Input		
Labor	5.3 hr	—
Machinery	1745 kg	33,403
Liquid fuel	90 ℓ	1,027,260
Electricity	—	—
Nitrogen	17 kg	97,300
Phosphorus (P_2O_5)	—	—
Potassium (K_20)	136 kg	249,900
Limestone[a]	—	—
Seed	13 kg alfalfa, 67 kg oats	1,074,000
Irrigation water	—	—
Insecticides	1.2 kg	29,000
Herbicides	—	—
Drying	—	—
Transportation	2027 kg	520,939
Total fossil energy input		**3,031,802**
Output		
Dry matter yield (10% H_2O)	2548 kg oat grain	
Protein yield (H_2O-free)	305 kg oat protein	
Digestible energy output/ fossil energy input		2.82

[a] Prorated over production years.

Table 2
AN ENERGY AUDIT OF ALFALFA ESTABLISHMENT WITH AN HERBICIDE IN SOUTHEASTERN MINNESOTA (ESTABLISHMENT YEAR, TWO BALED-HAY HARVESTS; RAINFED, TEMPERATE CLIMATE)

Item	Quantity/ha	kcal/ha
Input		
Labor	6.4 hr	—
Machinery	1465 kg	30,375
Liquid fuel	95 ℓ	1,084,330
Electricity	—	—
Nitrogen	—	—
Phosphorus (P_2O_5)	45 kg	49,500
Potassium (K_2O)	136 kg	217,600
Limestone[a]	—	—
Seed	14 kg	868,000
Irrigation water	—	—
Insecticides	—	—
Herbicides	5.7 kg	138,000
Drying	—	—
Transportation	1717 kg	441,269
Total fossil energy input		**2,829,074**
Output		
Dry matter yield (15% H_2O)	5682 kg	
Protein yield (H_2O-free)	889 kg	
Digestible energy output/ fossil energy input		4.28

[a] Prorated over production years.

Table 3
AN ENERGY AUDIT OF BALED ALFALFA HAY PRODUCTION IN SOUTHEASTERN MINNESOTA (SECOND OR THIRD PRODUCTION YEAR; RAINFED, TEMPERATE CLIMATE)

Item	Quantity/ha	kcal/ha
Input		
Labor	7.2 hr	—
Machinery	852 kg	19,906
Liquid fuel	92 ℓ	1,050,090
Electricity	—	—
Nitrogen	—	—
Phosphorus (P_2O_5)	34 kg	37,400
Potassium (K_2O)	68 kg	108,800
Limestone	2652 kg	835,390
Seed	—	—
Irrigation water	—	—
Insecticides	—	—
Herbicides	—	—
Drying	—	—
Transportation	3655 kg	939,335
Total fossil energy input		**2,990,921**
Output		
Dry matter yield (15% H_2O)	10,000 kg	
Protein yield (H_2O-free)	1,580 kg	
Digestible energy output/ fossil energy input		7.20

Table 4
AN ENERGY AUDIT OF BALED ALFALFA HAY PRODUCTION IN SOUTHEASTERN MINNESOTA (SECOND OR THIRD PRODUCTION YEAR; IRRIGATED, TEMPERATE CLIMATE)

Item	Quantity/ha	kcal/ha
Input		
Labor	11.8 hr	—
Machinery[a]	772 kg	20,285
Liquid fuel	76 ℓ	867,460
Electricity	—	—
Nitrogen	—	—
Phosphorus (P_2O_5)	34 kg	37,400
Potassium (K_2O)	135 kg	216,000
Limestone	2652 kg	835,390
Seed	—	—
Irrigation water[b]	10 cm	696,250
Insecticide	—	—
Herbicides	—	—
Drying	—	—
Transportation	3633 kg	933,681
Total fossil energy input		**3,606,466**
Output		
Dry matter yield (15% H_2O)	11,800 kg	
Protein yield (H_2O-free)	1,845 kg	
Digestible energy output/ fossil energy input		6.98

[a] Exclusive of irrigation system.
[b] Energy for pumping H_2O.

Table 5
AN ENERGY AUDIT OF OAT SILAGE — BALED ALFALFA HAY PRODUCTION IN SOUTHEASTERN MINNESOTA (ESTABLISHMENT YEAR, OAT SILAGE AND ONE CUT OF BALED HAY HARVESTED; IRRIGATED, TEMPERATE CLIMATE)

Item	Quantity/ha	kcal/ha
	Input	
Labor	13.7 hr	—
Machinery[a]	1736 kg	52,857
Liquid fuel	133 ℓ	1,518,060
Electricity	—	—
Nitrogen	56 kg	823,200
Phosphorus (P_2O_5)	34 kg	37,400
Potassium (K_2O)	67 kg	107,200
Limestone	—	—
Seed	17 kg alfalfa, 84 kg oats	1,390,000
Irrigation water[b]	10 cm	696,250
Insecticides	—	—
Herbicides	—	—
Drying	—	—
Transportation	2065 kg	530,705
Total fossil energy input		**5,155,682**
	Output	
Dry matter yield	20650 kg	
(68% H_2O oat silage)		
(15% H_2O hay)	4200 kg	
Protein yield (H_2O-free)	2003 kg oat protein 657 kg alfalfa protein	
Digestible energy output/ fossil energy input		5.07

[a] Exclusive of irrigation system.
[b] Energy for pumping H_2O.

Table 6
AN ENERGY AUDIT OF BALED ALFALFA HAY PRODUCTION IN EASTERN COLORADO (SECOND OR THIRD PRODUCTION YEAR; RAINFED, SEMIARID, TEMPERATE CLIMATE)

Item	Quantity/ha	kcal/ha
Input		
Labor	18.8 hr	—
Machinery	681 kg	14,300
Liquid fuel	142 ℓ	1,620,790
Electricity	—	—
Nitrogen	—	—
Phosphorus (P_2O_5)	91 kg	100,100
Potassium (K_2O)	—	—
Limestone	—	—
Seed	—	—
Irrigation water	—	—
Insecticides	1.2 kg	29,000
Herbicides	—	—
Drying	—	—
Transportation	845 kg	217,165
Total fossil energy input		**1,981,335**
Output		
Dry matter yield (15% H_2O)	1718 kg	
Protein yield (H_2O-free)	269 kg	
Digestible energy output/fossil energy input		1.85

Table 7
AN ENERGY AUDIT OF PRODUCING DEHYDRATED ALFALFA PELLETS IN COLORADO (SECOND OR THIRD PRODUCTION YEAR; RAINFED, SEMIARID, TEMPERATE CLIMATE)

Item	Quantity/ha	kcal/ha
Input		
Labor	20 hr	—
Machinery	1560 kg	32,760
Liquid fuel	208 ℓ	2,374,110
Electricity	—	—
Nitrogen	—	—
Phosphorus (P_2O_5)	227 kg	249,700
Potassium (K_2O)	—	—
Limestone	—	—
Seed	—	—
Irrigation water	—	—
Insecticides	1.2 kg	29,000
Herbicides	—	—
Drying	—	560,000
Transportation	1893 kg	486,500
Total fossil energy input		**3,732,070**
Output		
Dry matter yield (8% H_2O)	1594 kg	
Protein yield (H_2O-free)	306 kg	
Digestible energy output/fossil energy input		1.07

REFERENCES

1. **Lowe, C. C., Marble, V. L., and Rumbaugh, M. D.,** Adaptation, varieties, and usage, in *Alfalfa Science and Technology,* Hanson, C. H., Ed., American Society of Agronomy, Inc., Madison, Wis., 1972.
2. **Radcliffe, E. B.,** The Influence of Insects on Alfalfa Production, Proc. 8th Annual Alfalfa Symp., Bloomington, Minn., 1978.
3. **Smith, N. L.,** Opportunities for Energy Savings in Crop Production, Report to the Subcommittee on Advanced Energy Technologies and Energy Conservation Research, Development, and Demonstration, U.S. House of Representatives, U.S. Government Printing Office, Washington, D.C., 1978.
4. U.S. Department of Agriculture, Agricultural Statistics, Washington, D.C., 1977.

ENERGY INPUTS IN HAY PRODUCTION

Sterling Chick and John Krummel

Over 34.5 million acres of hay, other than alfalfa, is harvested yearly in the U.S.[12] Every state has significant acreage in hay production and only corn, wheat, and soybeans have more acreage harvested each year. The leading states in hay production include Texas, Missouri, Kentucky, New York, Nebraska, and Wisconsin. While most of the hay that is harvested is used directly on the farm, hay (excluding alfalfa) has an estimated market value of over $2.85 billion.[12]

Generally, hay is a generic term describing a myriad of plant species that have all of their above ground plant parts cut, dried, harvested, and stored for future use as livestock feed. While legumes yield the most nutritious hay, mixtures of legumes and grass often give an optimum return of stand durability, yield, and quality of forage. The plant species that comprise hay often grow in areas unsuitable for row crops, provide excellent protection against soil erosion, and offer a relatively cheap source of nutrients and energy for ruminants when pasture is not available.

Make hay while the sun shines. Indeed, the labor required and the uncertainty of weather conditions means that harvesting must be done efficiently and quickly. Also, the cutting, drying, and harvesting operation must be done during the times of the year when farmers have to deal with other important projects. Thus most farmers mechanize their harvest operations to reduce the labor input and shorten the time required to make hay. While there are a number of ways to make hay, the method of harvest we analyze represents average conditions (Table 8). Most farmers cut hay and rake it into rows to dry in the sun. To avoid serious wetting from a rainy period, farmers, especially in the East and South, try to shorten the field drying time. However, while hay has to be dry for storage, the leaves must not be so dry that they are shattered and lost during the harvest. The mower-conditioner greatly improves the field drying and forage quality. Hay is usually baled to increase storage capacity and improve transport and handling efficiency of this bulky crop. The PTO baler reduces labor, and presently over 90% of the hay harvested in the U.S. is baled.[3]

While hay is often harvested several times per year, the inputs we show in Tables 1-7 represent those that are necessary for one harvest operation. For the hay systems in states we analyze, fertilizer inputs are relatively small (Tables 1-7), while lime inputs, in soils of low pH, increase the hay yields. If water is pumped from groundwater supplies, irrigation represents the most energy intensive operation in hay production (Table 7). However, in the absence of irrigation, hay, compared to other crops,requires low energy inputs.

Table 1
ENERGY INPUTS PER HECTARE FOR HAY IN NEW YORK

Item	Quantity/ha	kcal/ha	Ref.
	Input		
Labor	7.0 hr	—	9
Machinery	22.94 kg[a]	404,250	
Gasoline	7.76 ℓ	78,446	1
L.P. gas	2.76 ℓ	21,266	1
Electricity	5.19 kWh	14,859	1
Nitrogen	3.18 kg	46,746	9
Phosphorus	3.18 kg	3,498	9
Potassium	13.62 kg	21,792	9
Herbicides	0.091 kg	9,092	
Seed	1.14 kg[b]	23,940	
Transportation	31.53 kg[c]	8,103	
Total		**631,992**	
	Output		
Hay yield	2,539 kg	5,458,850[d]	9
Protein yield	160.0 kg[d]		
kcal output/kcal input		8.6	
kcal output/labor hour		779,835	

[a] See Table 8.
[b] Assuming $5.00 per lb.[9]
[c] For machinery and fuels (other transportation figures included in kcal calculations).
[d] Timothy (sun-cured) as fed to cattle contains 2150 kcal/kg digestible energy and 6.3% protein.[6]

Table 2
ENERGY INPUTS PER HECTARE FOR HAY IN GEORGIA

Item	Quantity/ha	kcal/ha	Ref.
	Input		
Labor	7.0 hr[a]	—	
Machinery	22.94 kg[b]	404,250	
Gasoline	10.22 ℓ	103,314	1
L.P. gas	2.69 ℓ	20,726	1
Electricity	4.4 kWh	12,597	1
Nitrogen	42.68 kg	627,396	1
Phosphorus	11.35 kg	12,485	1
Potassium	11.35 kg	18,160	1
Lime	290.6 kg	91,670	8
Herbicides	0.0045 kg	450	1
Seed	1.14 kg	23,940	2
Transportation	33.33 kg[c]	8,566	
Total		**1,323,554**	
	Output		
Hay yield	2,177 kg	4,680,550[d]	1
Protein yield	137.2 kg[d]		
kcal output/kcal input		3.5	
kcal output/labor hour		668,650	

[a] Assumed similar to New York.[9]
[b] See Table 8.
[c] For machinery and tools.
[d] Timothy (sun-cured) as fed to cattle contains 2150 kcal/kg digestible energy and 6.3% protein.[6]

Table 3
ENERGY INPUTS PER HECTARE FOR HAY IN MISSOURI

Item	Quantity/ha	kcal/ha	Ref.
	Input		
Labor	4.7 hr[a]	—	
Machinery	22.94 kg[b]	404,250	
Gasoline	9.08 ℓ	91,790	1
L.P. gas	2.65 ℓ	20,418	1
Electricity	4.9 kWh	14,029	1
Nitrogen	36.32 kg	533,904	1
Phosphorus	11.35 kg	12,485	1
Potassium	2.27 kg	3,632	1
Lime	290.6 kg	91,670	8,13
Herbicides	0.0045 kg	450	1
Seed	1.14 kg	23,940	2
Transportation	32.45 kg[c]	8,340	
Total		**1,204,908**	
	Output		
Hay yield	1,451 kg	3,119,650[d]	1
Protein yield	91.4 kg[d]		
kcal output/kcal input		2.6	
kcal output/labor hour		663,755	

[a] Assumed similar to Arkansas.[4]
[b] See Table 8.
[c] For machinery and fuels.
[d] Timothy (sun-cured) as fed to cattle contains 2150 kcal/kg digestible energy and 6.3% protein.[6]

Table 4
ENERGY INPUTS PER HECTARE FOR HAY IN WISCONSIN

Item	Quantity/ha	kcal/ha	Ref.
	Input		
Labor	4.7 hr[a]	—	
Machinery	22.94 kg[b]	404,250	
Gasoline	14.76 ℓ	149,209	1
Diesel	0.15ℓ	1,712	1
L.P. gas	3.33 ℓ	25,658	1
Electricity	7.37 kWh	21,100	1
Nitrogen	9.5 kg	139,650	1
Lime	290.6kg	91,670	1
Herbicide	0.0045 kg	450	1
Seed	1.14 kg	23,940	2
Transportation	37.23 kg[c]	9,567	
Total		**876,206**	
	Output		
Hay yield	1,814 kg	3,900,100[d]	1
Protein yield	114.3 kg[d]		
kcal output/kcal input		4.5	
kcal output/labor hour		829,809	

[a] Assumed similar to Arkansas.[4]
[b] See Table 8.
[c] For machinery and fuels.
[d] Timothy (sun-cured) as fed to cattle contains 2150 kcal/kg digestible energy and 6.3% protein.[6]

Table 5
ENERGY INPUTS PER HECTARE FOR HAY IN SOUTH DAKOTA

Item	Quantity/ha	kcal/ha	Ref.
	Input		
Labor	2.0 hr[a]	—	
Machinery	22.94 kg[b]	404,250	
Gasoline	13.40 ℓ	135,461	1
Diesel	0.11 ℓ	1,256	1
L.P. gas	3.52 ℓ	27,122	1
Electricity	5.41 kWh	15,489	1
Nitrogen	0.005 kg	74	1
Herbicides	0.0045 kg	450	1
Seed	1.14 kg	23,940	2
Transportation	36.29 kg[c]	9,326	
Total		**617,368**	
	Output		
Hay yield	726 kg	1,560,900[d]	1
Protein yield	45.7 kg[d]		
kcal output/kcal input		2.5	
kcal output/labor input		780,450	

[a] After Colorado.[2]
[b] See Table 8.
[c] For machinery and fuels.
[d] Timothy (sun-cured) as fed to cattle contains 2150 kcal/kg digestible energy and 6.3% protein.[6]

Table 6
ENERGY INPUTS PER HECTARE FOR HAY IN COLORADO

Item	Quantity/ha	kcal/ha	Ref.
	Input		
Labor	1.99 hr	—	2
Machinery	22.94 kg[a]	404,250	
Gasoline	6.06 ℓ	61,261	2
Diesel	11.96 ℓ	136,511	2
Herbicides	0.1589 kg[b]	15,876	
Seed	1.14 kg	23,940	2
Transportation	37.60 kg[c]	9,662	
Total		**651,500**	
	Output		
Hay yield	907 kg	1,950,050[d]	2
Protein yield	57.1 kg[d]		
kcal output/kcal input		3.0	
kcal output/labor hour		979,925	

[a] See Table 8.
[b] Assumes $5/lb for insecticide.[2]
[c] For machinery and fuels.
[d] Timothy (sun-cured) as fed to cattle contains 2150 kcal/kg digestible energy and 6.3% protein.[6]

Table 7
ENERGY INPUTS PER HECTARE FOR HAY IN ARIZONA

Item	Quantity/ha	kcal/ha	Ref.
	Input		
Labor	2.35 hr	—	14
Machinery	22.94 kg[a]	404,250	
Gasoline	21.88 ℓ	221,185	1
Irrigation fuels			
Water (acre ft. applied)	2.7	—	1
Natural gas (m^3)	286	3,378,518	1
Electricity (kWh)	1,470	377,790	1
Nitrogen	0.23 kg	3,381	1
Herbicides	0.0045 kg	450	1
Seed	1.14 kg	23,940	2
Transportation	40.49 kg[b]	10,406	
Total		**4,419,920**	
	Output		
Hay yield	2,086 kg	4,484,90[c]	1
Protein yield	131.4 kg[c]		
kcal output/kcal input		1.0	
kcal output/labor hour		1,908,468	

[a] See Table 8.
[b] For machinery and fuels.
[c] Timothy (sun-cured) as fed to cattle contains 2150 kcal/kg digestible energy and 6.3% protein.[6]

Table 8
MACHINERY REQUIREMENTS FOR HAY IN COLORADO

Type[2]	HP[2]	Weight (kg)	Hours/hectare[2]	Lifetime (year)[5]	Ref.
Tractor	75	3,260	0.78	12	7
Tractor	105	3,710	0.98	12	7
Pickup ½ ton	250	1,900	1.18	12	7
Mower conditioner (or SP swather)		1,320	0.29	10	10
Tandem disks		2,090	0.20	12	10
Drill		900	0.15	14	10
Harrow-spike		450	0.07	12	7
PTO baler		1,300	0.69	8	7
Hay wagon		2,180	0.49	12	7

Average farm size	65 ha
Total machinery weight	17,110 kg
Weight per hectare per year	22.94 kg

REFERENCES

1. Energy and U.S. Agriculture: 1974 Data Base, Office of Energy Conservation and Environment, Federal Energy Administration, FEA/D-76/459, 1976.
2. U.S. Department of Agriculture, Economic Research Service, and Firm Enterprise Data System, 1975, Oklahoma State University, Stillwater, 1977.
3. **Fergunson, W. L., Strickler, P. E., and May, R. C.,** Hay harvesting practices and labor used, 1967, 48 States, Economic Research Service, U.S. Department of Agriculture, Washington, D.C., Rep. 460, 1971.
4. **Halbrook, W. A., Denton, D. C., Spooner, A. E., and Ray, M. L.,** Production items, estimated costs, and returns for beef cattle and selected forage crops, upland soils of the upper Arkansas river valley area, Report Series 177, Agricultural Experimental Station, University of Arkansas, Fayetteville, 1969.
5. **James, S. C. and Stoneberg, E.,** Farm Accounting and Business Analysis, Iowa State University Press, 1974.
6. National Academy of Sciences, Atlas of nutritional data on United States and Canadian feeds, National Academy of Sciences, Washington, D.C., 1971.
7. **McDonald, J. L., Ed.,** *Power Farming Technical Annual,* Pacific Publications, Sydney, Australia, 1975.
8. **Saunders, F. B. and Moss, R. B.,** Costs and returns for selected crop enterprises at the southwest Georgia branch experiment station 1972—1974, Research Report 207, Code Station, University of Georgia, Athens, 1975.
9. **Snyder, D. P.,** Field crops costs and returns from farm cost accounts, A.E. Res. 76-25, Department of Agricultural Economics, Cornell University, Ithaca, New York, 1976.
10. Range Seeding Equipment Handbook, U.S. Department of Agriculture, U.S. Department of the Interior, Forest Service Handbook, Washington, D.C., 1965.
11. Hay Harvesting Practices and Labor Used, 1967, 48 States, Economic Research Service and Statistical Report Service, U.S. Department of Agriculture Statistical Bulletin 460, U.S. Department of Agriculture, Washington, D.C., 1971.
12. Agricultural Statistics, 1977. U.S. Department of Agriculture, U.S. Government Printing Office, Washington, D.C., 1977.
13. **Wise, J. O.,** *Selected Crop and Livestock Budgets for the Piedmont Area of Georgia,* Research Report 188, Department Agricultural Economics, University of Georgia, Athens, 1974.
14. **Wright, N. G., Stubblefield, T. M., Ott, G. O., Gorman, W. D., Lansford, R. R., and Swope, D. A.,** Cost of Producing Crops in the Irrigated Southwest, Part II — New Mexico, Technical Bulletin 222, Agricultural Experimental Station, University of Arizona, Tucson, 1976.

ENERGY INPUT AND PRODUCTION FOR CORN SILAGE

Wayne R. Knapp

In areas where corn or maize can be grown, corn silage will produce more kilocalories of digestible energy per hectare than any other feed crop. Corn silage also has the features of consistently high digestibility and of being well suited to efficient harvesting, storing, and feeding systems.

During the past 10 years, corn harvested for silage in the U.S. has accounted for approximately 12% of the total area planted to corn each year. In 1977, this amounted to approximately 4,500,000 ha of corn silage. Silage yields for all this area have averaged approximately 8400 kg of dry matter per hectare.[2] Recent values for corn silage have been in the range of 5 to 7¢/kg of dry matter. This puts the annual value of corn silage in the U.S. at over $2¼ billion. Most corn silage is fed on the farm where it is grown, so except in cases such as a neighbor buying a silage crop, there is not a cash market for silage.

Since corn silage is a livestock feed, its production is centered in areas with large numbers of dairy or beef cattle. States that lead in corn silage production each year and their average annual number of hectares over the past 10 years are South Dakota (400,000), Wisconsin (378,000), Minnesota (303,000), Iowa (265,000), New York (232,000), Nebraska (220,000), Pennsylvania (161,000), Michigan (158,000), Kansas (128,000),North Dakota (126,000), Illinois (114,000), Colorado (105,000), Ohio (90,000), and Virginia (83,000). Corn silage is also a major crop in the New England states.[2]

Corn being grown for silage has the same basic environmental requirements as corn intended for grain. Since corn is harvested for silage at a less mature stage than a grain crop, however, it is possible to grow silage corn in regions where the growing season is too short or cool for satisfactory grain maturation. Water requirements are similar, regardless of how the crop is used. In areas receiving less than 15 cm of rainfall during the growing season, irrigation is necessary. In addition, irrigation will increase silage yields in areas receiving 15 to 50 cm of rainfall through the summer and also in regions where the rainfall is poorly distributed through the growing season.[8] Irrigation, of course, increases the energy requirements in production. In states such as the Dakotas, Minnesota, Iowa, and Missouri, corn intended for grain, but stricken by drought during the growing season, is frequently harvested for silage, since this does allow the crop to be salvaged.

For best performance, silage corn needs a soil pH of 6.0 or above.[3] This necessitates lime applications on acid soils. Adequate levels of nitrogen, phosphorus, and potassium must also be available in the soil. A corn silage crop yielding 10,500 kg of dry matter per hectare will contain approximately 200 kg of N, 40 kg of P (91 kg of P_2O_5), and 195 kg of K (236 kg of K_2O). Approximately 70% of the N, 80% of the P, and 20% of the K will be in the grain at maturity.[7] Therefore, a grain corn crop and a silage crop will remove similar amounts of nitrogen and phosphorus at harvest. However, the high potassium content of corn stover results in a much greater removal of K from the soil by a silage crop than by a grain crop. In most corn producing regions, at least part of these nutrients need to be supplied by chemical fertilizers if high yields are to be obtained. This in turn requires energy consumption. Some soils also require applications of other nutrients such as zinc for good corn silage production. However, these minor nutrients will not be considered in these energy budgets.

Cultural practices involving energy inputs are similar for both silage and grain corn.

In most regions, conventional tillage is used for silage production, but minimum tillage or no tillage is widely used in some areas. Highest silage yields generally result from somewhat higher plant population levels than for grain corn. Usually about 5000 additional plants per hectare are desirable for silage, so this means a slightly greater input in seed.

Diseases are a relatively minor concern with corn silage. In most production areas, the corn rootworm is the major insect affecting corn grown for silage. Soil insecticide applications are often a standard procedure for controlling this pest. In some situations, later spraying may be needed for controlling European corn borers or corn leaf aphids, but this is not a routine operation. Weed control in all areas is now based on herbicide applications. Most corn silage land now receives herbicides. However, the herbicide materials used vary with tillage methods and weed problems. Cultivation for additional weed control is still widely used in the central and western U.S.

Energy requirements for harvesting corn silage are greater than for grain corn because of the high power required for the chopper and the blower.

USING THE TABLES

The tables of energy inputs and returns for corn silage are based on geographical regions of production and different basic production procedures. Different regions of the country vary enough in their production operations and yields to permit a breakdown into five general areas. The northeastern U.S. includes New England, New York, Pennsylvania, and Maryland. The Cornbelt and Lake states consist of Michigan, Ohio, Indiana, Illinois, Iowa, Missouri, southern Minnesota and Wisconsin, and eastern Nebraska and Kansas. The states south of and including Kentucky and Virginia make up the southeastern region. The northern region consists of North and South Dakota and northern Minnesota and Wisconsin. The western region is all the area west of central Nebraska and Kansas. A budget for no tillage production in the Cornbelt is also given. Little corn is used for silage in developing nations, but a final budget is presented to give a comparison of a high labor system with the highly mechanized ones.

Machinery inputs are presented separately for each budget as equipment most commonly used in each region varies.[4] The average number of hectares on which the machinery is used differs for each region depending on the average area of corn silage produced on a farm.[1]

The fuel requirements given first are for the smaller tractor being gasoline powered and the larger one used for tillage and chopping being a diesel. In addition, however, if both are gasoline engines, the value for gasoline is shown in the parenthesis.[5]

The fertilizer rates given are for soils having medium phosphorus and potassium levels and for continuous corn production. Manure applications, cover crops, or a rotation with a legume are not considered in establishing these rates, so if one of these is used, the fertilizer rates should be adjusted accordingly. The energy requirements assume the nitrogen is being supplied by anhydrous ammonia (82-0-0), the phosphorus by triple superphosphate (0-46-0), and the potassium by muriate of potash (0-0-60). Two fertilizer rates are given. The first is the average rate applied by producers in the region in 1973.[3] The second in the parenthesis is the average rate recommended by agronomists in the region.[4] Phosphorus and potassium rates are in terms of P_20_5 and K_2O, respectively. The rates of lime are also suggested levels for best corn growth.[6]

Unless otherwise noted, the herbicide given is a combination of 1.7 kg/ha active ingredient of atrazine and 1.7 kg/ha active ingredient of alachlor or a similar material. The insecticide listed is for a 1.1 kg/ha active ingredient application of carbofuran for corn rootworm control. In rotations or where rootworms are not a problem, this would not be applied, so the percentage of corn land receiving it varies among regions.[1]

Transportation was computed for the total weight of machinery, fuel, fertilizers, and seed used per crop hectare.

Two values are also given for each of the output categories. The first in each is based on the average yields obtained over the past 10 years in the region.[2] This uses the average fertilizer rates applied by farmers. The second, in parenthesis, is based on production using the recommended rates of fertilizer.[6] Protein yield is valued as 8% of the dry matter production.

Table 1
ENERGY INPUTS AND RETURNS PER HECTARE OF CORN SILAGE IN THE NORTHEASTERN U.S. AND SOUTHEASTERN CANADA

Item	Quantity/ha	kcal/ha
	Input	
Labor	14.5 hr	—
Machinery	59.7 kg[a]	1,074,600
Gasoline	62.0 ℓ	626,800
	(145.0 ℓ)	(1,465,800)
Diesel	59.3 ℓ	676,850
Nitrogen	106 kg	1,272,000
	(110 kg)	(1,320,000)
Phosphorus	47 kg	141,000
	(75 kg)	(225,000)
Potassium	54 kg	86,400
	(65 kg)	(104,000)
Lime	960 kg	302,800
Seed	22 kg[b]	550,000
Insecticide	1.1 kg[c]	152,300
Herbicide	3.4 kg	277,000
Transportation	178.6 kg	45,900
		5,205,650
Total		**(5,355,250)**
	Output	
Silage dry matter yield	9,400 kg	29,070,700
	(11,425 kg)	(35,300,200)
kcal output/kcal input		5.58
		(6.59)
Protein yield	753 kg	
	(914 kg)	
kcal output/labor hour		2,004,900
		(2,434,500)

[a] Conventional tillage with moldboard plow, spring tooth harrow, planter, 125-hp tractor, 60-hp tractor, tank sprayer, fertilizer sidedress unit or manure spreader, chopper, two forage wagons, and pickup truck. Total weight of 17.4 metric tons of machinery used on an average of 25 ha and depreciated over 12 years.

[b] Final plant population of 61,800 plants per hectare.

[c] Applied to approximately 50% of the corn in this area.

Table 2
ENERGY INPUTS AND RETURNS PER HECTARE OF CORN SILAGE IN THE CORNBELT AND LAKE STATES

Item	Quantity/ha	kcal/ha
	Input	
Labor	14.7 hr	—
Machinery	59.8 kg[a]	1,076,400
Gasoline	65.7 ℓ	664,200
	(160.8 ℓ)	(1,625,500)
Diesel	67.4 ℓ	769,300
Nitrogen	128 kg	1,536,000
	(180 kg)	(2,160,000)
Phosphorus	49 kg	147,000
	(80 kg)	(240,000)
Potassium	84 kg	134,400
	(135 kg)	(216,000)
Lime	560 kg	176,400
Seed	19 kg[b]	475,000
Insecticide	1.1 kg[c]	152,300
Herbicide	3.4 kg	277,000
Transportation	185.4 kg	47,648
		5,455,648
Total		**(6,254,248)**
	Output	
Silage dry matter yield	9,140 kg	28,239,500
	(11,425 kg)	(35,300,200)
kcal output/kcal input		5.18
		(5.64)
Protein yield	731 kg	
	(914 kg)	
kcal output/labor hour		1,921,100
		(2,401,400)

[a] Conventional tillage with moldboard plow, disk, fertilizer spreader or anhydrous ammonia applicator, tank sprayer, spike tooth harrow, planter, rotary hoe, cultivator, 140-hp tractor, 80-hp tractor, fertilizer sidedress unit or manure spreader, chopper, tao forage wagons, and pickup truck. Total weight of 23.3 metric tons of machinery used on an average of 34 ha and depreciated over 12 years.

[b] Final plant population of 54,300 plants per hectare.

[c] Applied to approximately 65% of the corn in this area.

Table 3
ENERGY INPUTS AND RETURNS PER HECTARE OF CORN SILAGE IN THE SOUTHEASTERN U.S.

Item	Quantity/ha	kcal/ha
	Input	
Labor	14.5 hr	—
Machinery	39.0 kg[a]	702,000
Gasoline	59.5 ℓ	601,500
	140.5 ℓ	(1,420,300)
Diesel	57.8 ℓ	659,700
Nitrogen	151 kg	1,812,000
	(200 kg)	(2,400,000)
Phosphorus	62 kg	186,000
	(85 kg)	(255,000)
Potassium	81 kg	129,600
	(140 kg)	(224,000)
Lime	930 kg	292,950
Seed	19 kg[b]	475,000
Insecticide	1.1 kg[c]	152,300
Herbicide	3.4 kg	277,000
Transportation	151.8 kg	39,013
Total		**5,327,063**
		(6,078,463)
	Output	
Silage dry matter yield	8,530 kg	26,370,000
	(10,080 kg)	(31,147,200)
kcal output/kcal input		4.95
		(5.12)
Protein yield	683 kg	
	(806 kg)	
kcal output/labor hour		1,818,600
		(2,148,100)

[a] Conventional tillage with moldboard plow or chisel plow, fertilizer spreader, tank sprayer, harrow, planter, 125-hp tractor, 70-hp tractor, manure spreader, chopper, two forage wagons, and pickup truck. Total weight of 19.0 metric tons of machinery used on an average of 40 ha and depreciated over 12 years.

[b] Final plant population of 54,300 plants per hectare.

[c] Applied to approximately 20% of the corn in this area.

Table 4
ENERGY INPUTS AND RETURNS PER HECTARE OF CORN SILAGE IN THE NORTHERN U.S.

Item	Quantity/ha	kcal/ha
	Input	
Labor	13.3 hr	—
Machinery	61.5 kg[a]	1,107,000
Gasoline	60.0 ℓ	606,500
	(161.5 ℓ)	(1,632,600)
Diesel	72.4 ℓ	826,400
Nitrogen	73 kg	876,000
	(90 kg)	(1,080,000)
Phosphorus	27 kg	81,000
	(35 kg)	(105,000)
Potassium	19 kg	30,400
	(15 kg)	(24,000)
Lime	0 kg	—
Seed	14 kg[b]	350,000
Insecticide	1.1 kg[c]	152,300
Herbicide	2.2 kg[d]	124,300
Transportation	182.0 kg	46,774
		4,200,674
Total		**(4,422,274)**
	Output	
Silage dry matter yield	5240 kg	16,197,800
	(7390 kg)	(22,841,300)
kcal output/kcal input		3.86
		(5.17)
Protein yield	419 kg	
	(591 kg)	
kcal output/labor hour		1,217,900
		(1,717,400)

[a] Conventional tillage with moldboard plow, tandem disk, anhydrous ammonia applicator, spring tooth harrow, tank sprayer, planter 125-hp tractor, 90-hp tractor, cultivator, chopper, three 2-ton trucks, and pickup truck. Total weight of 27.0 metric tons of machinery used on an average of 36 ha and depreciated over 12 years.

[b] Final plant population of 39,500 plants per hectare.

[c] Applied to approximately 25% of the corn in this area.

[d] Consists of atrazine at 1.7 kg/ha active ingredient and 2,4-D at 0.5 kg/ha active ingredient.

Table 5
ENERGY INPUTS AND RETURNS PER HECTARE OF IRRIGATED CORN SILAGE IN THE WESTERN U.S.

Item	Quantity/ha	kcal/ha
	Input	
Labor	11.8 hr	—
Machinery	55.1 kg[a]	991,800
Gasoline	64.7 ℓ	654,100
	(158.5 ℓ)	(1,603,300)
Diesel	66.9 ℓ	763,600
Nitrogen	143 kg	1,608,000
	(180 kg)	(2,160,000)
Phosphorus	36 kg	108,000
	(65 kg)	(195,000)
Potassium	22 kg	35,200
	(0 kg)	(0)
Lime	0 kg	0
Seed	23 kg[b]	575,000
Irrigation (water applied)	50 cm[c]	6,175,800
Insecticide	1.1 kg[d]	152,300
Herbicide	3.4 kg	277,000
Transportation	183.5 kg	47,160
		11,387,960
Total		**(11,991,760)**
	Output	
Silage dry matter yield	12,360 kg	38,207,800
	(14,110 kg)	(43,606,100)
f1p08kcal output/kcal input		3.64
		(3.36)
Protein yield	986 kg	
	(1,129 kg)	
kcal output/labor hour		3,237,900
		(3,695,400)

[a] Conventional tillage with moldboard plow or offset disk, fertilizer spreader, anhydrous ammonia applicator, spring tooth mulcher or sweep, planter, cultivator, cultivator-sidedress unit, 150-hp tractor, 125-hp tractor, chopper, three 2-ton trucks, and pickup truck. Total weight of 29.5 metric tons of machinery used on an average of 45 ha and depreciated over 12 years.

[b] Final plant population of 66,700 plants per hectare.

[c] Average amount of water applied in western Great Plains region. Energy requirements based on a center pivot sprinkler powered by electricity and having a lift of 50 m.

[d] Applied to approximately 70% of the corn in this area.

Table 6
ENERGY INPUTS AND RETURNS PER HECTARE OF NO-TILLAGE CORN SILAGE IN THE CORNBELT AND LAKE STATES

Item	Quantity/ha	kcal/ha
	Input	
Labor	11.9 hr	—
Machinery	44.4 kg[a]	799,200
Gasoline	40.7 ℓ	411,400
	(100.9 ℓ)	(1,020,000)
Diesel	43.1 ℓ	491,900
Nitrogen	128 kg	1,536,000
	(180 kg)	(2,160,000)
Phosphorus	49 kg	147,000
	(80 kg)	(240,000)
Potassium	84 kg	134,400
	(135 kg)	(216,000)
Lime	560 kg	176,400
Seed	21 kg[b]	525,000
Insecticide	1.1 kg[c]	152,300
Herbicide	2.3 kg[d]	172,100
Transportation	132.6 kg	34,078
Total		**4,579,778**
		(5,378,378)
	Output	
Silage dry matter yield	9,140 kg	28,239,500
	(11,425 kg)	(35,300,200)
kcal output/kcal input		6.17
		(6.56)
Protein yield	731 kg	
	(914 kg)	
kcal output/labor hour		2,373,100
		(2,966,400)

[a] No-tillage system with a no-till planter, fertilizer spreader or manure spreader, tank sprayer, 140-hp tractor, 80-hp tractor, chopper, two forage wago s, and pickup truck. Total weight of 17.3 metric tons of machinery used on an average of 32 ha and depreciated over 12 years.

[b] Final plant population of 54,300 plants per hectare.

[c] Applied to essentially all of the corn planted no-till.

[d] Consists of atrazine at 1.7 kg/ha active ingredient and paraquat at 0.5 kg/ha active ingredient.

Table 7
ENERGY INPUTS AND RETURNS PER HECTARE OF CORN SILAGE PRODUCED USING ONLY MANPOWER[6]

Item	Quantity/ha	kcal/ha
Input		
Labor	620 hr	—
Ax, hoe, and corn knife	0.6 kg[a]	12,500
Nitrogen	11 kg[b]	161,700
Phosphorus	4 kg	10,800
Potassium	6 kg	9,600
Seed	10.4 kg	36,600
Total		**231,200**
Output		
Silage dry matter yield	2000 kg[c]	6,204,700
kcal output/kcal input		26.8
Protein yield	160 kg	
kcal output/labor hour		10,000

[a] Assumes implements weigh 4 kg, last 5 years, and are used on 1.3 ha.
[b] Provided by ammonium nitrate.
[c] Based on grain yield and assuming the grain makes up 50% of the total weight.

REFERENCES

1. Bureau of the Census, 1969 Census of Agriculture, Special Reports, Vol. 5, Parts 1 and 8, U.S. Department of Commerce, Washington, D.C.
2. Crop Reporting Board, Statistical Reporting Service, U.S. Department of Agriculture Statistical Bulletin, 582, 1977.
3. Crop Reporting Board, Statistical Reporting Service, Crop Production, U.S. Department of Agriculture, 1974.
4. Extension agronomists at respective Land Grant Universities, personal communication.
5. **Gunkel, W. W., Price, D. R., Lucas, G. M., Murray, D. L., Casler, G. L., and Supper, S.,** *Energy Requirements for New York State Agriculture,* Part 1, Cornell University Agricultural Engineering Extension Bulletin 405, Cornell University, Ithaca, N.Y.
6. **Akinwumi, J. A.,** Cost and return of commercial maize production in Savannah Belt of western Nigeria, *Bull. Rural Econ. Sociol.,* 6(2), 219, 1971.
7. **Larson, W. E. and Hanway, J. J.,** *Corn and Corn Improvement,* Am. Soc. Agronomy, Madison, Wis., 1976, 625.
8. **Shaw, R. H.,** *Corn and Corn Improvement,* Am. Soc. Agronomy, Madison, Wis., 1976, 591.

Energy Inputs and Outputs for Crop Systems — Vegetables

CABBAGE

R. Brian How

Currently about 40,500 ha of cabbage is growing in the U.S. annually. In 1976 an average yield of 10.9 tons per acre resulted in total production of 1.1 million metric tons with a total farm value of $113 million. Cabbage hectarage for harvest is fairly evenly distributed throughout the year, but yields and consequently production are higher in the summer and the fall.

Florida leads other states in total cabbage production, with harvest spread over the fall, winter, and spring seasons amounting to about 20% of total U.S. output. New York ranks second with harvest concentrated in the late summer and fall equal to about 16% of the total. Texas and Wisconsin follow Florida and New York in order of production, and together the four leading states account for about 60% of U.S. cabbage output.

In the U.S., most cabbage is marketed in fresh form, either as whole heads or chopped as cole slaw. Many different varieties of cabbage are grown for fresh sale, including some with purple as well as the customary green leaves. Fresh market varieties are generally of the Danish type with small compact heads. Customarily, fresh cabbage has been prepared for serving by boiling, but increasingly is served raw as a salad vegetable. Consumption of cole slaw has increased with the addition of this item to the offerings at fast-food restaurants. The increased demand for cabbage for cole slaw has in turn expanded the market opportunities for fall harvested storage cabbage. Methods of storing cabbage have also been improved, including limited use of refrigerated and controlled-atmosphere systems. As a result of these changes in market demands and storage techniques, the proportion of the total crop harvested in the northern states of New York, Wisconsin, and Ohio in the fall has tended to increase.

About 20% of the cabbage grown commercially in the U.S. is processed into sauerkraut. Varieties grown for sauerkraut typically are of the domestic type, with larger and looser leaves than those grown for fresh market. Processing cabbage for sauerkraut involves coring and chopping cabbage, adding salt, and then storing the chopped cabbage in large vats to permit the pickling process to take place. About two thirds of the cabbage made into sauerkraut is grown and processed in New York and Wisconsin, with Ohio and several other states accounting for the remainder. Most commercially produced sauerkraut is packed in cans for distribution with canned fruits and vegetables. A small proportion of the product is packaged in film bags and marketed and merchandized under refrigeration.

The traditional method of growing cabbage has been to start plants in a seed bed or a greenhouse and then to transplant 4- to 6-week-old plants into open fields. In northern states, young plants are sometimes obtained from growers in southern regions. In recent years, an increasing acreage of cabbage has been grown from seed planted directly and then thinned later. Direct seeding avoids the labor required to transplant from seed bed to field, but requires special attention to thinning, increased use of seed, and a possible delay in the maturity of the crop. Thinning of direct seeded cabbage has generally been done by hand, but mechanical equipment is being developed to do this job. In the meantime, many growers feel the saving in transplanting more than offsets the added costs of seed and labor for thinning.

Cabbage is fertilized by broadcasting before plowing, by deep drilling after plowing, and by banding at time of planting. The growing crop is side dressed with nitrogen, the amount depending on rainfall since preplant fertilizer was applied.

Weed control is achieved by cultivation and chemical applications. Cabbage is susceptible to numerous diseases and insects. Chief among the diseases are club root, blackleg, damping-off, and downy mildew. Serious insect pests include root maggots, aphids, worms, and loopers. Biological materials as well as chemicals are used to control cabbage loopers.

Cabbage for sauerkraut is generally delivered direct to the processor from the field. Cabbage for fresh market harvested in the fall may be sold from the field or placed in common storage for 2 to 3 months, in refrigerated storage for 5 to 6 months, or in limited quantities in controlled atmosphere storage for 5 to 8 months. At other seasons of the year, cabbage for fresh sale is largely marketed directly from the field.

Table 1
CABBAGE: FLORIDA, FRESH MARKET, TRANSPLANTED, HAND HARVESTED

Item	Quantity/ha	kcal/ha
	Input	
Labor	240 hr	—
Gasoline	460 ℓ	4,650,140
Diesel	158 ℓ	1,803,412
Machinery	170 kg	3,060,000
Nitrogen	146 kg	1,752,000
Phosphorus	140 kg	420,000
Potassium	138 kg	220,800
Seed	0.2 kg	6,000
Herbicide	2.2 kg	219,802
Insecticide	6.8 kg	590,988
Fungicide	4.5 kg	292,095
Containers	10 kg	120,000
Electricity	5 kWh	14,315
Buildings	2 m^2	6,840
Transportation	659.5 kg	169,492
Total		**13,325,884**
	Output	
Cabbage yield	26,945 kg	6,466,800
Protein yield	539 kg	
kcal output/kcal input		0.49
kcal output/labor hour		26,945

Note: 257 kcal/kg transported.

Table 2
CABBAGE: NEW YORK, SAUERKRAUT, DIRECT SEEDED, MACHINE HARVESTED

Item	Quantity/ha	kcal/ha
Input		
Labor	95 hr	—
Gasoline	420 ℓ	4,245,780
Diesel	355 ℓ	4,051,970
Machinery	300 kg	5,400,000
Nitrogen	198 kg	2,376,000
Phosphorus	185 kg	555,000
Potassium	182 kg	291,200
Seed	0.2 kg	6,000
Herbicide	2.2 kg	219,802
Insecticide	5.6 kg	486,696
Fungicide	3.4 kg	220,694
Electricity	3 kWh	8,589
Buildings	2 m²	6,840
Transportation	917.0 kg	235,669
Total		**18,104,240**
Output		
Cabbage yield	67,363 kg	16,167,120
Protein yield	1,347 kg	
kcal output/kcal input		0.89
kcal output/labor hour		170,180

Table 3
CABBAGE: NEW YORK, FRESH MARKET, TRANSPLANTED, HAND HARVESTED

Item	Quantity/ha	kcal/ha
Input		
Labor	289 hr	—
Gasoline	562 ℓ	5,681,258
Diesel	198 ℓ	2,259,972
Machinery	225 kg	4,050,000
Nitrogen	198 kg	2,376,000
Phosphorus	185 kg	555,000
Potassium	182 kg	291,200
Seed	0.2 kg	6,000
Herbicide	2.2 kg	219,802
Insecticide	5.6 kg	486,696
Fungicide	3.4 kg	220,694
Containers	30 kg	360,000
Electricity	20 kWh	57,260
Buildings	10 m²	34,200
Transportation	815 kg	209,455
Total		**16,807,537**
Output		
Cabbage yield	53,000 kg	12,720,000
Protein yield	1,060 kg	
kcal output/kcal input		0.76
kcal output/labor hour		44,014

REFERENCES

1. Vegetables — Fresh Market, Annual Summary, Crop Reporting Board, Statistical Reporting Service, U.S. Department of Agriculture, 1977.
2. **Snyder, P.,** *Fruit and Vegetable Crops Costs and Returns From Farm Cost Accounts 33 Farmers 1975,* A. E. Res. 76-26, Department of Agricultural Economics, Cornell University, Ithaca, N.Y., 1976.
3. **Brooke, D. L.,** Costs and Returns from Vegetable Crops in Florida, Economics Mimeo Report, Department of Agricultural Economics, University of Florida, Gainesville.
4. *Cornell Recommendations For Commercial Vegetable Production,* New York State College of Agriculture and Life Sciences, Cornell University, Ithaca, N.Y., 1978.

ENERGY USE IN FLORIDA CELERY PRODUCTION

J. W. Mishoe and Jose Alvarez

Celery is one of the most important vegetables in the U.S. Over 33,000 acres of celery are harvested every year for a total production close to 17 million cwt and a value between $130 million and $150 million. Celery has ranked fourth and sixth, by value, among all U.S. vegetables during the past 3 years (Table 1).

The main celery producing states in the U.S. are California and Florida. They dominate the market during the winter and spring seasons. Other states (New York, Ohio, Michigan, and Washington) contribute small quantities during the summer and fall seasons (Table 2).

This distribution of production is the result of the climate requirements. Celery cannot tolerate prolonged and/or severe freezing temperatures. The best growing period is during the dry, cool winter months with daily temperatures ranging from 50 to 73°F.[2]

In many cases, the harvesting and farm processing equipment is one of a kind with much on-farm fabrication. The production of the crop begins in the seedbed with 1 ha of seedbed, in Florida, producing enough transplants for 25 ha of celery in the field.

The following tables use values taken from the Florida industry; therefore, they should be considered only representative of other areas. The major differences in production techniques in different areas occur mainly in seedbed production. In Florida, the seedbeds must be shaded during most of the growing season. Pest control is quite extensive, including the practice of flooding and drying cycles of the land prior to seeding which is designed to destroy most soil-borne pests. In some of the other areas of the U.S., seedlings are produced in greenhouses, thereby reducing production cost but increasing capital outlay cost.

The crop is grown as an extensive, highly mechanized operation. Specialized equipment for harvesting and preparation for shipment make celery production and marketing highly dependent on energy.

Table 1
U.S. CELERY COMMERCIAL AREA, PRODUCTION AND VALUE, 1975, 1976, and 1977

Year	Harvested area (1000 ha)	Production (1000 kg)	Value: Per 1000 kg ($)	Value: Total ($ 1000)	Rank by value
1975	12.72	7194	16.42	118,122	6
1976	13.48	7684	17.35	133,336	4
1977	13.44	7549	19.63	148,193	6

From Vegetable Situation, TVS-207, Economics, Statistics, and Cooperative Services, U.S. Department of Agriculture, February 1978.

Table 2
AVERAGE U.S. CELERY PRODUCTION BY SEASON AND REGION, 1973 TO 1975

Season and region	Average production 1973 to 1975 (1000 cwt)
Winter	
Florida	2052
California	2207
Spring	
Florida	1501
California	2771
Summer	
New York	270
Ohio	99
Michigan	861
California	2005
Fall	
New York	74
Michigan	146
Florida	613
Washington	112
California	3650

From Usual Planting and Harvesting Dates for Fresh Market and Processing Vegetables, Agricultural Handbook No. 507, Statistical Research Service, U.S. Department of Agriculture, February, 1977.

Table 3
ENERGY INPUTS PER HECTARE FOR CELERY SEEDBED IN THE EVERGLADES[a]

Inputs	Quantity/ha	kcal/ha	Ref.
Labor	1,617 hr	—	1
Machinery	3 kg	54,000	1
Diesel	32 ℓ	365,248	4
Nitrogen	337 kg[b]	4,951,000	1
Phosphorus	112 kg[c]	123,500	1
Potassium	225 kg[d]	359,000	1
Seeds	3.5 kg	10,500	1
Irrigation (water applied)	23.9 ha-cm[e]	293,114	2
Insecticides	135 kg[f]	8,281,685	1
Herbicides	467 ℓ[g]	1,018,568	1
Transportation	2,329 kg[h]	599,000	
Soil fumigant	224.5 kg[i]	6,234,365	1
Shading	1,509,000 kcal	1,509,000	1
Total		**23,798,980**	

[a] All input requirements are for 1 ha of seedbed which will produce enough plants to set out 25 ha of field-set celery. They include all activities performed on the seedbed from plowing until plants are ready to be transplanted.
[b] Sodium nitrate. 14.7 Mcal/kg N.
[c] 1.1 Mcal/kg P.
[d] 1.6 Mcal/kg K_2O.
[e] 23.9 (NIR + 20″).
[f] Inorganic coppers, organic fungicides, organic phosphate insecticides.
[g] Mineral spirits. Assumes a production input of 2179 kcal/ℓ.
[h] Includes seed, machinery, fuel, and shading materials.
[i] Methyl bromide, chloropicrin.

Table 4
ENERGY INPUTS PER HECTARE FOR CELERY SEEDBED IN CENTRAL FLORIDA[a]

Inputs	Quantity/ha	kcal/ha	Ref.
Labor	1,601 hr	—	1
Machinery	3 kg	54,000	1
Diesel	32 ℓ	365,248	4
Nitrogen	123.5 kg[b]	1,815,000	1
Phosphorus	179.6 kg[c]	197,600	1
Potassium	190.9 kg[d]	305,400	1
Lime	336.8 kg	106,243	1
Seed	3.8 kg	11,400	1
Irrigation (water applied)	23.2 ha-cm[e]	287,789	2
Insecticides	84 kg[f]	5,176,053	1
Herbicides	467 ℓ[g]	1,018,568	1
Transportation	2,206 kg	567,000	
Soil fumigant	552 kg[h]	15,329,040	1
Shading	1,509,000 kcal	1,509,000	1
Soil amendments	1,123 kg[i]	—	1
Total		**26,742,341**	

[a] All input requirements are for 1 ha of seedbed which will produce enough plants to set out 25 ha of field-set celery. They include all activities performed on the seedbed from plowing until plants are ready to be transplanted.
[b] Actual N. 14.7 Mcal/kg N.
[c] 1.1 Mcal/kg P.
[d] 1.6 Mcal/kg K_2O.
[e] 23.2 (NIR + 20'').
[f] Inorganic coppers, organic fungicides, organic phosphate insecticides.
[g] Mineral spirits. Assumes a production input of 2,179 kcal/ℓ.
[h] Vapam®.
[i] Castor pomace, sludge, tankage.

Table 5
ENERGY INPUTS PER HECTARE FOR CELERY FIELD IN THE EVERGLADES

Item	Quantity/ha	kcal/ha	Ref.
	Inputs		
Labor	674 hr	—	1
Machinery	2 kg	36,000	1
Diesel	373 ℓ	4,257,422	4
Nitrogen	64.5 kg[a]	948,200	1
Phosphorus	360 kg[b]	396,000	1
Potassium	598 kg[c]	956,800	1
Irrigation	3.9 ha-cm[d]	140,988	2
Insecticides	135 kg[e]	8,281,685	1
Transportation	1,868 kg	480,000	
Precooling and refrigeration	12.4 kWh	35,500	
Seedbed energy[f]	—	951,959	
Total		**16,484,554**	
	Outputs		
Celery field yield	51,820 kg/ha	8,809,000	5,7
Protein yield	466.4		7
kcal output/kcal input		0.534	
kcal output/labor hours		13,070	

[a] 14.7 Mcal/kg.
[b] 1.1 Mcal/kg.
[c] 1.6 Mcal/kg.
[d] 3.9 NIR ha-cm.
[e] Inorganic coppers, organic fungicides, organic phosphate insecticides.
[f] Total inputs for 1-ha seedbed per 25 ha.

Table 6
ENERGY INPUTS PER HECTARE FOR CELERY FIELD IN CENTRAL FLORIDA

Item	Quantity/ha	kcal/ha	Ref.
	Inputs		
Labor	746 hr	—	1
Machinery	2 kg	36,000	1
Diesel	373 ℓ	4,257,422	4
Nitrogen	168.4 kg[a]	2,475,500	1
Phosphorus	168.4 kg[b]	185,200	1
Potassium	404.2 kg[c]	646,700	1
Lime	1,684 kg[d]	531,246	1
Irrigation	3.2 ha-cm[e]	135,663	2
Insecticides	78 kg[f]	4,830,983	1
Transportation	1,941 kg	499,000	
Precooling and refrigeration	12.4 kWh	35,500	1
Seedbed[g]	—	1,069,693	
Total		**14,702,907**	
	Outputs		
Celery field yield	51,820 kg/ha	8,809,000	5,7
Protein yield	466.4 kg		7
kcal output/kcal input		0.599	
kcal output/labor hour		11,808	

[a] 14.7 Mcal/kg.
[b] 1.1 Mcal/kg.
[c] 1.6 Mcal/kg.
[d] Crushed and ground lime. 315.45 kcal/kg.
[e] 3.2 NIR ha-cm.
[f] Inorganic coppers, organic fungicides, organic phosphate insecticides.
[g] Total inputs for 1-ha seedbed per 25 ha.

Table 7
MATERIALS USED IN A-FRAME SHADING OF SEEDBEDS ON A PER ACRE BASIS

4200 ft of 1 × 2 in. wood (used 12 times)
18,000 ft of 12-gauge galvanized wire (used 12 times)
40 wooden posts, 4 in. diameter × 3 ft long (used 12 times)
6000 ft muslin in 5-ft-wide strips (used eight times)
1500 nails and staples

REFERENCES

1. **Brooke, D. L.,** Labor and Material Requirements, Costs and Returns for Celery by Areas in Florida, Economics Mimeo Rep. EC70-2, Department of Agricultural Economics, Florida Agricultural Experiment Stations, Institute of Food and Agricultural Sciences, University of Florida, Gainesville, August 1969.
2. **Guzman, V. L., Burdine, H. W., Harris, E. D., Jr., Orsenigo, J. R., Sholwater, R. K., Thayer, P. L., Winchester, J. A., Wolf, E. A., Berger, R. D., Genung, W. G., and Zitter, T. Z.,** Celery Production on Organic Soils of South Florida, Bulletin 757, Florida Agricultural Experiment Stations, Institute of Food and Agricultural Sciences, University of Florida, Gainesville, September 1973.
3. **Rogers, J. S. and Marlowe, G. A., Jr.,** Water Needs of Florida Vegetable Crops, WRC-2, Water Resources Council, Institute of Food and Agricultural Sciences, University of Florida, Gainesville.
4. **Smerdon, E. T.,** Fuel Use Estimates for Florida Agriultural Production. Agricultural Growth in an Urban Age, Institute of Food and Agricultural Sciences, University of Florida, Gainesville; By IFAS Energy Committee, E. T. Smerdon, Chm., Undated.
5. Usual Planting and Harvesting Dates for Fresh Market and Processing Vegetables, Agricultural Handbook No. 507, Statistical Research Service, U.S. Department of Agriculture, February, 1977.
6. Vegetable Situation, TVS-207, Economics, Statistics, and Cooperative Services, U.S. Department of Agriculture, February, 1978.
7. Composition of Foods, Agricultural Handbook No. 8, Agricultural Research Service, U.S. Department of Agriculture, Washington, D.C., 1963.

LETTUCE

Edward J. Ryder*

Lettuce is the third most important vegetable crop grown in the U.S., behind potatoes and tomatoes. It is important also in Western Europe, parts of Africa and South America, and Australia.

Crisphead lettuce, by far the most important type, was grown on 90,085 ha in 1976 in the U.S. Total production was 2,416,500 metric tons, and the total value was $473,837,000. About 70% of U.S. production is in California, nearly 20% in Arizona and the rest is grown in Florida, Texas, Colorado, New Mexico, New York, New Jersey, Washington, Michigan, and several other states. The two largest production areas in the United States are the Salinas and Imperial Valleys of California.

Two factors strongly influence the concentration of production in the arid valleys of the West. Lettuce performs best under cool conditions. It is grown for its foliage and is less subject to foliar diseases in dry conditions. The coastal valleys in spring, summer, and early fall, and the desert valleys in late fall, winter, and early spring satisfy these requirements. These areas are generally too cool in those seasons for crops like beans, tomatoes, and cucurbits.

Because it is a leafy vegetable, lettuce must be succulent and green, which necessitates a high water and nitrogen requirement. These are furnished by irrigation and chemical fertilizer, respectively.

The western soils of the U.S. are mineral or upland type and usually have a pH around 7. Lime is, therefore, not a major requirement. Eastern lettuce is grown mostly on organic soils, which are usually acid, and lime is usually required.

Much northern European lettuce is grown under glass, and additional energy inputs, primarily for heating, are necessary. In general, most energy inputs for lettuce in developed countries are high; the low inputs are for gasoline, seeds, seed coatings, insecticides, and herbicides (Tables 1 and 2).

The primary inputs for lettuce in less-developed countries is hand labor and water, the latter because most are in arid, warm zones (Table 3).

The energy output of lettuce is low. It is, however, a moderate source of calcium, Vitamins A and C, and several other minerals. Lettuce is a staple food, consumed regularly (daily in most U.S. homes), and the amounts of these nutrients supplied on a yearly basis in the average diet is significant.

* Western Region, U.S. Agricultural Research Station, Salinas, California

Table 1
ENERGY INPUTS PER HECTARE FOR CRISPHEAD LETTUCE IN THE SALINAS VALLEY, CALIFORNIA, SUMMER PRODUCTION PERIOD[a]

Item	Quantity/ha	kcal/ha	Ref.
Input			
Labor	170.9 hr	—	1
Machinery	30.0 kg[b]	540,000	
Gasoline	37.1 ℓ[c]	375,044	
Diesel	617.0 ℓ[d]	7,042,438	
Electricity	989.3 kWh[e]	2,832,366	
Nitrogen	280.1 kg	3,361,200	1
Phosphorus	78.4 kg	235,200	1
Potassium	78.4 kg	125,440	1
Seeds	.3 kg[f]	1,200	
Seed coating	1.5 kg[g]	52,683	
Irrigation	150.6 cm[h]	2,279,927	1
Insecticide	29.0 kg[i]	2,520,390	
Herbicide	1.7 kg	169,847	1
Transportation	582.6 kg[j]	149,728	
Total		**19,685,463**	
Output			
Total yield	31,595 kg[k]	4,107,350	
Protein yield	284 kg[l]	1,136,000	
kcal output/kcal input		0.21	
kcal output/labor hour		24,034	

[a] These tables were prepared with the substantial aid of G. I. Shalaby, Richard Lindsey, and Keith Mayberry.
[b] Average value, obtained from other vegetable crops.
[c] Minimal use for gasoline except for one flyover per crop by crop dusting airplane (Stearman with 600 hp motor); information furnished by grower.
[d] Diesel use for farm machinery, trucks and crew busses; information furnished by grower.
[e] Primarily for irrigation pumps; information furnished by grower.
[f] Most common planting rate.
[g] Minicoat seed coating material.
[h] Six applications — 1 pre-irrigation, 2 sprinkler, 3 furrow.
[i] Average of 2 minimum, 8 maximum applications; information furnished by grower.
[j] Transportation of all on-farm materials except machinery and nitrogen.
[k] From average production in Salinas Valley — 600 cartons/acre.
[l] Calculated on basis of 0.9% protein for crisphead lettuce.

Table 2
ENERGY INPUTS PER HECTARE FOR CRISPHEAD LETTUCE IN THE IMPERIAL VALLEY, CALIFORNIA, WINTER PRODUCTION PERIOD

Item	Quantity/ha	kcal/ha	Ref.
	Input		
Labor	243.8 hr	—	2
Machinery	30.0 kg[a]	540,000	
Gasoline	283.9 ℓ[b]	2,869,945	
Diesel	576.5 ℓ[c]	6,580,171	
Electricity	0[d]	0	
Nitrogen	224.1 kg	2,689,200	2
Phosphorus	201.7 kg	605,100	2
Potassium	0	0	2
Seeds	.3 kg[e]	1,200	
Seed coating	1.5 kg[f]	52,683	
Irrigation	357.6 cm[g]	5,428,792	2
Insecticide	53.8 kg[h]	4,675,758	
Herbicide	1.7 kg	169,847	2
Transportation	733.3 kg[i]	188,458	
Total		**23,801,154**	
	Output		
Total yield	26,217 kg[j]	3,408,210	
Protein yield	236 kg[k]	944,000	
kcal output/kcal input		0.14	
kcal output/labor hour		13,980	

[a] Average value, obtained from other vegetable crops.
[b] Minimal use, except 5 flyovers by crop dusting airplane (Stearman 600 hp motor).
[c] From figures supplied by K. Mayberry, Farm Advisor.
[d] Essentially no electricity used in lettuce production.
[e] Most common planting rate.
[f] Minicoat seed coating material.
[g] Eight applications: 1 flooding, 1 sprinkler, 6 furrow.
[h] Average 8 applications.
[i] Transportation of all on-farm materials except machinery and nitrogen.
[j] From average production in Imperial Valley — 500 cartons/acre.
[k] Calculated on basis of 0.9% protein for crisphead lettuce.

Table 3
ENERGY INPUTS FOR LETTUCE (NON-HEADING TYPE) IN EGYPT, USING ONLY MANPOWER[a]

Item	Quantity/ha	kcal/ha
Input		
Labor	952 hr	—
Knife and hoe	7.1 kg[b]	29,411
Nitrogen	71.4 kg	856,800
Phosphorus	35.7 kg	39,270
Potassium	35.7 kg	57,120
Seeds	1.2 kg	4,800
Water (irrigation)	90.7 cm	1,377,852
Total		**2,365,253**
Output		
Total yield	23,800 (7140)kg[c]	4,284,000 (1,285,200)
Protein yield	309 (93) kg[d]	1,236,000 (372,000)
kcal output/kcal input		1.81 (0.54)
kcal output/labor hour		4,500 (1,350)

[a] Courtesy of G. I. Shalaby, University of Assiut, Assiut, Egypt.
[b] Assume 2 knives and 8 hoes, weighing 15 kg, lasting 5 years and used on one feddan; one feddan = 4201 m^2.
[c] First number — assumed to be maximum yield under experiment station condition; second number — estimate of probable average yield.
[d] Calculated on basis of 1.3% protein for romaine lettuce.

REFERENCES

1. Agricultural Statistics 1977, U.S. Department of Agriculture, U.S. Government Printing Office, Washington, D.C., 1977.
2. **Cudney, D. W., et al.,** Guidelines to Production Costs and Practices 1976—1977, Imperial County Crops Circ. 104, Imperial County, California, 1977.
3. **Huffman, J. W. and Yeary, E. A.,** Sample Costs to Produce Lettuce in Monterey County — 1977, Mimeograph, Monterey County (California) Extension Service, 1977.

ENERGY INPUTS FOR POTATO PRODUCTION

Ilse H. Schreiner and Donald M. Nafus

Potatoes, valued at $1.5 billion per year, are the eighth most important crop in the U.S. About 14.5 million tonnes are produced annually on approximately 510,000 hectares. Yields average about 28,000 kg/ha, but can be as high as 45,000 kg/ha in Oregon and Washington.[7]

The leading area of potato production is the Columbia plateau region, with Idaho producing between 34 and 36 million kg, Washington 16—22 million kg, and Oregon 7—10 million kg annually. About 40% of the acreage harvested is in these three states. Other major producers are Maine with 12 million kg on 24,000 hectares, California with 9 million kg on 24,000 hectares, and North Dakota with 8 million kg on 45,000 hectares.[7]

Types and sizes of machinery vary markedly with soil types. On sandy soil, relatively smaller tractors are used for tillage. For example, on Long Island, farmers commonly use 75 hp tractors, but on the upland mineral soils of New York, 100 hp tractors are more common.[9] The size of the tillage equipment also varies with farm size. In Washington, for example, 4.3 m chisels and 4.6 m spiketooth harrows are most commonly used. However, large-scale farmers may use 8.5 m wide chisels and 9.1 m spiketooth harrows.[10]

The number of insecticide and fungicide sprays varies considerably between the drier and the wetter areas of production. The pesticides may be applied by ground rigs, commonly using 9.7 m booms, or by air.[10] The vines may be killed with flailcutters, draggers, or with herbicides.

The kind of equipment used to harvest the crop will vary with soil type. Where the soil is rocky, an air separator, frequently with auxiliary power, is required to sort the potatoes from the rocks. In less rocky soils, this is not needed. The kind of machinery used will also vary with the yields expected. In areas with relatively low yields per hectare, such as Idaho, 4-row cross-over diggers are most commonly used. In states with very high yields, such as Washington, 2-row diggers are more commonly used to enable the trucks removing the potatoes from the field to keep up with the combine.

In areas where irrigation is required, one quarter-mile-center-pivot systems are the most frequent.[10]

Table 1
ENERGY INPUTS AND OUTPUTS PER HECTARE OF IRRIGATED POTATOES IN CALIFORNIA

Item	Quantity/ha	kcal[a]	Ref.
	Input		
Labor	62 hr[b]	—	
Machinery	14 kg[b]	252,000	
Gasoline	76 ℓ	768,284	2
Diesel	326ℓ	3,720,964	2
L.P. gas	42ℓ	323,610	2
Electricity	212 kWh	606,956	2
Nitrogen	254 kg	3,048,000	1
Phosphorus	212 kg	636,000	1
Potassium	103 kg	164,800	1
Seeds	2805 kg	1,722,825	3
Irrigation	$24	134,158	8
Insecticides	4.2 kg	365,022	1
Herbicides	11.8 kg	1,178,938	1
Transportation	3177.9 kg[b]	816,715	
Total		**13,738,272**	
	Output		
Total yield	36,736 kg/ha	22,548,333	
Protein yield	622.2 kg/ha	2,488,772	
kcal output/kcal input		1.64	
kcal output/labor hour		363,683	

[a] See appropriate table for conversion factors.
[b] Estimate.

Table 2
ENERGY INPUTS AND OUTPUTS PER HECTARE OF POTATOES IN MAINE

Item	Quantity/ha	Kcal	Ref.
	Input		
Labor	89 hr	—	4
Machinery	14 kg	252,000	
Gasoline	171 ℓ	1,728,639	2
Diesel	104 ℓ	1,187,056	2
Electricity	17.6 kWh	50,389	2
Nitrogen	185 kg	2,220,000	4
Phosphorus	249 kg	747,000	4
Potassium	247 kg	395,200	4
Seeds	2408 kg	1,478,184	4
Insecticides	3.4 kg	295,494	4
Herbicides	10.6 kg	1,059,046	4
Transportation	2638.7 kg[a]	678,146	
Total		**10,091,154**	
	Output		
Total yield	23,689 kg/ha	14,540,643	
Protein yield	399 kg/ha	1,597,596	
kcal output/kcal input		1.44	
kcal output/labor hour		163,378	

[a] Estimate.

Table 3
ENERGY INPUTS AND OUTPUTS PER HECTARE FOR IRRIGATED POTATOES IN IDAHO

Item	Quantity/ha	kcal	Ref.
	Input		
Labor	62 hr	—	2
Machinery	14 kg[a]	252,000	
Gasoline	170 ℓ	1,718,530	2
Diesel	509 ℓ	5,809,726	2
L.P. gas	6.5 ℓ	50,083	2
Natural gas	9.1 m^3	107,498	2
Nitrogen	281 kg	4,130,700	2
Phosphorus	257 kg	771,000	2
Potassium	202 kg	323,200	2
Seeds	2247 kg	884,260	4
Irrigation	$25	134,158	8
Electricity	1282 kWh	3,670,366	2
Insecticide	8.6 kg	747,426	2
Herbicide	3.4 kg	339,694	2
Transportation	2825.0 kg	726,025	
Total		**19,664,666**	
	Output		
Total yield	26,720 kg	16,401,294	
Protein yield	452 kg	1,811,004	
kcal output/kcal input		0.83	
kcal output/labor hour		265,608	

[a] Estimate.

Table 4
ENERGY INPUTS AND OUTPUTS PER HECTARE FOR POTATOES IN NEW YORK

Item	Quantity/ha	kcal	Ref.
	Input		
Labor	35 hrs	—	6
Machinery	14 kg[a]	252,000	
Gasoline	261 ℓ	2,638,449	2
Diesel	152 ℓ	1,734,928	2
Electricity	45.7 kWh	130,839	2
Nitrogen	229 kg	2,748,000	6
Phosphorus	390 kg	1,170,000	6
Potassium	222 kg	355,200	6
Seed	2134 kg	1,309,347	6
Insecticides	31.4 kg[b]	2,678,20	6
Herbicides	18.0 kg[b]	1,798,380	6
Transportation	2473 kg	635,561	
Total		**15,451,124**	
	Output		
Total yield	34,468 kg	21,156,291	
Protein yield	538.9 kg	2,335,632	
kcal output/kcal input		1.37	
kcal output/labor hour		611,807	

[a] Estimate.

[b] $126/acre for spray material assumed $3/lb and ⅔ for insecticides and ⅓ for herbicide.

Table 5
ENERGY INPUTS AND OUTPUTS PER HECTARE FOR POTATOES IN SOUTH CAROLINA

Item	Quantity/ha	kcal	Ref.
	Input		
Labor	42 hrs	—	5
Machinery	14 kg[a]	252,000	
Gasoline	178 ℓ	1,799,402	2
Diesel	101 ℓ	1,152,814	2
Nitrogen	188 kg	2,256,000	5
Phosphorus	225 kg	675,000	5
Potassium	225 kg	360,000	5
Lime	560 kg	176,652	5
Seeds	2020 kg	1,240,434	5
Insecticides	32 kg	2,781,120	5
Herbicides	6.7 kg	669,397	5
Transportation	2253.4 kg	579,124	
Total		**11,941,543**	
	Output		
Total yield	19,648 kg	12,059,898	
Protein yield	332.9 kg	1,331,824	
kcal output/kcal input		1.01	
kcal output/labor hour		287,209	

[a] Estimate.

Table 6
ENERGY INPUTS AND OUTPUTS PER HECTARE OF POTATOES IN MICHIGAN

Item	Quantity/ha	kcal	Ref.
	Input		
Labor	65 hrs	—	3
Machinery	14 kg[a]	252,000	
Gasoline	285 ℓ	2,881,065	2
Diesel	223 ℓ	2,545,322	2
Electricity	1000 kWh	2,863,000	2
Nitrogen	336 kg	4,032,000	3
Phosphorus	136 kg	408,000	3
Potassium	336 kg	537,600	3
Seeds	2805 kg	1,722,825	3
Insecticides	6.2 kg	538,842	3
Herbicides	6.7 kg	669,397	3
Transportation	3222.3 kg	828,131	
Total		**17,278,182**	
	Output		
Total yield	31,435 kg	19,295,640	
Protein yield	531 kg	349,149	
kcal output/kcal input		1.12	
kcal output/labor hour		297,034	

[a] Estimate.

Table 7
ENERGY INPUTS AND OUTPUTS PER HECTARE FOR IRRIGATED POTATOES IN COLORADO

Item	Quantity/ha	kcal	Ref.
	Input		
Labor	34 hrs	—	8
Machinery	14 kg[a]	252,000	
Gasoline	207 ℓ	2,092,563	2
Diesel	175 ℓ	1,997,450	2
L.P. gas	1.08	7,705	2
Electricity	60 kWh	171,780	2
Nitrogen	390 kg	4,680,000	8
Phosphorus	383 kg	1,149,000	8
Potassium	333 kg	532,800	8
Seeds	2805 kg	1,722,825	8
Irrigation	$25	134,158	8
Natural gas	18.7 m^3	220,903	2
Herbicides	2.2 kg[b]	219,802	8
Transportation	3123.6 kg	802,765	
Total		**13,983,751**	
	Output		
Total yield	28,069 kg	17,228,250	
Protein yield	475.5 kg	1,901,900	
kcal output/kcal input		1.23	
kcal output/labor hour		501,799	

[a] 1976.
[b] $5 for herbicides at $3/lb.

Table 8
WEIGHTS AND USE OF MACHINERY FOR POTATO PRODUCTION

Machine	Weight (kg)	Width (m)	Speed (m/hr)	Lifetime (yr)	Actual working lifetime (hr)	Hr work/hr	Ha/yr	Kg/ha/yr
Tractor[a]								2.40
Plow	550	4.3	7200	10	1600	160	496	.11
Disc	1600	3.7	2400	12	1660	138	124	1.07
Harrow	700	6.1	8500	12	1600	133	690	.08
Cultivator	200	3.7	6100	12	1520	126	277	.06
Planter	2300	3.7	6400	10	780	78	187	1.23
Spreader	350	7.4	8000	8	4489	561	3321	.01
Sprayer	110	9.7	6400	10	780	78	484	.02
Flailcutter	900	3.7	5900	10	1340	134	281	.30
Seed cutter	700	—	2 ha/hr	10	2000	200	400	.20
Harvester	5500	1.85	3200	10	1750	175	103	5.34
Trucks								3.59
Total								**14.41**

[a] Kg/ha for tractor and trucks taken from sugar beet paper; other data derived from figures provided by N. A. Wynn, personal communication.

REFERENCES

1. **Cervinka, V., Chancellor, W. J., Coffelt, R. J., Curley, R. G., and Dobie, J. B.,** Energy Requirements for Agriculture in California, California Department of Food and Agriculture, University of California, Davis, 1974.
2. Energy and U.S. Agriculture: 1974 Data Base, FEA/D-76/459, Office of Energy Conservation and Environment, Federal Energy Administration, 1976.
3. **Knoblauch, W., Nott, S., Schwab, G., Harsh, S., and Black, J.,** Enterprise Budgets, Rep. No. 295, Agricultural Economics Reports, Michigan State University, East Lansing, 1976.
4. **Krofta, R. N. and Harlan, R. K.,** Costs, Returns, and Capital Requirements for Producing Potatoes in Maine, Bull. 730, Life Science Agricultural Experiment Station, University of Maine, Orono, 1976.
5. **Smith, D. B., King, G. A., and Ezell, D. O.,** Selected Vegetable Budgets: Crop Production, Cost Estimates, 1977, Spec. Rep. 43, Agricultural Experiment Station, University of Arkansas, Fayetteville, 1976.
6. **Snyder, D. P.,** Cost of Production — Update for 1976 on Muck Onions, Potatoes, Sweet Corn, Dry Beans, Apples, A.E. Res. 77-11, Cornell University Agricultural Experiment Station, Ithaca, 1977.
7. Agricultural Statistics 1976, U.S. Department of Agriculture, Washington, D.C., 1976.
8. **Wright, N. G., Stubblefield, T. M., Gronewoller, H. P., Crim, H. E., and Swope, D. A.,** Cost of producing crops in the irrigated Southwest, Part IV, *Colorado Agricultural Technical Bulletin,* Experimental Station, University of Arizona, Tucson, 1975.
9. **Sieczka, J. B.,** Cornell University, personal communication.
10. **Wynn, N. A.,** U.S. Department of Agriculture, personal communication.

PICKLING CUCUMBER PRODUCTION — HARVESTING

L. R. Baker

The importance of pickling cucumber production in the U.S. has increased dramatically over the last 25 years. This is evidenced by the per capita consumption that has increased from some 2 lb in 1932 to the present 8.6 lb in 1976 (Figure 1). The acreage of pickling cucumber production in the U.S. has remained constant at about 120,000 acres, but the total production has increased to some 800,000 ton from a previous production of 200,000 ton during the 1950s (Figure 1).

Michigan is the leading state in pickling cucumber production (Tables 1 and 2). Some 80 to 90% of the state's production is mechanically harvested (Figure 2). Growing of pickling cucumbers for mechanical harvest differs greatly from that for hand-harvest. Maximum fruit numbers must be marketable at harvest time for the once-over destructive harvest system. Proper cultural practices are necessary to obtain maximum uniformity in cucumber plants and fruit growth and development.

Numerous factors affect the yield, quality, and return of pickling cucumbers grown for once-over mechanical harvest. These include: proper field selection, soil preparation, fertilization, variety, seeding method, plant population, soil moisture, weed-insect-disease control, bee pollination, and timing of harvest.

FERTILIZATION

Production of pickling cucumbers at high plant populations (above 50,000 plants per acre) requires good levels of soil fertility. Many sandy soils offer desirable moisture characteristics, but do not have a large nutrient-holding capacity. Nitrogen applied to soils may be taken up by plants, lost by leaching, or lost by denitrification under excessively wet conditions. Hence, application of nitrogen close to the time of plant uptake is essential for efficient use of nitrogen fertilizer. Hybrid cucumber varieties that mature in 50 to 60 days, require less nitrogen than monoecious varieties, which are hand-harvested. About 75 lb of actual nitrogen are removed from the field, with yields of 300 to 350 bu of pickles per acre.

For soils containing less than 2% organic matter, 50 to 75 lb of nitrogen per acre is recommended. On soils with greater than 2% organic matter, 50 lb of nitrogen per acre is adequate. On sandy soils where leaching occurs, one half of the nitrogen should be applied close to planting, and the remainder applied as the vines begin to fill the rows.

Lime, phosphorus, potassium, and micronutrient needs should be based on a good soil test that truly represents a given field. Soil pH is one of the most important properties of soil. The most desirable pH range is between 6.0 and 7.0. A 300 to 350 bu/acre pickle yield requires from 25 to 40 lb of phosphate (P_2O_5) per acre. Only 10 to 15 lb of phosphate per acre are removed from the field in the cucumbers. With close row spacings (9 to 10 in.), phosphate can be broadcast and disked-in prior to planting, and band placement offers no particular advantage. Early seedling growth may be somewhat improved by row application, but usually there is no difference in maturity or yield compared to complete broadcast application of phosphorus.

Potassium recommendations depend upon concentrations already present in the soil. From 100 to 140 lb of potash (K_2O) is taken up by a 300 to 350 bu/acre pickle crop, with 40 to 50 lb K_2 being removed from the field in the pickles. Broadcast and disk-in potassium fertilizer prior to planting.

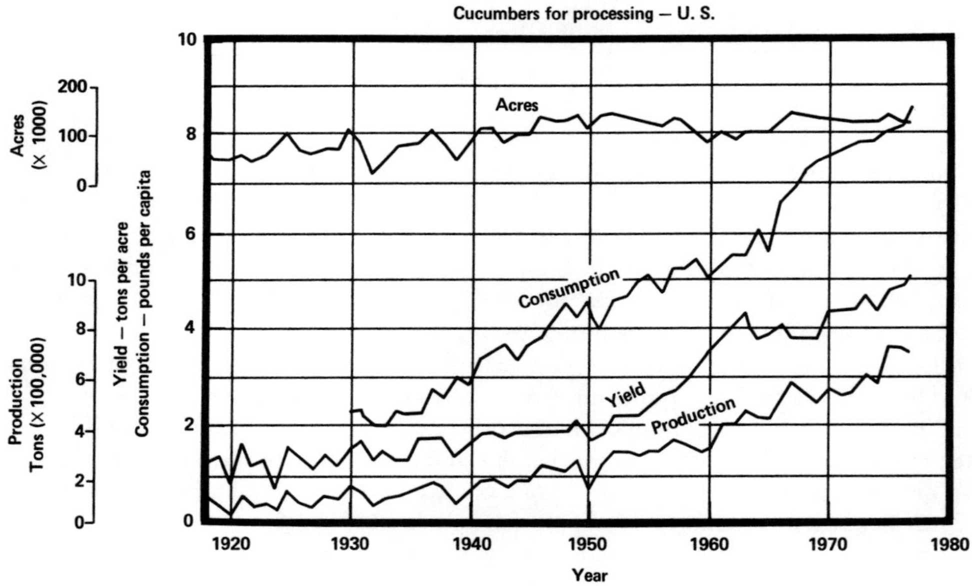

FIGURE 1. Importance of pickling cucumbers in the U.S.

Zinc (Zn) and manganese (Mn) are the most important micronutrients in the production of pickling cucumbers. Zn and Mn needs can be met by soil or foliar application. Where additional Zn and Mn are required, rates for band application range from 2 to 5 lb Zn per acre and from 4 to 8 lb Mn per acre. Broadcast application of 25 lb Zn per acre usually supplies sufficient zinc for 5 years. Manganese is readily tied up in soil, so broadcast applications are not recommended.

Foliar application of Zn and Mn should be made 2 to 3 weeks after emergence. Since Mn and Zn are not readily translocated, a second spray application will be needed 2 weeks later to cover new vine growth. Recommended rates are 1 to 2 lb of Mn per acre and 0.3 to 0.7 lb of Zn per acre. When using chelated materials, use the rates recommended on the label.

WEED CONTROL

For maximum yields of pickling cucumbers that can be successfully mechanically harvested, weeds must be effectively controlled from planting until harvest time.

The proper herbicide program depends on the soil type and the weed species present in the field. Usually, better results are obtained by using combinations of herbicides. Application of ¼ to ½ acre-in. of water as soon as possible after herbicide application is the key to effective weed control and also helps to provide uniform emergence of cucumber plants.

Where row spacing allows cultivation, cucumbers may be cultivated until the three to four leaf stage with a minimal amount of damage. Rolling cultivators or finger weeders that stir only the top inch of soil are preferred. Deeper cultivation can cause serious root injury, as well as expose many new weed seeds.

MECHANICAL HARVESTING

Once-over mechanical harvesting should be done when cucumbers have attained a size distribution that produces the greatest dollar return. Planting schedules, growing

Table 1
INCREASING IMPORTANCE OF PICKLING CUCUMBERS IN THE U.S. AND MICHIGAN, WITH DRAMATIC CONVERSION TO MECHANICAL HARVEST IN MICHIGAN

	Acreage		Yield[a]		Production		Value		Mechanical harvesting
Year	Planted	Harvested	Tons per acre	Bushels per acre	1,000 tons	% of U.S.	1,000 $	Dollars per bushel	(% of acreage)[b]
1950	39,600	33,600	0.72	30	24.19	14	2,218	2.20	
1951	47,600	44,800	1.54	64	68.81	25	4,730	1.65	
1952	48,800	45,800	1.85	77	84.65	26	5,290	1.50	
1953	41,700	39,700	2.04	85	80.98	25	5,230	1.55	
1954	38,000	36,200	2.06	86	74.71	25	4,203	1.35	
1955	37,500	35,500	2.45	102	86.90	28	4,345	1.20	
1956	38,000	35,800	2.54	106	91.03	28	4,744	1.25	
1957	40,000	38,000	3.05	127	115.82	31	5,791	1.20	
1958	31,000	28,200	3.53	147	99.48	28	4,974	1.20	
1959	25,000	22,700	4.61	184	104.65	32	5,013	1.20	
1960	23,500	20,600	4.82	193	99.29	30	5,173	1.30	
1961	26,500	24,400	5.23	209	127.61	31	6,916	1.36	
1962	23,800	22,100	5.30	212	117.13	29	5,997	1.28	
1963	26,000	24,900	5.60	224	139.44	30	7,335	1.32	
1964	25,500	23,700	5.05	202	119.68	28	7,971	1.67	0
1965	17,800	16,500	5.75[c]	230[c]	94.83[c]	22[c]	7,584	2.33	2.5
1966	20,900	19,600	6.23	249	122.11	23	11,161	2.29	13
1967	29,600	26,600	4.45	178	118.37	20	13,139	2.77	20
1968	22,900	21,200	4.37	175	92.64	16	9,208	2.48	20
1969	23,100	22,200	4.08	163	90.60	18	9,332	2.58	35
1970	24,700	23,200	4.48	179	103.95	18	10,353	2.49	65
1971	25,000	24,500	3.38	135	82.80	15	7,618	2.30	85
1972	28,700	26,000	3.76	150	97.80	17	8,391	2.14	90
1973	27,200	26,400	4.07	163	107.45	18	9,477	2.20	90+
1974	29,400	27,400	4.23	169	115.90	19	14,488	3.13	90+
1975	30,400	27,600	4.68	187	129.20	19	16,021	3.10	90+
1976	27,000	25,500	4.05	162	103.30	16	11,570	2.80	90
1977[d]	25,200	24,500	4.65	186	114.00	18	12,426	2.73	80

[a] Prior to 1959 the production reporting unit was bushels (48 lb). From 1959 on, the production reporting unit has been tons (50 lb/bu used for calculations).
[b] Industry Advisory Committee Estimate.
[c] Includes 13,630 ton harvested but not marketed.
[d] Preliminary.

From U.S. Department of Agriculture Statistical Reporting Service.

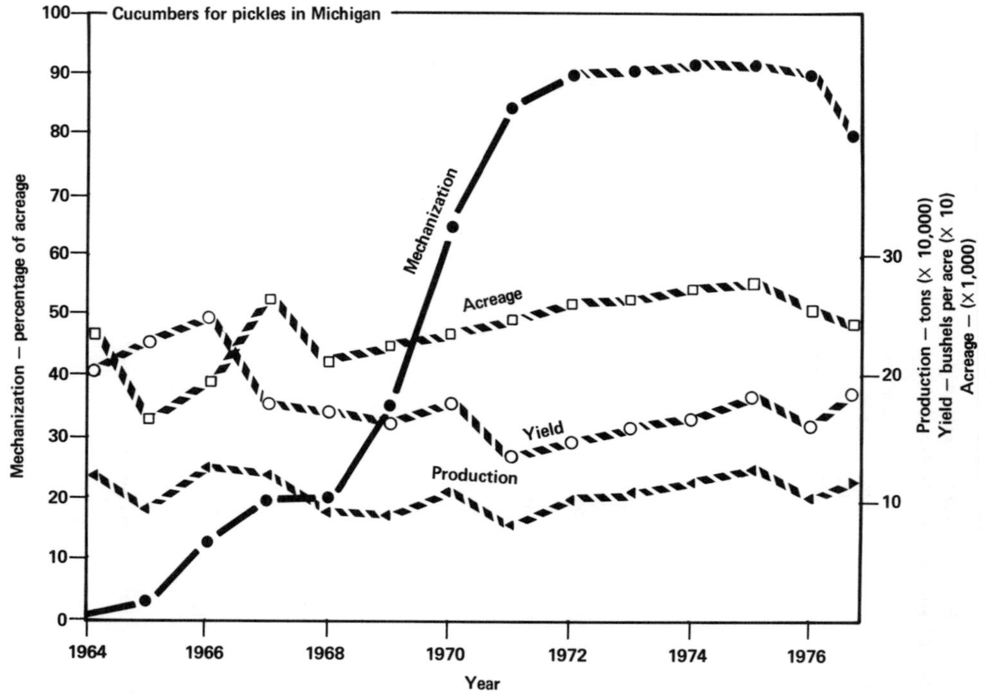

FIGURE 2. Rapid mechanization of pickling cucumber harvest in Michigan.

FIGURE 3. Mechanical harvester.

conditions, acres to be harvested, weather, and fruit size should all be considered when determining the optimum time for harvest. Pickling cucumber fruit can have a 40% weight increase in 24 hr. Hot, humid weather with adequate soil moisture promotes rapid increase in fruit size. The value of a field of cucumbers changes as the fruit size

Table 2
SCAB-RESISTANT VARIETIES SUGGESTED FOR ONCE-OVER MECHANICAL HARVEST IN MICHIGAN[a]

Hybrid	Spine color	Maturity
Premier	White	Early
Score	White	Mid
Bounty	Black	Mid
Tally	Black	Mid
Carolina	White	Mid
Pioneer	Black	Early
Calypso	White	Early
Compass	White	Early
Green Star	White	Mid
Earlipik 14	White	Mid
Green Spear 14	White	Mid
Perfecto Verde 14	White	Mid

[a] University releases are available from most seed companies.

Table 3
ENERGY INPUTS FOR PRODUCING PICKLING CUCUMBERS

Item	Quantity/ha	kcal
	Input	
Labor	48 hr	—
Machinery	30 kg	540,000
Diesel fuel	150 ℓ	1,712,000
Nitrogen	100 kg	1,451,520
Phosphorus	56 kg	168,000
Potassium	84 kg	134,400
Seeds	7 kg	28,000
Herbicides	6.7 kg	420,559
Irrigation	10 cm	159,000
Bee pollinators	1 hive (70 kg)	Transport costs only
Transport	164.5 kg	42,277
Total		**4,655,856**
	Output	
Total yield	10,935 kg	1,640,250
Protein yield	98.4 kg	
kcal output/kcal input		0.35
kcal output/labor hour	34,172 kcal	

changes; typically, depreciation in value of 5 to 10%, and sometimes 20%, occurs in a 24-hr period. Thus time of once-over mechanical harvesting is very critical.

Sampling should be started before the field appears to be ready for harvest and repeated daily. Size grade these cucumber fruits by hand, and consult the pickle processor's field man for a recommendation on optimum harvest time.

Follow the manufacturers' instructions for recommended harvester operation. Although several machines were made by different manufacturers in the 1960s, two manufacturers remain at the present (Wilde Mfg. and Cukes, Inc.). The harvester unit requires a 75 to 100 hp tractor for operation. The relationship of recovery to ground speed and pickup reel speed, pinch-roll speeds, and pinch-roll pressure is very critical. Proper harvester adjustments will result in increased field recovery and fewer damaged cucumbers. Adjust components of the harvester as often as necessary. The pickup of dirt and mud is influenced by soil moisture and the depth of the cut-off blade below the soil surface. For conservation of power requirements and reduction of dirt, operate the cut-off blade as near the surface as possible, and yet maintain a smooth flow of dirt and vines over the blade.

The energy inputs for producing pickling cucumbers are listed in Table 3. Note that machinery, diesel fuel, herbicides, and nitrogen include about 89% of the total fossil energy going into the system.

CANTALOUPES

Hunter Johnson, Jr. and William J. Chancellor

Cantaloupe production in the U.S. is largely centered in the southwestern states of California, Texas, and Arizona. Growers in these states provide 90% of the U.S. production of this crop on 85% of the total acreage. California is the largest producer of cantaloupes, with 15,060 ha in 1976 (51% of the U.S. total and 66% of the U.S. production). Total U.S. value of cantaloupes in 1976 was $108 million, with the three leading states accounting for $100 million (93%).

Peak production occurs from June through September. The earliest U.S. production begins in early May in the Rio Grande Valley of Texas, and harvest from summer plantings in California's Imperial Valley often continues into December. Table 1 shows production and value for the leading states for the 1976 season. Cantaloupes are transported to market from the producing districts by rail and truck, but the great majority of shipments are by diesel-powered refrigerator trucks. Table 2 provides data for monthly shipments in the U.S. for the 1976 season. One carlot equivalent of cantaloupes, including containers in which they are packed, will weigh 22,800 to 25,650 kg.

In the southwestern states, cantaloupes are grown as an irrigated crop on raised beds. The initial energy input to the crop is for various tillage operations, including final preparation of the beds and seeding, all of which is performed with a wide range of tools (such as landplanes, discs, subsoilers, or plows) drawn by diesel tractors. Either gasoline or diesel-powered tractors may be used for planting, cultivation to control weeds, and to apply fertilizers or pesticides. From the time that the vines cover the beds until harvest begins, fungicides and insecticides are applied by fixed-wing aircraft, since it is no longer practical to drive tractors through the fields.

Water for the great majority of all California and Arizona acreage is applied by the furrow method from irrigation district canals. On some farms, water is pumped from deep wells using electric power, but diesel, gasoline, or propane engines are also used. Powered booster pumps are used where preplant irrigation is applied by sprinklers or sometimes to pump water from a canal source to a distant field for furrow irrigation. In some areas, the land is irrigated before planting to supply soil moisture for germination (typical of the San Joaquin Valley in California) or to leach salt from the land (typical of Arizona and California desert valley plantings). After the planting beds have been prepared and seeded, the standard practice is to apply water by the furrow method for the balance of the growing period. Five to six irrigations, totaling 9,144 to 12,192 M^3/ha are applied per crop.

Cantaloupes respond to applications of nitrogen fertilizer in all growing locations and to phosphorus in some locations. Growers generally apply 68 to 90 kg of nitrogen and 36 to 45 kg of phosphorus per acre. All of the phosphorus and about 23 kg of nitrogen are applied prior to seeding as ammoniated phosphates, usually as broadcast application. The balance of the nitrogen is applied after thinning in one or two applications as a side-dressing by tractor or sometimes in the irrigation water. Other fertilizer nutrients, such as potassium or minor elements, or amendments such as lime or gypsum, are rarely required and thus generally not applied.

Herbicides are commonly used in the San Joaquin Valley for the control of seedling weeds with excellent results. A mixture of naptalam and bensulide are incorporated in the surface soil on the planting beds prior to seeding with the use of a powered tiller. Later, at the 3- to 5-leaf stage, after final bed shaping has occurred, trifluralin is applied and incorporated with rolling cultivators. Herbicides are not as effective under

Table 1
CANTALOUPE YIELD, PRODUCTION, AND VALUE — 1976

State	Hectars harvested	Yield/ha (metric tons)	Production 1000 (metric tons)	Value 1000 ($)
California	15,066	19.5	293.8	68,831
Texas	6,439	10.1	65.0	17,729
Arizona	3,483	13.8	48.1	13,673
Other	4,669		40.3	7,842
Total U.S.	**29,657**		**547.2**	**108,075**

[a] Ranges from 4.5-20.2

cultural conditions in the desert districts and therefore are rarely used. Weeds are controlled in those areas by tractor cultivation and with hand labor using long-handled hoes.

An unusual and important energy input to cantaloupe production in the southwest U.S. is the use of an asphalt petroleum emulsion to enhance seed germination. After seeding, a 15-cm-wide band of asphalt emulsion (58% asphalt) is sprayed over the seed row by tractor ground sprayer at the rate of 150 to 190 ℓ per acre. This black surface increases the soil temperature by several degrees at seed zone depth (approximately 2.5 cm) and accelerates germination. The asphalt band also serves to reduce surface evaporation, thus maintaining good soil moisture around the seed.

Almost all insecticides and fungicides are applied by fixed-wing aircraft because the period of greatest need for these materials occurs after the vine growth has covered the beds. During earlier growth stages, insecticides are applied by ground dusters or sprayers.

The selection and picking of mature cantaloupes is entirely by hand labor, although self-propelled labor-aid devices and trucks are used in the collection of harvested fruit. Field workers select mature fruit and either place it in a sack, which is carried on the worker's back, or on the cross-conveyor of a belt-loader labor-aid device. In the case of sack crews, the worker carries his full sack up an inclined plank at the rear of a truck that travels through the field with the crew. Sack crews are generally composed of 12 pickers, and 12 rows are harvested during each pass through the field. Where belt loaders are used, the pickers follow closely behind the loader which is traveling perpendicular to and across 14 to 15 rows. Fruit is placed on the moving cross-conveyor belt as it is harvested, and the belt is then elevated at one end to load a truck which is moving through the field with the loader. The trucks (sometimes with trailer) carry 5.5 to 10.9 metric tons of fruit when loaded, depending upon truck size, and transport the fruit to packing sheds (frequently 33 to 48 km from the field) where it is graded and packed.

At the packing shed, several operations are performed prior to shipment, using various mechanical devices for conveying, cooling, and packaging. Removal of defective or over-mature fruit, sizing, and packaging are most frequently accomplished by hand labor with the aid of conveying equipment, although automatic sizing and packing devices have been developed and are in operation in some of the large sheds. Both wood crates and fiberboard cartons are in use, but the use of cartons predominates. The fully-packed fiberboard half-carton weighs about 18 kg, and the wood crate about 36 kg. Since cantaloupes are generally harvested during hot weather, from late spring to early fall, the field heat must be removed as soon as possible after harvest in order to preserve quality. This may be accomplished by hydrocooling, package icing, or

Table 2
U.S. TRUCK AND RAIL SHIPMENTS OF CANTALOUPES — 1976

	Carlot equivalents[a]								
State	**May**	**June**	**July**	**August**	**September**	**October**	**November**	**December**	**Total**
California									
Truck	17	1703	2907	2783	1191	456	300	22	9379
Rail	1	570	1210	832	328	60	60	—	3061
Total									**12,440**
Arizona									
Truck	1	828	531	1	22	173	5	—	1561
Rail	—	293	208	—	—	3	—	—	504
Total									**2,065**
Texas									
Truck	814	272	221	38	1	—	—	2400	1054
Rail	33	19	—	—	—	—	—	—	52
Total									**2,452**
Other states									
Truck	31	172	221	389	138	13			964
Total U.S.									
Truck									14,304
Rail									3,617

[a] Carlot equivalent = 600 Jumbo crates (@ 36 to 38 kg each) or 1350 half cartons (@ 18 to 19 kg each).

forced-air cooling. Hydrocooling for carton packing is the predominant system used in California. Top-icing is used when melons are packed in wood crates. Forced air or pressure cooling of carton-packed cantaloupes is a relatively new practice adopted by only a few shippers. During transit to market, continued cooling is provided for carton loads by mechanical refrigeration equipment in trucks and in rail cars. Wood crate shipments are refrigerated in transit by crushed ice on top of the load.

Energy inputs to the production of cantaloupes in two major California districts are shown in Tables 3 and 4. The input range from 20 to 25 million kcal/ha includes major contributions for mobile equipment, electricity, nitrogen fertilizer, and packing cartons. In the desert valleys (Table 3), the asphalt emulsion used on the seed rows provides another important contribution for those areas. The kilocalorie output-input ratio is very low in both districts, about 0.10.

Table 3
ENERGY INPUTS TO CANTALOUPES — IMPERIAL COUNTY, CALIFORNIA

Item	Quantity/ha	kcal/ha
Input		
Labor	521.0 hr	—
Machinery (farm equipment)	18.46 kg[a]	3,067,343
Machinery (vehicles)	5.03 kg	751,094
Gasoline (vehicles only)	220.0 ℓ[b]	2,223,980
Gasoline (aircraft)	77.0 ℓ	778,393
Diesel	292.0 ℓ[c]	3,332,888
Electricity (packing shed)	407.0 kWh[d]	1,165,241
Natural gas (packing shed)	1.93 thm[d]	48,550
Nitrogen (N)	162.0 kg[e]	3,348,311
Phosphorus (P_2O_5)	94.0 kg[e]	186,150
Potassium	0	
Lime	0	
Seeds	2.25 kg[f]	112,500
Irrigation (water applied)	137.0 cm[g]	317,851
Insecticides	3.65 kg[h]	91,028
Herbicides	0	
Asphalt emulsion spray (50% asphalt)	421.0 ℓ[i]	1,681,775
Packing shed (repair and depreciation)	$30.38[j]	205,732
Packing cartons[k] (0.907 kg/carton)	742.7 cartons	3,042,149
Transportation (fertilizer)[l] (other items are accounted for elsewhere)	—	70,144
Beehive service (without transportation)	$18.52[m]	76,135
Spray aircraft (repair and depreciation)	$ 5.51[n]	31,756
Total		**20,150,087**
Output		
Melon yield (total)	13,473.0 kg	4,041,818
Melon yield (51% edible)	6,871.0 kg	2,061,327
kcal output/kcal input	—	0.102
protein yield (edible)	481.0 kg @ 7%	—

[a] Tractor weights (1): 235 hp (wheel) = 11,127 kg (24,530 lb)
185 hp (wheel) = 8,480 kg (18,696 lb)
70 hp (wheel) = 3,230 kg (7,120 lb)

Tractor life: 12,000 hr (2)

Subsoiler weight = 2381 kg (5250 lb), life = 2500 hr (2)

Disc harroweight = 2722 kg (6000 lb), life = 2500 hr (2)

Planter sled weight = 2722 kg (6000 lb), life = 1200 hr (2)

Furrower weight = 363 kg (800 lb), life = 2500 hr (2)

Cultivator weight = 1134 kg (2500 lb), life = 2500 hr (2)

Table 3 (continued)
ENERGY INPUTS TO CANTALOUPES — IMPERIAL COUNTY, CALIFORNIA

Item	Quantity/ha	kcal/ha

Fertilizer spreader weight	= 340 kg (750 lb), life = 2000 hr (2)
250-hp truck weight	= 3629 kg (8000 lb), life = 5000 hr
Crew bus weight	= 3175 kg (7000 lb), life = 4000 hr
Pickup truck weight	= 1814 kg (4000 lb), life = 4000 hr
Harvest conveyor weight (used on 20% of acreage; capacity is 7.2 metric tons (8 U.S. tons) per hour)	= 5443 kg (12,000 lb, life = 4000 hr
Farm machinery	@ \$2.2/kg (\$1/lb), 1967; 19,033 kcal/\$1 (75,528 Btu/\$1), 1967 (3)
Motor vehicles	@ \$2.2/kg (\$1/lb), 1967; 17,104 kcal/\$1 (67,874 Btu/\$1), 1967 (3)

[b] A 250-hp truck consumes 34.1 ℓ gasoline per hour; a crew bus consumes 26.5 ℓ (7 gal) gasoline per hour; a 600-hp agricultural spray aircraft consumes 155 ℓ (41 gal) gasoline per hour; a pickup truck consumes 15.1 ℓ (4 gal) gasoline per hour, 10,109 kcal/ℓ total.

[c] Diesel tractors (wheel), when operating at 75% of rated power, consume 0.065 U.S. gal of fuel per horsepower-hour (2); 11,414 kcal/ℓ total.

[d] Packing shed energy needs are 27.44 kWh electricity and 3620 kcal (0.14 thm) of natural gas per metric ton packed.[4] Yield is 13.4 metric tons/ha (6 U.S. tons/acre).

[e] Nitrogen fertilizer @ 4261 kcal/kg N (37,281 Btu/lb N).[5] phosphate fertilizer @ 408 kcal/kg P_2O_5 (3,572 Btu/lb P_2O_5).[5]

[f] Assumed 50,000 kcal/kg seed.

[g] 81 kWh per hectar-meter (10 kWh per acre-foot)[6] and 2863 kcal/kWh.

[h] 0.56 kg Benomyl approx. Manet + $Captan_2$ = 25,475 kcal/kg; 1.68 kg Methomyl. Diazanon approx. Methyl Parathion; 13,810 kcal/kg; 1.40 kg Sevin approx. Toxaphene; 38,100 kcal/kg.

[i] For an asphalt energy content of 1714 kcal/kg (15,000 Btu/lb), a weight of 0.96 kg/ℓ (8.0 lb/gal), and an asphalt content of 50% by weight, the resulting energy content of the asphalt emulsion spray is 3,994 kcal/ℓ.

[j] Shed cost of \$172,000 handles 7600 metric tons (8400 U.S. tons) per year. A 15-year life equals \$1.36 per U.S. ton depreciation plus 50% for repairs equals \$2.05 per U.S. ton. For \$2 (1977) = \$1 (1967) and 13,543 kcal/\$1 (53,744) Btu/\$1), 1967, for repair and maintenance construction,[3] the result is 13,882 kcal/ton.

[k] Fiberboard carton material approximately equal to softwood = 4515 kcal/kg.

[l] Transport and handling of fertilizers at 274 kcal/kg of primary nutrient[4].

[m] Half of beehive service assumed as truck transport, balance was computed at \$2 (1976)/\$1(1967) and 8,220 kcal/\$1 (32,618 Btu/\$1), 1967, for agricultural services.[3]

[n] Repairs \$7/hr and depreciation \$4.16/hr in 1977 at \$2 (1977) per \$1 (1967) and 12,032 kcal/\$1 (47,747 Btu/\$1), 1967, for repairs and 10,667 kcal/\$1 (42,331 Btu/\$1), 1967, for aircraft and parts;[3] a total of 64,283 kcal/hr results (0.494 hr/ha used).

Table 4
ENERGY INPUTS TO CANTALOUPES — SAN JOAQUIN VALLEY, CALIFORNIA

Item	Quantity/ha	kcal/ha
Input		
Labor	310.0 hr	—
Machinery (farm equipment)	16.43 kg[a]	2,730,035
Machinery (vehicles)	5.13 kg	766,026
Gasoline (vehicles)	220.0 ℓ[b]	2,223,980
Gasoline (tractor)	61.0 ℓ[c]	616,649
Gasoline (aircraft)	38.0 ℓ[c]	384,142
Diesel	177.0 liters[d]	2,020,278
Electricity (packing shed)	542.0 kWh[e]	1,551,746
Natural gas (packing shed)	2.57 thm[e]	64,764
Nitrogen (N)	194.0 kg[f]	4,009,706
Phosphorus (P_2O_5)	96.0 kg[f]	190,110
Potassium	0	
Lime	0	
Seeds	3.37 kg[g]	168,500
Irrigation	66.0 cm[h]	4,955,853
Insecticides	1.30 kg[i]	57,499
Herbicides	9.82 kg[j]	574,811
Packing shed (repair and depreciation)[k]	$40.51	274,310
Packing cartons[l] (0.907 kg/carton)	990.3 cartons	4,055,382
Transportation (fertilizer)[m]	—	79,460
Beehive service (without transportation)	$18.52[n]	76,135
Spray aircraft (repair and depreciation)	$ 2.70[o]	15,878
Total		**24,815,264**
Output		
Melon yield (total)	17,964.0 kg	5,389,090
Melon yield (edible)	9,161.0 kg	2,748,436
kcal output/kcal input	—	0.111
protein yield (edible)	641.0 kg @ 7%	

[a]Tractor weights (1):

90-hp diesel (track) = 9,418 kg (20,765 lb)
120-hp diesel (wheel) = 6,189 kg (13,645 lb)
70-hp gasoline (wheel) = 3,549 kg (7,835 lb)

Tractor life: 12,000 hr (2)
Subsoiler weight = 2381 kg (5250 lb), life = 2500 hr[2]
Disc harrow weight = 2721 kg (6000 lb), life = 2500 hr[2]
Planter sled weight = 2721 kg (6000 lb), l fe = 1200 hr[2]
Furrower weight = 363 kg (800 lb), life = 2500 hr[2]
Landplane weight = 4014 kg (8850 lb), life = 2500 hr[2]
Cultivator weight = 1134 kg (2500 lb), life = 2500 hr[2]

Table 4 (continued)
ENERGY INPUTS TO CANTALOUPES — SAN JOAQUIN VALLEY, CALIFORNIA

Item	Quantity/ha	kcal/ha

Fertilizer spreader weight = 340 kg (750 lb), life = 2000 hr[2]
250-hp truck weight = 3629 kg (8000 lb), life = 5000 hr[2]
Crew bus weight = 3175 kg (7000 lb), life = 4000 hr[2]
Pickup truck weight = 1814 kg (4000 lb), life = 4000 hr[2]

Farm machinery @ $2.2/kg ($1/lb), 1967; 19,033 kcal/$1 (75,528 Btu/$1), 1967;[3] motor vehicles @ $2.2/kg ($1/lb), 1967; 17,104 kcal/$1 (67,874 Btu/$1), 1967.[3]

[b] A 250-hp truck consumes 34 ℓ (9 gal) per hour; a crew bus consumes 26 ℓ (7 gal) per hour; a pickup truck consumes 15 ℓ (4 gal) per hour.

[c] Gasoline tractors (wheel), when operating at 75% of rated power, consume 0.32 ℓ (0.085 gal) per horsepower-hour. A 600-hp agricultural aircraft consumes 155 ℓ (41 gal) per hour, 0.1 hr required for three applications.

[d] Diesel tractors, when operating at 75% of rated power, consume 0.25 ℓ (0.065 gal) per horsepower-hour (wheel) and 0.28 ℓ (0.075 gal) per horsepower-hour (track).

[e] Packing shed energy needs are 27.44 kWh electricity and 3620 kcal (0.14 thm) of natural gas per metric ton packed.[4] Yield is 21.7 tonnes/ha (8 U.S. tons/acre).

[f] Nitrogen fertilizer at 20,711 kcal/kg N (37,281 Btu/lb N).[5] Phosphorus fertilizer at 1984 kcal/kg P_2O_5 (3,572 Btu/lb P_2O_5).[5]

[g] Assumed 50,000 kcal/kg seed.

[h] Preplant irrigation (20 cm) is applied by sprinkler, the balance (46 cm) is applied by surface methods. Water (20%) for cantaloupe is from deep wells, the balance is from gravity sources. A total of 2592 kWh/ha-m (320 kWh/acre-foot) is required for pumping from well sources and 2187 kWh/ha-m (270 kWh/acre-foot) from gravity sources (6). The mean value is 2568 kWh/ha-m (280 kWh/acre-foot). An additional 1021 kwh/ha-m (126 kWh/acre-foot) is added for that water applied by sprinklers.[6] A total of 2863 kcal/kWh is used.

[i] An average value of 44,128 kcal/kg of active ingredient was used.

[j] An average value of 60,700 kcal/kg of active ingredient was used for Alnap and Prefar [each 4.5 kg/ha (4 lb/acre) active ingredient]. A total of 0.84 kg/ha (0.75 lb/acre) of Treflan (trifluralin) was used at 35,170 kcal/kg.

[k] Shed cost of $172,000 handles 7600 metric tons (8400 U.S. tons) per year. Assuming a 15-year life, depreciation is $1.36 per U.S. ton, and allowing 50% for repairs, total cost is $2.05 per U.S. ton. For $2 (1977) equal to $1 (1967) and 13,543 kcal/$1 (53,744 Btu/$1), 1967, for repair and maintenance construction,[3] the resulting energy cost is $13,882 kcal/ton.

[l] Fiberboard carton material approximately equal to softwood — 4515 kcal/kg.

[m] Transport and handling of fertilizers at 274 kcal/kg of primary nutrient (4).

[n] Half of beehive service assumed as truck transport, balance was computed at $2 (1976) per $1 (1967) and 8220 kcal/$1 (32,618 Btu/$1), 1967 for agricultural services (3).

Repairs at $7/hr, and depreciation at $4.16/hr in 1977. At $2 (1977) per $1 (1967) and 12,032 kcal/$1 (47,747 Btu/$1), 1967, for repairs and 10,667 kcal/$1 (42,331 Btu/$1), 1967, for aircraft and parts (3), a total of 64,283 kcal/hr results (0.247 hr/ha used).

REFERENCES

1. *Implement and Tractor, Red Book,* Intertec, Overland Park, Kan., 1977.
2. **Reed, A. D. and Horel, L. A.,** Farm Machinery Costs, Leaflet No. 2263, University of California Division of Agricultural Sciences, Berkeley, 1975.
3. **Bullard, C., Hannon, B., and Hevendeen, R.,** Energy Flow Through the United States Economy, a wall chart, University of Illinois Press, Urbana, 1975.
4. **Cervinka, V., Chancellor, W. J., Coffelt, R. J., Curley, R. G., and Dobie, J. B.,** Energy Requirements for Agriculture in California, California Department of Food and Agriculture and Agricultural Engineering Department, University of California, Davis, 1974.
5. **Avlani, P. K.,** Energy Requirements for Production and Use of Wheat in California, M.S. thesis, University of California, Davis, 1977.
6. **Knutson, G. D., Curley, R. G., Roberts, E. B., Hagan, R. M., and Cervinka, V.,** Pumping Energy Requirements for Irrigation in California, Special Publication No. 3215, Division of Agricultural Sciences, University of California, Berkeley, 1977.
7. **Matthews, R. H. and Garrison, Y. J.,** Food Yields Summarized by Different Stages of Preparation, Agriculture Handbook No. 102, Agricultural Research Service, U.S. Department of Agriculture, Washington, D.C., 1975.
8. **Van Maren, A.,** Guidelines to Production Costs and Practices 1976-77, Circular 104, Imperial County Cooperative Extension Service, California, 1976.
9. **Yeary, E.,** Cost Analysis Work Sheet — Cantaloupes, Southern San Joaquin Valley — 1975, Special Publication No. 3038, University of California, Division of Agricultural Sciences, Berkeley, 1975.
10. **Kasmire, R.,** Muskmelon Production in California, University of California Division of Agricultural Sciences Leaflet, University of California, Berkeley, in preparation.
11. Marketing California Melons — 1976 Season, Federal-State Market News Service, Sacramento, Calif., 1977.
12. Shipments of California Fruits and Vegetables — 1976, Federal-State Market News Service, Sacramento, Calif., 1977.

WATERMELON

Norman C. Glaze*

In 1976, 1.13 million t of watermelons were produced and marketed in the U.S. The value of the crop to the producers amounted to $85,000,000. The majority of the crop is produced in the southern U.S. The six leading states with the percent of the total production in parentheses are Florida (38), Texas (17.3), Georgia (9.3), California (7.5), South Carolina (4.5), and Alabama (4.3). Other states producing watermelon in smaller amounts include Mississippi, Oklahoma, Missouri, Indiana, Arizona, North Carolina, Virginia, Maryland, and Delaware. The southeastern and southwestern states production peaks about the fourth of July, and the production moves northward to provide continuity of supply on the major markets during the summer. The majority of the crop is seeded, but some transplanting is done in certain areas for earlier production.

The energy use for disease control is low in semiarid areas because the environmental conditions are less conducive to disease proliferation. In the humid areas of the Southeast and Atlantic Coast, more energy is expended for disease control and pesticides. In contrast, growing crops in drier climates necessitates more irrigation than is needed in humid areas, where irrigation is normally limited to supplemental irrigation applied during short periods of inadequate rainfall. Watermelons are frequently grown without irrigation in the Southeast and Atlantic Coast regions. Watermelon harvest is labor intensive as fruits are often picked in two or three hand harvests. Melons are frequently taken to sorting tables for sizing before shipment. Most of the crop is shipped by truck; although, a small portion of the southeastern melons is shipped by rail.

Table 1
AN EXAMPLE OF ENERGY INPUTS PER HECTARE FOR WATERMELON IN CALIFORNIA

Item	Quantity/ha	kcal/ha	Ref.
	Input		
Labor	134 hr	—	1
Machinery	47 kg[a]	846,000	1,6
Diesel	600 ℓ[a]	6,848,400	
Nitrogen	112 kg	1,601,600	1
Phosphorus	56 kg	168,000	1
Potassium	56 kg	89,600	1
Seeds	3.4 kg	12,240	1
Irrigation	76.2 cm	1,210,253	1
Insecticides	9 kg	782,190	1
Herbicides	6.7 kg	669,397	1
Transportation	560.4 kg	144,023	
Total	**12,371,703**		
	Output		
Watermelon yield	22,176 kg	2,661,120	2
kcal output/kcal input		0.22	
protein yield	51 kg		
kcal output/labor hour	19,859		

[a] Assumed from data in Georgia.

* Coastal Plain Station, Science and Education Administration, U.S. Department of Agriculture, Tifton, Ga.

Table 2
AN EXAMPLE OF ENERGY INPUTS PER HECTARE FOR WATERMELON IN FLORIDA

Item	Quantity/ha	kcal/ha	Ref.
	Input		
Labor	150 hr	—	
Machinery	47kg[a]	846,000	
Diesel	600 ℓ	6,48,400	
Nitrogen	135 kg	1,930,500	
Phosphorus	180 kg	540,000	
Potassium	180 kg	288,000	
Lime	4,480 kg	8,480,640	
Seeds	2.2 kg	7,920	
Irrigation	20 cm	2,803,621	
Insecticides	0.6 kg	52,146	
Fungicides	11 kg	714,010	
Transportation	559.2 kg	143,714	
Total		**22,654,951**	
	Output		
Watermelon yield	19,264 kg	2,311,680	1
kcal output/kcal input		0.10	
protein yield	44 kg		
kcal output/labor hour	15,411		

[a] Employed a similar value to that for machinery in Georgia.

Table 3
AN EXAMPLE OF ENERGY INPUTS PER HECTARE FOR WATERMELON IN GEORGIA

Item	Quantity/ha	kcal/ha	Ref.
	Input		
Labor	138 hr	—	3, 4
Machinery	47 kg[a]	850,935	
Diesel	600 ℓ	6,848,400	3, 4
Nitrogen	101 kg	1,444,300	3, 4
Phosphorus	112 kg	336,000	3, 4
Potassium	168 kg	268,800	3, 4
Lime	2,240 kg	4,240,320	3, 4
Seeds	2.2 kg	7,920	3, 4
Irrigation	7.5 cm	119,119	3, 4
Insecticides	3.4 kg	295,494	3, 4
Fungicides	11.2 kg	729,992	3, 4
Herbicides	5 kg	499,550	3, 4
Transportation	559.2 kg	143,714	
Fumigant	9 kg	584,190	3, 4
Total		**16,363,799**	
	Output		
Watermelon yield	9,259 kg	1,111,080	2
kcal output/kcal input		0.07	
protein yield	21 kg		
kcal output/labor hour	8,051		

[a] Calculated based on dollar costs of machinery use and employing a value 18,105 kcal per 1967 dollar in the conversion.[6]

Table 4
AN EXAMPLE OF ENERGY INPUTS PER HECTARE FOR WATERMELON IN INDIA[5]

Item	Quantity/ha	kcal/ha
	Input	
Labor	800 hr	—
Machinery	3 kg	54,000
Diesel	60 ℓ	684,840
Nitrogen	120 kg	1,716,000
Phosphorus	25 kg	75,000
Organic	5,100 kg	—
Seeds	4 kg	14,400
Irrigation	60 cm	836,656
Insecticides	0.1 kg	8,691
Transportation	58 kg	14,906
Total		**3,404,493**
	Output	
Watermelon yield	15,273 kg	1,832,760
kcal output/kcal input		0.54
protein yield	35 kg	
kcal output/labor hour	2,291	

REFERENCES

1. Watermelon Production, Leaflet 2672, Division of Agricultural Sciences, University of California, Davis.
2. Marketing Watermelons — 1976 Crop, Federal-State Market News, U.S. Department of Agriculture, North Palm Beach, 1977.
3. Selected Horticultural Crop Budgets for South Georgia, Research Report 39, College of Agricultural Experiment Station, University of Georgia, 1969.
4. Vegetable Enterprise Cost Analysis, Farm Management Department, Georgia Cooperative Extension Service, 1977.
5. **Seshadri, U.S.,** Division of Vegetable and Ornamental Crops, Indian Agricultural Research Institute, New Delhi, India, personal communication, 1978.
6. **Herendeen, R. A. and Bullard, C. W.,** Energy Cost of Goods and Services, 1963 and 1967, Doc. No. 140, Center for Advance Computation, University of Illinois, Urbana, 1974.

ENERGY USE IN PEPPERS

Philip A. Minges

Peppers belong to the Solanaceae family with two species being of commercial importance, *Capsicum annum* and *C. frutescens.* The tabasco type (very hot), which is processed for pickles or sauces, is in the latter species. The several remaining types are in *C. annum*. The pungent type with a wide range of shapes is more popular in warmer areas of the world while the sweet type, also with varying shapes and colors, is more commonly grown in northerly latitudes, although there is overlapping. Both types are used fresh and for processing, including dehydration, pickling, freezing, and canning.

In the U.S. the principal pepper producing states are Florida, California, Texas, New Jersey, and North Carolina with smaller hectarages in a number of other states for a total of around 22,267 ha. About one third of pepper production in California is hot peppers for dehydration. Most of the other processing (canning and pickling) centers are in the Southern states. The bell type is the principal sweet pepper grown for fresh market, and well over 50% of the U.S. acreage is devoted to this type. Average yields range from 1591 to over 9091 kg depending on the season (winter, spring, summer), the type, and locality. The U.S. average is about 100 cwt for the bell type. Yields double these figures are obtained by the better growers.

Estimates (from Food and Agriculture Organization and others) of hectares harvested in major areas of production in the world are as follows:

Continent	Number of hectares	Principal countries with hectares (ha)
North America	77,000	Mexico (53), U.S. (22)
South America	20,000	Argentina (9), Peru (5)
Europe	141,000	Yugoslavia (32), Spain (29), Italy (22), Bulgaria (19), Hungary (17), Romania (13)
Asia	346,000	China (126), Sri Lanka (57), Pakistan (51), Korea (50), Turkey (57)
Africa	106,000	
World total	691,000	

INPUTS

Inputs of labor and energy to produce an acre of peppers will vary greatly depending on the area, the type of pepper, the use of the crop, whether irrigated or not, the amount and type of equipment used, the soil fertility level, and the sophistication of culture. Labor inputs are far higher for fresh market peppers than for those used for processing, especially dehydration where mechanical harvesting is now practiced. Cultural practices within a country may vary greatly. For example, in Mexico the northwest coast district practices are similar to those in California while in other parts of Mexico culture may range down to very primitive.

Since peppers are a relatively minor crop, even in states like California and Florida, studies on inputs are rare, and of those available few are up-to-date. Fertilizer usage, which may account for about one third of the total energy input (manufacture, transportation, and application), is probably the easiest to estimate with some degree of accuracy. Some estimates on other inputs have been gleaned from cost input studies (mostly out-of-date), crop publications, and questionnaires to a few growers. Unfortunately, few of the studies or surveys followed similar patterns, which makes summarizing still more difficult.

Generally, the inputs for peppers are similar to unstaked tomatoes and eggplant. Most pepper fields are established using transplants, but some acreage for processing, especially for dehydration, is direct (field) seeded, which saves some on costs even though more seed is used.

Table 1
ENERGY INPUTS FOR BELL PEPPER PRODUCTION IN NEW YORK STATE

Item	Quantity/ha	kcal/ha
	Input	
Labor	225 hr	—
Machinery	31 kg	558,000
Diesel fuel	420 ℓ	4,793,880
Nitrogen	150 kg	2,205,000
Phosphorus	85 kg	255,000
Potassium	85 kg	136,000
Seeds	0.4 kg	1,440
Electricity	8 kWh	22,904
Fuel oil	470 ℓ	5,364,580
Pesticides	15 kg	1,303,650
Transportation	788 kg	202,516
Total		**14,842,970**
	Output	
Pepper yield	11,200 kg	2,016,000
Protein yield	11 kg	
kcal output/kcal input		0.14
kcal output/labor hour		8,960

Table 2
ENERGY INPUTS FOR IRRIGATED BELL PEPPER PRODUCTION IN NEW YORK STATE

Item	Quantity/ha	kcal/ha
	Input	
Labor	225 hr	—
Machinery	31 kg	558,000
Diesel fuel	420 ℓ	4,793,880
Nitrogen	175 kg	2,572,500
Phosphorus	85 kg	255,000
Potassium	85 kg	136,000
Seeds	0.4 kg	1,440
Electricity	8 kWh	22,904
Fuel oil	470 ℓ	5,364,580
Irrigation	15 cm	238,238
Pesticides	15 kg	1,303,650
Transportation	788 kg	202,516
Total		**15,448,618**
	Output	
Pepper yield	11,200 kg	2,016,000
Protein yield	11 kg	
kcal output/kcal input		0.13
kcal output/labor hour		8,960

Table 3
ENERGY INPUTS FOR IRRIGATED CHILI-TYPE PROCESSING PEPPERS IN CALIFORNIA

Item	Quantity/ha	kcal/ha
	Input	
Labor	100 hr	—
Machinery	31 kg	558,000
Diesel fuel	225 ℓ	2,568,150
Nitrogen	200 kg	2,940,000
Phosphorus	25 kg	75,000
Seeds	0.75 kg	2,700
Irrigation	45 cm	6,700,000
Pesticides	12 kg	1,042,920
Dehydration	6,360 kg	9,540,000
Transportation	202 kg	51,914
Total		**23,478,684**
	Output	
Pepper yield	11,200 kg	2,240,000
Protein yield	11 kg	
kcal output/kcal input		0.10
kcal output/labor hour		20,160

ENERGY USE IN SPINACH *(SPINACIA OLERACEA)* PRODUCTION

G. A. Bradley

IMPORTANCE OF SPINACH ECONOMICALLY

Spinach is a crop of some legend in the U.S. due to its association with renewed energy in the comic character Popeye.

In composition, spinach is above 90% water regardless of the form in which it is consumed; raw, canned, or frozen.[1] Spinach is a good-to-excellent source of iron, ascorbic acid, vitamin A potential, calcium, potassium, and sodium. Spinach is also high in oxalate salts, and under rarely encountered environmental conditions may accumulate high levels of nitrates.[2,3] New Zealand spinach *(Tetragonia expansa)*, a somewhat similar plant, but not closely related to spinach, is similar to spinach in nutritional value. It can be grown under warmer conditions than spinach.

Spinach is produced in many home gardens, where it can be grown easily during the cooler parts of the year. It is produced commercially in the U.S. and is sold in the fresh, canned, and frozen forms.

The average production area in spinach harvested for fresh market from 1975 to 1977 was approximately 3,700 ha, yielding about 32 million kg, with a value to producers of about $12 million annually. Leading states in the production of spinach for fresh market are Texas, California, Maryland, and Virginia.[4]

Harvested spinach for processing in the U.S. averaged about 8,900 ha from 1975 to 1977, production about 145,000 tonnes, and grower value $10.5 million. About half is canned and half is frozen. Spinach is seventh in value to growers of processing vegetables produced in the U.S., exceeding beets and cabbage for kraut in value and land used. California leads in the production of processing spinach; Arkansas and Oklahoma follow.

ENVIRONMENTAL CONDITIONS INFLUENCING PRODUCTION AND ENERGY USE

Environment affects the production of spinach greatly because spinach is sensitive to temperature, to light intensity, and to daylength.[5,6] Spinach grows most rapidly under sunny conditions, with fairly long days, and temperatures that do not exceed about 22°C in the daytime or drop below 10°C at night. Under certain conditions of daylength, temperature, or a combination of daylength and temperature, the spinach plant will produce a seedstalk (bolt), rendering it undesirable for culinary use, but of course an essential process for seed production in the crop.[6] Conditions most favorable for such seedstalk appearance include exposure to temperatures in the range of 4 to 10°C during early stages of vegetative growth, especially if the daylength is 12 hr or longer, then continuing growth under increasing daylength and increasing temperatures. Thus the so called spring crop of spinach grown in the Arkansas-Oklahoma area is more subject to bolting than the fall or the overwinter crop.[5] There is genetic variation in susceptibility to bolting and cultivars have been selected for the long-standing character; that is, they are less likely to bolt when subjected to conditions of daylength and temperature that favor bolting. Once a seedstalk has been initiated in spinach, growing conditions that favor vegetative growth also tend to favor seedstalk extension. Therefore cultivars selected for their long-standing ability tend to yield less when compared to cultivars without such ability, if grown under conditions unfavorable for seedstalk elongation.[6]

Table 1
ENERGY INPUTS PER HECTARE FOR SPINACH GROWN AS A SPRING CROP IN OZARKS REGION OF U.S.

Item	Quantity/ha	kcal/ha	Ref.
	Input		
Labor	31 hr	—	
Machinery	27 kg[a]	486,000	
Gasoline	47 ℓ[b]	378,500	
Diesel	79 ℓ[b]	736,227	
Nitrogen	79 kg	1,213,124	9
Lime	560 kg	1,060,000	9
Seeds	28 kg	100,800	9
Irrigation (water applied)	15 cm[c]	1,832,000	
Insecticides	3.7 kg[d]	321,567	
Fungicides	7.2 kg[e]	169,704	
Herbicides	2.2 kg[f]	219,802	
Transportation	157.4 kg[g]	40,452	
Total		**6,558,176**	
	Output		
Spinach yield	13,209 kg	3,434,340	
Protein yield	422 kg		
kcal output/kcal input		0.52	
kcal output/labor hour		110,785	

[a] Assumes 15 metric tons of machinery used for production (tractor, harvester, truck, etc.) and the machinery is depreciated over 15 years and serves an average of 36 ha.[9]
[b] Gasoline and diesel used for operation of tractor, harvester, trucks, sprayer, etc.[9]
[c] 15 cm water per hectare applied by side roll sprinkle pumping against an average 100 m head.[9]
[d] 2.2 ℓ 15% EC Phosdrin, 2.2 ℓ 25% EC Parathion.[9]
[e] 7.2 kg Maneb.[9]
[f] 2.2 kg Cycloate.[9]
[g] Total weight of machinery, fuel, fertilizer, seeds, etc. transported to crop hectare.

Energy requirements for production of spinach may vary somewhat with the season of the year in which it is produced because of different number of days to maturity and possible differences in fertilizer amounts and schedules, and in the programs needed for control of insects, diseases, and weeds. Several energy budgets are given that reflect to some extent such differences (Tables 1 to 4).

Spinach plants do have considerable tolerance to cold. With proper conditioning, spinach may tolerate temperatures as low as −15°C with only minor damage to leaves of the plant. Under snow cover, spinach may survive temperatures of −25°C. This cold tolerance makes it possible to produce spinach during the winter in areas such as Arkansas, Oklahoma, and Tennessee.[5] A crop may be planted from mid-October until near Christmas in these states with harvest from early January until mid-March, depending upon the nature of the growing season and demand of the processing industry for spinach at any particular time when weather is conducive to harvest. California also produces much of its spinach in the winter season, planting mainly in January,

Table 2
ENERGY INPUTS PER HECTARE FOR SPINACH GROWN AS A OVERWINTER CROP IN OZARKS REGION U.S.

Item	Quantity/ha	kcal/ha	Ref.
	Input		
Labor	35 hr	—	
Machinery	27 kg[a]	486,000	
Gasoline	47 ℓ	378,500	9
Diesel	83 ℓ	773,504	9
Nitrogen	105 kg	1,543,000	9
Lime	560 kg	1,060,000	9
Seeds	28 kg	100,800	9
Irrigation (water applied)	10 cm[b]	1,220,000	
Insecticides	5.6 kg	486,696	
Fungicides	3.6 kg[c]	84,850	
Herbicides	2.2 kg[c]	219,802	
Transportation	160.8 kg[d]	41,326	
Total		**6,394,478**	
	Output		
Spinach yield	14,0000 kg	3,640,000	
Protein yield	447 kg		
kcal output/kcal input		0.57	
kcal output/labor hour		104,000	

[a] Assumes 15 metric tons of machinery used for production. Price production.[9]

[b] 10 cm water applied by side roll sprinkler pumping against an average 100 m head.

[c] 3.3 ℓ 15% E. C. Phosdrin, 3.3 ℓ 25% EC Parathion, 3.6 kg Maneb, 2.2 kg Cycloate.[9]

[d] Total weight of machinery, fuel, fertilizers, seeds, etc. transported to crop hectare.

Table 3
ENERGY INPUTS PER HECTARE FOR PROCESSING SPINACH GROWN AS WINTER CROP IN THE SALINAS VALLEY OF CALIFORNIA

Item	Quantity/ha	kcal/ha	Ref.
	Input		
Labor	45 hr	—	
Machinery	40 kg[a]	720,000	
Gasoline	55 ℓ	442,925	8-10
Diesel	100 ℓ	931,932	8-10
Nitrogen	202 kg	2,424,000	7
Phosphorus	60 kg	180,000	7
Seeds	17 kg	61,200	7
Irrigation (water applied)	15 cm	738,300	8, 10
Insecticides	5.6 kg	486,696	
Fungicides	3.6 kg[b]	84,850	
Herbicides	2.2 kg[b]	219,802	
Transportation	183.3 kg[c]	47,108	
Total		**6,336,813**	
	Output		
Spinach yield	22,400 kg	5,824,000	
Protein yield	716 kg		
kcal output/kcal input		0.92	
kcal output/labor hour		129,422	

[a] Assume 22.2 metric tons of machinery used for production and the machinery is depreciated over 15 years and serves an average of 36 ha.[8-10]

[b] 3.6 Maneb, 2.2 kg Cycloate, 3.3 ℓ 15^{70} E.C. Phosdrin and 3.3 ℓ 25^{70} Parathion.[8, 10]

[c] Total weight of machinery, fuel, fertilizers, seeds, etc. transported to crop hectare.

Table 4
ENERGY INPUTS PER HECTARE TO PRODUCE FRESH MARKET SPINACH. WINTER GARDEN REGION OF TEXAS

Item	Quantity/ha	kcal/ha	Ref.
	Input		
Labor	270 hr	—	
Machinery	40 kg[a]	722,000	
Gasoline	50 ℓ	402,600	11
Diesel	90 ℓ	838,700	11
Nitrogen	67 kg	806,400	11
Phosphorus	90 kg	270,000	11
Seeds	20 kg	72,000	11
Irrigation (water applied)	11.4 cm	979,000	11
Insecticides	4.5 kg[b]	391,095	
Fungicides	7.8 kg[b]	183,840	
Herbicides	2.2 kg[b]	219,802	
Transportation	174 kg	44,718	11
Total		**4,930,155**	
	Output		
Spinach yield	20,160 kg	5,241,600	
Protein yield	645 kg		
kcal output/kcal input		1.06	
kcal output/labor hour		19,413	

[a] Assumes 22.2 metric tons of machinery used for production and the machinery is depreciated over 15 years and serves an average of 36 ha.[11]

[b] 2.0 kg Phosdrin, 2.5 kg Parathion, 7.8 kg Maneb, 2.2 kg Cycloate.[11]

but temperatures do not go low enough to cause damage.[7] The Salinas Valley of California has one of the most favorable climates for spinach production in the winter season, with temperatures in a favorable growing range most of the time.[7,8] Bolting can be a problem and does limit the potential yield to some extent. Spinach grown in the overwinter season grows in cycles, depending upon the temperature patterns. If a crop is in the field for a long growing period (sometimes as much as 135 to 150 days in the Ozarks region), rain may leach out nitrogen and result in an extra application being needed. Sometimes insect infestations develop or disease outbreaks occur, but usually in the Ozarks area such problems are less in the overwinter crop than in the spring or fall crop because of shorter periods of favorable temperatures. It is also more difficult to predict the probability of bolting in the overwinter crop because each planting is exposed to unique temperature patterns. The likelihood of bolting is generally less than for the spring crop. Slow bolting varieties are usually grown for the overwinter and spring crops, but some fall-adapted varieties may be planted for the overwinter crop.

REFERENCES

1. **Watt, B. K. and Merrill, A. L.,** Composition of foods, Agricultural Handbook No. 8, Agricultural Research Service, U.S. Department of Agriculture, 1963.
2. **Brown, J. R. and Smith, C. E.,** Nitrate accumulation in vegetable crops as influenced by soil fertility practices, *Mo. Agric. Exp. Stn. Bull.*, 920, 1967.
3. Cantliffe, Div. Nitrate accumulation in spinach grown under different light intensities, *J. Am. Soc. Hortic. Sci.*, 97, 152, 1972.
4. **Anon.,** Vegetables, 1977 Annual Summary. Crop Reporting Board. Statistical Reporting Service, U.S. Department of Agriculture, Vg. 2-2(77). 1977.
5. **Bradley, G. A., Sistrunk, W. A., Baker, E. C., and Cash, J. N.,** Effect of plant spacing, nitrogen and cultivar on spinach *(Spinacia oleracea L.)* yield and quality, *J. Am. Soc. Hortic. Sci.*, 100, 45, 1975.
6. **Knott, J. E.,** The effect of temperature on the photoperiodic response of spinach, *N. Y. Agric. Exp. Sta., Cornell, Mem.*, 218, 1939.
7. **Zink, F. W.,** Growth and nutrients absorption in spring spinach, *Proc. Am. Soc. Hortic. Sci.*, 87, 380, 1965.
8. **Sims, W. L.,** Spinach production in California, *Proc. Arkansas State Hortic. Soc. 93rd Ann. Meet.*, 50, 1972.
9. **Price, C., Bowers, J. L., Bradley, G. A., Kattan, A. A., and McFerran, J.,** Guidelines for production of vegetables for processing in Arkansas, Spec. Rep. 40, Arkansas Agric. Exp. Sta., 1977.
10. **Sims, W. L.,** personal communication.
11. **Fisher, G.,** Economics of profitable spinach production, L-1076A., Texas Agricultural Extension Service.

Energy Inputs and Outputs for Crop Systems—Fruits and Tree Crops

ENERGY USE IN LOW, MEDIUM, AND HIGH DENSITY APPLE ORCHARDS — EASTERN U.S.

Richard C. Funt

During the 1975-1977 period the utilized commercial apple crop was 2.8 million metric tons.*[2] Washington, New York, Michigan, and Pennsylvania are the leading apple producing states. Washington produces more than one third of the U.S. total and ships apples to eastern markets, especially New York City. In the eastern U.S., New York, Pennsylvania, Virginia, and North Carolina are the leading producers.[5]

Apples are second (bananas are first) among noncitrus purchases of fresh fruit in the U.S. About 56% of all apples sold over the past 4 years were sold on the fresh market.[6] The processed market utilized almost 44% of the crop with apple juice and apple sauce as the chief items. Apple juice amounted to 43% of apples processed (16% of total production) while sauce averaged 26.5%.[6]

Generally, 5 to 6% of the apples have been exported.[2] However, in the 1976 to 1977 season, U.S. apple exports of 6.3 million bushels (120,054 metric tons) set a post-World War II record.[2] Exports to Europe increased due to the lowest yield in 15 years and to the Mideast (Saudi Arabia) and Far East (Hong Kong).

Generally apples are grown on sites that are protected from freezing temperatures during bloom. Nationally, only 10% of the deciduous fruit and nut crops require frost protection.[1] Orchard heaters and wind machines utilize LP and natural gas, gasoline, diesel, and electric motors. In the Pacific Northwest, irrigation is depended upon for apple production. Supplying irrigation water can account for 40% or more of energy used in production.[1] However yields are generally higher in the Pacific Northwest than in the East due to controlled watering. Generally few growers in the East use frost protection or irrigation in their orchards.

Nationally, large-sized apple trees require larger equipment and use larger amounts of fuel than smaller sized trees.[4] Some growers in the East are using large mechanical harvesters rather than hand labor.

Pesticide usage can account for 11% of the total energy expended in orchards while field operations (tillage, pesticide application, etc.) account for 12% of the total energy use.[1] There is very little difference in energy usage between high-quality apples grown for processing or fresh market.[4]

Apples, whether they are marketed processed or fresh, are stored in common or controlled atmosphere refrigerated storage at 32°F (0°C) and 90% relative humidity. Generally, 46% of all apples are stored.[3] In the East, 40 to 55% (6 year average = 44.1%) of the stored apples are sold between October 31 and January 31 and less than 6.4% of all apples are left in storage after April 30.[3]

The energy tables for eastern apple production were developed from an economic doctoral thesis.[4] The thesis data is based on a typical Pennsylvania fruit farm having 44.4 acres (17.8 ha) of apple trees, 20 acres (8.0 ha) of other fruit crops, 85.6 acres (34.2 ha) of open land, and 25 acres (10.0 ha) of farmstead and nonproductive areas. The author was born and raised on a similar farm in the leading fruit production area of the state and utilized this operation for developing the data.

The data presented in Tables 1 through 9 represent low, medium, and high density apple orchards with selected ages of 1, 10, and 20 years. Only the high-density orchard

* This converts to 6.2 billion lb or 147.6 million bushels (42 lb equivalent).

is irrigated and no orchards utilize frost control. The high-density orchard has a trellis system for support.

The low- and medium-density orchards show comparisons using hand or a mechanical harvester which is a shake-and-catch system moving and stopping at each tree. In the high-density system, an over-the-row harvester moves continuously down the row at 1.7 km per hour and is capable of harvesting 28,600 kg of fruit per hour. All fruit is transported by truck 17.6 km to a packing plant. No refrigeration or grading energy values are included.

The equipment utilized in the low- and medium-density systems are larger and have higher horsepower requirements than the equipment used in the high-density system. For example, large air-blast sprayers with over 100 horsepower motors are necessary for low density while small PTO driven air blasts are used in the high-density system. However, high-density systems require greater travel time through the orchard due to a greater number of rows than the low- or medium-density systems.

The results in Tables 1 through 9 indicate that the largest use of energy (kcal/ha) is fuel (gasoline and diesel). Generally 60 to 70% of the energy use can be attributed to fuel. Pesticide energy use ranged from 7 to 20% with an average of 14% over all densities and years of production. This is in general agreement with a U.S. Department of Agriculture publication where 44% of total direct energy went to fuel for orchard crops and pesticides represented 15.3% for apples.[1] Differences between these studies are due to specific densities and tree age here and general orchard conditions plus frost protection in the USDA publication.

Generally, it takes the same amount of pesticide per hectare regardless of yield. Therefore increasing yield utilizes pesticides more efficiently. Also greater knowledge and timing of pesticides and smaller droplet size will decrease pesticide usage.

Fuel usage can be reduced with small-size trees because small equipment can be used. However, fuel usage for transport increases as yield increases per hectare, but not per kg.

The input-output ratio was the lowest in the low-density system. The medium- and high-density systems were higher, but nearly equal to each other. High-density systems could be more efficient in labor and energy if the support system was eliminated. New rootstocks and interstem trees could be used to accomplish this goal.

Table 1
ENERGY INPUT PER HECTARE FOR LOW DENSITY (165 TREES/HA), HAND OR MECHANICALLY HARVESTED NONIRRIGATED APPLES FOR THE FIRST-YEAR (ESTABLISHMENT), EASTERN REGION U.S.[a]

Item	Quantity/ha	kcal/ha
	Input	
Labor		
Hand	70 hr	—
Mechanical[b]	70 hr	—
Machinery		
Hand	88 kg	481,606
Mechanical[b]	—	—
Gasoline		
Hand	325 ℓ	3,285,425
Mechanical[b]	—	—
Diesel	202 ℓ	2,305,628
Electricity	20 kwh	57,260
Nitrogen[c]	3 kg	44,100
Phosphorus[d]	114 kg	627,000
Potassium[d]	114 kg	231,408
Lime[d]	682 kg	1,437,656
Insecticide	4 kg	176,000
Fungicide	27 kg	594,000
Herbicide	2 kg	114,000
Transportation		
Hand	1,561 kg	401,177
Mechanical[b]	—	—
Building (450 sq ft)[e]		14,250
Total		
Hand		**9,769,510**
Mechanical[b]		—
	Output	
No yield first year		

[a] There are no differences in energy input between apples grown for fresh or processing.

[b] Mechanical harvester not used in nonbearing year.

[c] Ammonium nitrate applied annually at an increasing rate until age 16.

[d] Lime (calcium), phosphorus, and potassium applied every fifth year. Amount shown in average annual application.

[e] Service building only, with 20-year life and used for 60 ha.

From Funt, R. C., Ph.D. thesis, Pennsylvania State University, University Park, 1974. (With permission.)

Table 2
ENERGY INPUT PER HECTARE FOR LOW DENSITY (165 TREES/HA) HAND OR MECHANICALLY HARVESTED NONIRRIGATED APPLES FOR THE TENTH YEAR AFTER ESTABLISHMENT, EASTERN REGION U.S.[a]

Item	Quantity/ha	kcal/ha
Input		
Labor		
Hand	179 hr	—
Mechanical[b]	91 hr	—
Machinery		
Hand	88 kg	566,510
Mechanical[b]	151 kg	1,117,117
Gasoline		
Hand	530 ℓ	5,357,770
Mechanical	715 ℓ	7,227,935
Diesel	298 ℓ	3,401,372
Electricity	20 kwh	57,260
Nitrogen[c]	30 kg	441,000
Phosphorus[d]	114 kg	627,000
Potassium[d]	114 kg	231,408
Lime[d]	682 kg	1,437,656
Insecticides	47 kg	2,889,090
Fungicides	49 kg	1,360,730
Herbicides	6 kg	599,460
Transportation		
Hand	1958 kg	518,626
Mechanical	2206 kg	582, 362
Building (450 sq ft)[e]	2206	14,250
Total		
Hand		17,502,132
Mechanical		19,986,640
Output		
Yield		
Hand	22,861 kg	12,802,160
Mechanical	22,861 kg	12,802,160
kcal output/kcal input		
Hand		0.731
Mechanical		0.690
Protein yield	46 kg	

[a] Apples are transported 17.6 km to packing or processing plant. There are no differences in energy input between apples grown for fresh or processing markets.
[b] Mechanical harvester is shake and catch type and travels at same rate of speed regardless of yield. It has a 5-year life.
[c] Ammonium nitrate applied annually at an increasing rate until age 16.
[d] Lime (calcium), phosphorus, and potassium applied every fifth year. Amount shown is average annual application.
[e] Service building only with 20-year life and used for 60 ha.

From Funt, R. C., Ph.D. thesis, Pennsylvania State University, University Park, 1974.

Table 3
ENERGY INPUT PER HECTARE FOR LOW DENSITY (165 TREES/HA) HAND OR MECHANICALLY HARVESTED NONIRRIGATED APPLES FOR THE TWENTIETH YEAR AFTER ESTABLISHMENT, EASTERN REGION U.S.[a]

Item	Quantity/ha	kcal/ha
	Input	
Labor		
Hand	385 hr	—
Mechanical[b]	138 hr	—
Machinery		
Hand	88 kg	1,029,385
Mechanical[b]	151 kg	1,730,557
Gasoline		
Hand	1346 ℓ	13,606,714
Mechanical[b]	1531 ℓ	15,768,879
Diesel	483 ℓ	5,512,962
Electricity	20 kwh	57,260
Nitrogen[c]	45 kg	661,500
Phosphorus[d]	114 kg	627,000
Potassium[d]	114 kg	231,408
Lime[d]	682 kg	1,437,656
Insecticide	47 kg	2,889,090
Fungicide	49 kg	1,360,730
Herbicide	6 kg	599,460
Transportation		
Hand	2974 kg	787,448
Mechanical	3222 kg	851,184
Building (450 sq ft)[e]		14,250
Total		
Hand		28,814,773
Mechanical		31,741,846
	Output	
Yield		
Hand	54,743 kg	30,656,080
Mechanical	54,743 kg	30,656,080
kcal output/kcal input		
Hand		1.063
Mechanical		0.966
Protein yield	109 kg	

[a] Apples are transported 17.6 km to packing or processing plant. There are no differences in energy input between apples grown for fresh or processing markets.

[b] Mechanical harvester is shake and catch type and travels at same rate of speed regardless of yield. It has a 5-year life.

[c] Ammonium nitrate applied annually at an increasing rate until age 16.

[d] Lime (calcium), phosphorus, and potassium applied every fifth year. Amount shown in average annual application.

[e] Service building only with 20-year life and used for 60 ha.

From Funt, R. C., Ph.D. thesis, Pennsylvania State University, University Park, 1974.

Table 4
ENERGY INPUT PER HECTARE FOR MEDIUM DENSITY (453 TREES/HA), HAND OR MECHANICALLY HARVESTED NONIRRIGATED APPLES FOR THE FIRST YEAR (ESTABLISHMENT), EASTERN REGION U.S.[a]

Item	Quantity/ha	kcal/ha
Input		
Labor		
Hand	95hr	—
Mechanical[b]	—	
Machinery		
Hand	82 kg	449,744
Mechanical[b]	—	—
Gasoline		
Hand	325 ℓ	3,285,425
Mechanical[b]	—	
Diesel	202 ℓ	2,305,628
Electricity	20 kwh	57,260
Nitrogen[c]	8 kg	117,600
Phosphorus[d]	114 kg	627,000
Potassium[d]	114 kg	231,408
Lime[d]	682 kg	1,437,656
Insecticide	4 kg	176,000
Fungicide	27 kg	594,000
Herbicide	2 kg	114,000
Transportation		
Hand	1560 kg	400,920
Mechanical[b]	—	—
Building (450 sq ft)[e]		14,250
Total		
Hand		**9,810,891**
Mechanical[b]		—
Output		
No yield first year		

[a] There are no differences in energy input between apples grown for fresh or processing.
[b] Mechanical harvesters not used in nonbearing year.
[c] Ammonium nitrate applied annually at an increasing rate until age 10.
[d] Lime (calcium), phosphorus, and potassium applied every fifth year. Amount shown is average annual application.
[e] Service building only with 20-year life and used for 60 ha.

From Funt, R. C., Ph.D. thesis, Pennsylvania State University, University Park, 1974.

Table 5
ENERGY INPUT PER HECTARE FOR MEDIUM DENSITY (453 TREES/HA), HAND OR MECHANICALLY HARVESTED NONIRRIGATED APPLES FOR THE TENTH YEAR AFTER ESTABLISHMENT, EASTERN REGION U.S.[a]

Item	Quantity/ha	kcal/ha
Input		
Labor		
Hand	176 hr	—
Mechanical[b]	126 hr	—
Machinery		
Hand	82 kg	898,100
Mechanical[b]	139 kg	1,502,766
Gasoline		
Hand	1101 ℓ	11,130,009
Mechanical	1303 ℓ	13,172,027
Diesel	439 ℓ	5,010,746
Electricity	20 kwh	57,260
Nitrogen[c]	82 kg	1,205,400
Phosphorus[d]	114 kg	627,000
Potassium[d]	114 kg	231,408
Lime[d]	682 kg	1,437,656
Insecticide	47 kg	2,889,090
Fungicide	49 kg	1,360,730
Herbicide	6 kg	599,460
Transportation		
Hand	2716 kg	698,012
Mechanical	2975 kg	764,575
Building (450 sq ft)[e]		14,250
Total		
Hand		26,159,121
Mechanical		28,872,368
Output		
Yield		
Hand	41,546 kg	23,265,760
Mechanical	41,546 kg	23,265,760
kcal output/kcal input		
Hand		0.889
Mechanical		0.806
Protein yield	83 kg	

[a] There are no differences in energy input between apples grown for fresh or processing. Apples are transported 17.6 km to packing or processing plant.

[b] Mechanical harvester is a shake and catch type and travels at the same speed regardless of yield. It has a 5-year life.

[c] Ammonium nitrate applied annually at an increasing rate until age 10.

[d] Lime (calcium), phosphorus, and potassium applied every fifth year. Amount shown is average annual application.

[e] Service building only with 20-year life and used for 60 ha.

From Funt, R. C., Ph.D. thesis, Pennsylvania State University, University Park, 1974.

Table 6
ENERGY INPUT PER HECTARE FOR MEDIUM DENSITY (453 TREES/HA), HAND OR MECHANICALLY HARVESTED NONIRRIGATED APPLES FOR THE 20th YEAR AFTER ESTABLISHMENT, EASTERN REGION U.S.[a]

Item	Quantity/ha	kcal/ha
Input		
Labor		
Hand	246 hr	—
Mechanical[b]	144 hr	—
Machinery		
Hand	82 kg	1,167,088
Mechanical[b]	139 kg	1,771,754
Gasoline		
Hand	1685 ℓ	17,033,665
Mechanical[b]	1887 ℓ	19,075,683
Diesel	563 ℓ	6,426,082
Electricity	20kwh	57,260
Nitrogen[c]	82 kg	1,205,400
Phosphorus[d]	114 kg	627,000
Potassium[d]	114 kg	231,408
Lime[d]	682 kg	1,437,656
Insecticide	47 kg	2,889,090
Fungicide	49 kg	1,360,730
Herbicide	6 kg	599,460
Transportation		
Hand	3438 kg	883,566
Mechanical[b]	3683 kg	946,531
Building(450 sq ft)[e]		14,250
Total		
Hand		**33,932,655**
Mechanical		**36,642,304**
Output		
Yield		
Hand	70,961 kg	39,738,160
Mechanical	70,961 kg	39,738,160
kcal output/kcal input		
Hand		1.171
Mechanical		1.084
Protein yield	142 kg	

[a] There are no differences in energy input between apples grown for fresh or processing. Apples are transported 17.6 km to packing or processing plant.
[b] Mechanical harvester is a shake and catch type and travels at the same speed regardless of yield. It has a 5-year life.
[c] Ammonium nitrate applied annually at an increasing rate until age 10.
[d] Lime (calcium), phosphorus, and potassium applied every fifth year. Amount shown is average annual application.
[e] Service building only with 20-year life and used for 60 ha.

From Funt, R. C., Ph.D. thesis, Pennsylvania State University, University Park, 1974.

Table 7
ENERGY INPUT PER HECTARE FOR HIGH DENSITY (1512 TREES/HA) (TRELLIS) SUPPORTED, HAND OR MECHANICALLY HARVESTED, IRRIGATED (TRICKLE SYSTEM) APPLES FOR THE FIRST YEAR (ESTABLISHMENT), EASTERN REGION U.S.[a]

Item	Quantity/ha	kcal/ha
Input		
Labor		
Hand	206 hr	—
Mechanical[b]	—	—
Machine		
Hand[c]	72 kg	510,469
Mechanical[c]	—	—
Gasoline		
Hand	345 ℓ	3,487,605
Mechanical	—	
Diesel	95 ℓ	1,084,330
Electricity	95 kwh	271,985
Trellis[d]		
Posts	62 kg	127,782
Wire	252 kg	4,299,372
Irrigation (trickle)[e]		
PVC	10 kg	83
Polyethylene	26 kg	179
Nitrogen[f]	27 kg	396,900
Phosphorus[g]	114 kg	627,000
Potassium[g]	114 kg	231,408
Lime[g]	682 kg	1,437,656
Insecticide	4 kg	176,000
Fungicide	27 kg	594,000
Herbicide	2 kg	114,000
Transportation		
Hand	1832 kg	470,824
Mechanical[b]	—	
Building[h]		14,250
Total		
Hand		**13,843,843**
Mechanical		—
Output		
No yield in establishment year		

[a] There are no differences in energy inputs between apples produced for fresh or processing.
[b] Mechanical harvesters are not used in nonbearing years.
[c] Hand harvest has 12.8 tons per 17.8 ha while mechanical harvest has 13.1 tons.
[d] Trellis consists of 4'' × 8' treated end posts, 2½'' × 8' treated row post and four strands of #9 galvanized wire.
[e] Trickle system utilizes 4'' PVC main line and ½'' lateral polyethylene tubing. Pump uses an electric motor.
[f] Nitrogen applied annually at an increasing rate until 5 years of age.
[g] Lime (calcium), phosphorus, and potassium applied every 5 years and is shown as an annual amount.
[h] Service building only with 20-year life and used for 60 ha.

From Funt, R. C., Ph.D. thesis, Pennsylvania State University, University Park, 1974.

Table 8
ENERGY INPUT PER HECTARE FOR HIGH DENSITY (1512 TREES/HA) (TRELLIS) SUPPORTED, HAND OR MECHANICALLY HARVESTED, IRRIGATED (TRICKLE SYSTEM) APPLES FOR THE TENTH YEAR AFTER ESTABLISHMENT, EASTERN REGION U.S.[a]

Item	Quantity/ha	kcal/ha
	Input	
Labor		
Hand	362 hr	—
Mechanical[b]	158 hr	—
Machine		
Hand[c]	72 kg	1,138,285
Mechanical	118 kg	1,208,736
Gasoline		
Hand	1781 ℓ	18,004,129
Mechanical	1821 ℓ	18,408,489
Diesel	127 ℓ	1,381,094
Electricity	150 kwh	429,450
Trellis[d]		
Posts	62 kg	127,782
Wire	252 kg	4,299,372
Irrigation (trickle)[e]		
PVC	10 kg	83
Polyethylene	26 kg	179
Nitrogen[f]	137 kg	2,013,900
Phosphorus	114 kg	627,000
Potassium	114 kg	231,408
Lime[g]	682 kg	1,437,656
Insecticide	47 kg	2,889,090
Fungicide	49 kg	1,360,730
Herbicide	6 kg	599,460
Transportation		
Hand	3479 kg	894,103
Mechanical	3565 kg	916,205
Building[h]		14,250
Total		
Hand		**35,447,971**
Mechanical		**35,944,844**
	Output	
Yield: Hand	59,124 kg	33,109,745
Mechanical	59,124 kg	33,109,745
kcal output/kcal input: Hand		0.934
Mechanical		0.921
Protein yield	118 kg	

[a] There are no differences in energy inputs between apples produced for fresh or processing. Apples are transported 17.6 km to packing or processing plant.

[b] Mechanical harvester is an over-row continuous moving machine and has a 5-year life.

[c] Hand harvest has 12.8 tons per 17.8 ha while mechanical harvest has 13.1 tons.

[d] Trellis consists of 4'' × 8' treated end posts, 2½'' × 8' treated row post and four strands of #9 galvanized wire.

[e] Trickle system utilizes 4'' PVC main line and ½'' lateral polyethylene tubing. Pump uses an electric motor.

[f] Nitrogen applied annually at an increasing rate until 5 years of age.

[g] Lime (calcium), phosphorus, and potassium applied every 5 years and is shown as an annual amount.

[h] Service building only with 20-year life and used for 60 ha.

Table 9
ENERGY INPUT PER HECTARE FOR HIGH DENSITY (1512 TREES/HA) (TRELLIS) SUPPORTED, HAND OR MECHANICALLY HARVESTED, IRRIGATED (TRICKLE SYSTEM) APPLES FOR THE TWENTIETH YEAR AFTER ESTABLISHMENT, EASTERN REGION U.S.[a]

Item	Quantity/ha	kcal/ha
	Input	
Labor		
Hand	446 hr	—
Mechanical[b]	158 hr	—
Machine		
Hand[c]	72 kg	1,345,079
Mechanical	118 kg	1,415,530
Gasoline		
Hand	2252 ℓ	22,765,468
Mechanical	2292 ℓ	23,169,828
Diesel	127 ℓ	1,449,578
Electricity	150 kwh	429,450
Trellis[d]		
Posts	62 kg	127,782
Wire	252 kg	4,299,372
Irrigation (trickle)[e]		
PVC	10 kg	83
Polyethylene	26 kg	179
Nitrogen[f]	137 kg	2,013,900
Phosphorus	114 kg	627,000
Potassium	114 kg	231,408
Lime[g]	682 kg	1,437,656
Insecticide	47 kg	2,889,090
Fungicide	49 kg	1,360,730
Herbicide	6 kg	599,460
Transportation		
Hand	3,950kg	1,015,150
Mechanical	4,036 kg	1,037,252
Building[h]		14,250
Total		
Hand		**40,605,635**
Mechanical		**41,099,548**
	Output	
Yield		
Hand	79,748 kg	44,658,880
Mechanical	79,784kg	44,658,880
kcal output/kcal input		
Hand		1.100
Mechanical		1.087
Protein yield	159 kg	

[a] There are no differences in energy inputs between apples produced for fresh or processing. Apples are transported 17.6 km to packing or processing plant.

[b] Mechanical harvester is an over-row continuous moving machine and has a 5-year life.

From Funt, R. C., Ph.D. thesis, Pennsylvania State University, University Park, 1974.

REFERENCES

1. A Guide to Energy Saving — For the Orchard Grower, U.S. Department of Agriculture, Federal Energy Administration, 1977.
2. **Burrows, F.,** Apple Exports and the Middle East, *Am. Fruit Grow.,* 97(11), 4, 1977.
3. **Cadwallader, C. L.,** Apple Cold Storage Report, Pennsylvania Crop Reporting Service, Nov. 1977 Report, 1977.
4. **Funt, R. C.,** An Economic Analysis of Several Apple Orchard Systems in Pennsylvania, Ph.D. thesis, Pennsylvania State University, University Park, 1974.
5. Pennsylvania Orchard and Vineyard Survey, Pennsylvania Department of Agriculture Crop Reporting Service, 1972.
6. **Tukey, L. D.,** Pa. State Hortic. Rev., 26(4), 4, 1977.

ENERGY INPUTS IN APRICOT PRODUCTION

James A. Beutel and Kiyoto Uriu

Apricot production in the U.S. in 1977 was estimated to be about 134,000 t, with a value of $32,000,000.[1] Production, however, was concentrated nearly entirely in California, which produced 130,000 t — 97.0% of U.S. production. Utah produced 1.2% of the national production and Washington, 1.8%. The acreage in California was about 12,000 ha, with 6% of it nonbearing.[2]

Commercial production of apricots is restricted to areas of little to no summer rainfall. Regions of heavy summer rainfall are not suitable, because in those areas the fungus disease brown rot can cause very severe loss of flowers and fruit.

All apricots produced commercially are irrigated. Summer irrigations are essential for annual production of large commercial crops. Trees may survive dry summers, but the number of flower buds for the following year's crop is greatly reduced under dry farming conditions.

In California, about 6% of the production is normally consumed in the fresh state (half consumed within California and half shipped out of state), 7% frozen, 28% dried, and 59% canned.[1] Two main varieties of apricot dominate the industry in California.[2] The Tilton variety (46% of the state acreage) is the main processing variety, while the Royal, also known as Blenheim (43% of the state acreage), is consumed fresh, used for drying, and processed into nectar (juice). New and miscellaneous varieties (11% of the state acreage) such as Modesto, Patterson, and Tracy are used for freezing, canning, and canning and drying, respectively.

In California apricots are grown on deep alluvial soils.

The Tilton variety is grown almost entirely in the hot interior valleys, while the Royal is grown in cooler areas such as the coastal valleys and the areas of the interior valley that receive marine influence from the San Francisco Bay. Due to their geographical locations, the Tiltons receive irrigation water from canals originating from mountain reservoirs, while the Royals are mostly irrigated with ground water lifted by pumps.

The labor input in raising apricots for either processing or the fresh market ranges from 411 to 465 hr/ha (Table 1). Energy inputs were 6.2 million kcal/ha for Tiltons and 11.9 million kcal for Royals (Table 2 and 3). Tiltons yield about 22,000 kg/ha, whereas Royals were half or 11,000 kg/ha.

Table 1
HOURS OF LABOR REQUIRED PER HECTARE PER YEAR FOR APRICOT PRODUCTION UNDER IRRIGATION IN CALIFORNIA

Operation		Tilton variety for processing		Royal variety for fresh consumption
Dormant pruning		106.2		123.5
Remove large limbs from orchard		4.9		4.9
Rake and shred prunings in orchard		4.9		4.9
Spray (three times)		9.9		9.9
Cultivation		12.4[a]		9.9[b]
Furrowing		4.9[c]		4.9[d]
Irrigation		12.5[c]		20.0[d]
Fertilize		2.5		2.5
Fruit thinning		38.0		75.0
Harvest[e]	22t/ha	200.0	11t/ha	190.0
Haul fruit containers in and out of orchard		14.8		19.8
Total hours labor per hectare		411.0		465.3

[a] Ten times.
[b] Eight times.
[c] Five times.
[d] Four times.
[e] Tiltons — two times, Royals — three times.

Table 2
ENERGY INPUTS FOR TILTON APRICOT (PROCESSING) PRODUCTION UNDER IRRIGATION IN CALIFORNIA

Item	Quantity/ha	kcal/ha
	Input	
Labor	411 hr	—
Machinery	46 kg	828,000
Diesel fuel	151 ℓ	1,723,510
Nitrogen	160 kg	2,288,000
Zinc sulfate	30 kg	45,000
Irrigation (canals)	120 cm	50,000
Fungicides		
Benlate	2 kg	129,820
Captan	4 kg	281,160
Insecticides		
Parathion	2 kg	133,820
Neutral copper	15 kg	22,500
Supreme oil	26 ℓ	296,764
Diazinon	2 kg	173,820
Transportation	921 kg	253,275
Total		**6,225,669**
	Output	
Apricot yield	22,000 kg	10,515,418
Protein yield	946 kg	
kcal output/kcal input ratio		1.69

Table 3
ENERGY INPUTS FOR ROYAL APRICOT FRESH SHIPPING PRODUCTION UNDER IRRIGATION IN CALIFORNIA

Item	Quantity/ha	kcal/ha
Input		
Labor	465 hr	—
Machinery	46 kg	828,000
Diesel fuel	151 ℓ	1,723,510
Nitrogen	80 kg	1,144,000
Irrigation (30 m lift)	90 cm	1,020,492
Fungicides		
Benlate	2 kg	129,820
Captan	4 kg	281,160
Insecticides		
Parathion	2 kg	133,820
Neutral copper	15 kg	22,500
Supreme oil	26 ℓ	296,764
Diazinon	2 kg	173,820
Transportation	811 kg	223,025
Total		**5,976,911**
Output		
Apricot yield	11,000 kg	5,257,709
Protein yield	473 kg	
kcal output/kcal input ratio		0.88

Table 4
EQUIPMENT DESCRIPTION FOR A 30 HA FARM

Item	Approximate cost	Years life
60 hp diesel wheel tractor	$14,000	12
30 hp gasoline wheel tractor (used)	4,500	6
12′ disc	2,400	10
20′ springtooth	2,600	10
18′ furrower	800	10
7½′ shredder	1,450	10
18′ brush rake	750	10
Sprayer-air blast	14,000	15
PTO forklift	1,200	10
Fertilizer spreader	1,500	10
Pickup truck	6,000	10
2-ton flat-bed truck	8,000	8

REFERENCES

1. Crop Reporting Board, Noncitrus fruits and nuts, 1977 annual summary, Fr Nt 1-3, Economics, Statistics and Cooperatives Service, U.S. Department of Agriculture, 1978.
2. 1976 California fruit and nut acreage, California Crop and Livestock Reporting Service, 1977.
3. **Coates, W. W. and Benson, L. C.,** Walnuts and Apricots — Production Costs - 1978, Mimeo, University of California Cooperative Extension, San Benito County, Jan., 1978.
4. **Rough, D., Ross, N., and Yeary, E.,** Sample Costs to Produce Apricots in the Stanislaus and San Joaquin Counties — 1976, Mimeo, University of California Agricultural Extension, Mar., 1976.
5. **DeTar, J. E. and Reed, A. D.,** Sample Cost of Apricot Production, University of California Agricultural Extension, Mimeo, Solano County, Jan., 1963.

ENERGY INPUTS IN CHERRY PRODUCTION

E. L. Proebsting

Commercial cherry production in the U.S. comes from two species of *Prunus,* the red tart cherries (*Prunus cerasus*) in the Great Lakes region, and the sweet cherries (*Prunus avium*), mostly grown in the Pacific Coast states.

Production of red tart cherries averages 100,000 metric tons annually.[1] These are nearly all processed, about ⅔ frozen and ⅓ canned.

Sweet cherry production averages 140,909 metric tons, of which about half is shipped for fresh consumption. Most of the remainder, 50,000 metric tons, is brined for manufacture of maraschino cherries. About 10,000 metric tons are canned and 8,182 metric tons go into miscellaneous cherry products.

Nationwide, the red tart cherry crop is valued at about $41,000,000 annually. The sweet cherry crop averages $65,000,000 annually.

Red tart cherries are perhaps grown under the least variable conditions of any of our major fruit crops. They are nearly all of one cultivar, Montmorency; they are nearly all grown in one reasonably homogeneous region; and they are nearly all grown for processing. Their culture is highly mechanized and is quite standard throughout the producing region.

Sweet cherry culture is more heterogeneous. The cherries are mostly grown in the Pacific Coast states where climatic conditions vary widely. Precipitation and temperature are much different among sweet cherry districts. The cherries are utilized in several forms whose maturity requirements differ. Pest control problems differ with region.

Climate influences energy use patterns directly where frost-control systems are used. Where rain frequently cracks fruit prior to harvest, the fruit probably will be harvested immature for brining instead of for fresh consumption. The form of utilization presently determines whether the fruit is harvested mechanically or by hand. Processed fruit is mostly harvested mechanically.

Deciduous fruit operations are similar, regardless of crop. Orchards are commonly diversified, with equipment shared among several crops. This analysis assumes a 40-acre cherry orchard. The basic figures were compiled from economic analyses of sweet cherry production costs in the Yakima Valley of Washington and for tart cherry in western Michigan.[2,3] The basic growing inputs, exclusive of harvest, irrigation, and frost protection, for the two crops are similar (Tables 2 and 3). Washington reports the use of more labor, most of it going into pruning and brush removal. The complement of machinery for the two areas is similar enough that they were treated identically. Michigan cherry culture requires greater use of pesticides. This is reflected in the gasoline input as well as in the chemicals.

Hand harvest is a massive consumer of labor but there is a considerable input of energy as well (Table 4). Fuel is used to bring fruit in from the orchard, and there is an energy component in the picking equipment. Mechanical harvest minimizes the labor but greatly increases fuel inputs.

Irrigation, as computed here, consumes more energy than growing and harvesting the crop (Table 5). If irrigation is to be used for frost protection, an additional 50% can be added to the energy input from this source.

Frost protection is a highly variable input. Many cherry orchards do not practice frost protection. Those that do may protect only part of the orchard, and in many years the protection systems do not have to be activated. When they are activated, the old systems utilizing oil heat are prodigious consumers of energy (Table 6). Trends

Table 1
EQUIPMENT USED IN CHERRY CULTURE

Implement[a]	Gross weight (kg)	kg/acre/yr
Tractors[3]	5,864	14.66
Disc	364	0.91
Mower	682	1.71
Sprayer PTO	864	2.16
Herbicide sprayer	136	0.34
Forklift 3-point	114	0.29
Pruning aid	818	2.05
Trailer	227	0.57
Pickup ½-ton	2,045	5.11
Truck 1-½-ton	3,182	7.96
Brush rake	136	0.34
Chain saw	5	0.01
Planting auger	114	0.29
Spreader	114	0.29
Total	**14,665**	**36.69**

Note: 36.69 kg/acre/yr × 2.47 acre/ha × 18,000 kcal/kg = 1,630,800 kcal/ha.

[a] Equipment used on 40 acres for 10 years.

Table 2
AN EXAMPLE OF ANNUAL ENERGY INPUTS PER HECTARE FOR SWEET CHERRY CULTURE IN WASHINGTON STATE WITHOUT HARVEST, IRRIGATION, OR FROST PROTECTION

Item	Quantity/ha	kcal/ha
	Input	
Labor	155 hr	—
Machinery	90.6 kg	1,630,800[a]
Gasoline	90 ℓ	909,810
Nitrogen	168 kg	2,335,200
Insecticides	14 kg	1,120,000
Herbicides	1.3 kg	130,000
Transportation	158.1 kg	40,632
Total		**6,166,442**
	Output	
Sweet cherry yield	13,452 kg	9,416,400
Protein yield	175 kg	
kcal output/kcal input		1.527
kcal output/labor hr		60,751

[a] See *Note* Table 1.

Table 3
AN EXAMPLE OF ANNUAL ENERGY INPUTS PER HECTARE FOR RED TART CHERRY CULTURE IN MICHIGAN WITHOUT HARVEST, IRRIGATION, OR FROST PROTECTION

Item	Quantity/ha	kcal/ha
Input		
Labor	150 hr	—
Machinery	90.6 kg	1,630,800
Gasoline	116 ℓ	1,173,000
Nitrogen	168 kg	2,335,000
Potassium	112 kg	190,000
Insecticides	25 kg	1,516,000
Herbicides	0.85 kg	85,000
Fungicides	20 kg	833,000
Transportation	177.6 kg	45,644
Total		**7,808,444**
Output		
Tart cherry yield	6,726 kg	3,901,080
Protein yield	897 kg	
kcal output/kcal input		0.500
kcal output/labor hour		97,527

Table 4
EXAMPLE OF ANNUAL ENERGY INPUTS PER HECTARE FOR HARVESTING CHERRIES BY HAND AND MECHANICALLY

Input	Quantity/ha	kcal/ha
Hand Harvest		
Labor	650 hr	—
Equipment	17.54 kg	
Buckets 2500 × 2-½		
Ladders 280 × 30	23.58 kg	
Trailer 2 × 500	2.81 kg	
Equipment total	43.93 kg	790,740
Gasoline	100 ℓ	1,011,000
Total		**1,801,740**
Machine Harvest		
Labor	50 hr	—
Equipment	—	
Harvester	28.07 kg	
Cooling pads and tanks	14.04 kg	
Equipment total	42.11 kg	757,980
Gasoline (for swamping, shaker, truck, and pumps)	400 ℓ	4,044,000
Total		**4,801,980**

Table 5
EXAMPLES OF ANNUAL ENERGY PER HECTARE FOR IRRIGATING CHERRIES

Input	Kcal/ha
Labor	15 hr/ha
Hand-moved sprinkler (100 ha-cm applied, 50 m lift)	17,603,000
Solid set sprinkler (100 ha-cm applied, 50 m lift)	17,760,000
For frost protection, an additional 50 cm irrigation	8,161,000

Table 6
EXAMPLES OF ANNUAL ENERGY INPUTS PER HECTARE FOR FROST PROTECTION IN CHERRIES

Input	Quantity/ha	kcal/ha
Labor	20 hr	—
Equipment		
Heaters 99/ha × 13.64	3.37 kg	
Oil tank-trailer	8.42 kg	
Oil tank	14.03 kg	
Total equipment	25.82 kg	464,760
Fuel		
26 hr 280.84 ℓ/ha/hr 7350 ℓ		84,244,000
Total		**84,708,760**

toward use of water (Table 5) or other methods are resulting in substantial savings of energy.

Minimum technology cherry culture is most likely to be in units as small as one tree in a farm garden. Energy inputs other than labor will be mostly in pest control, which will be determined by the pests of the area and the willingness of the consumer to accept infested fruit.

REFERENCES

1. **Powell, J. V. and Huang, B. W.,** The Fruit Situation, Economic Research Service, U.S. Department of Agriculture, TFS-205, 1977.
2. **Doran, S. M. and Anderson, F. A.,** Estimated Sweet Cherry Production Costs in the Yakima Valley, Washington State University, E. M. 3941, 1975.
3. **Kelsey, M., Ricks, D., and Harsh, S.,** Economics of Tart Cherry Production in Western Michigan, Michigan State University Agric. Economics Rep. No. 188, 1971.

ENERGY INPUTS IN PEACH PRODUCTION

Norman F. Childers and Robert T. Kelly

Peaches are part of a variable specific complex. This variability is reflected in many commercially important characters: fruit conformation; fruit tissue color, texture, and adherence to pit; skin characteristics; soluble solids content; seasonal timing of bloom and ripening; and overall tolerance of environmental extremes, especially low temperature. Each cultivar exhibits different combinations of these and other characters.

Yet, despite this inherent variability, most peach cultivars share general traits, which account for the significant energy expenditures in their culture:

- Good vegetative growth is necessary for commercial fruiting.
- All parts of the plants are susceptible to many pests and pathogens.
- The blooming period occurs relatively early each spring.

Therefore, economically viable peach culture involves judicious fertilizing, pruning, thinning, herbicide application and/or cultivation, irrigation, disease/pest control measures, and, in some areas, frost injury protection during bloom period.

In the U.S., the peach is an important fruit crop, ranking behind only citrus, grapes, and apples in tonnage production. Worldwide, the U.S. and Italy each produce more than 1 million metric tons of peaches annually; these two countries constitute roughly 50% of the entire world production.

It is emphasized that Tables 1 and 2 are only approximate guidelines, for there is

Table 1
WEST COAST, U.S., PEACH CULTURE

Item	Quantity/ha	kcal/ha	Ref.
	Input		
Labor	277 hr	Variable	
Machinery/tools	3200 kg	1,100,000	
Gasoline	16.7	168,820	3
Diesel	28 ℓ	319,592	3
Electricity	695 kWh	1,989,785	3
Nitrogen	153 kg	2,142,000	1
Phosphorus	370 kg	1,110,000	1
Potassium	—	—	1
Lime	—	—	1
Irrigation	61 cm	7,560,000[a]	
Pesticides (assumed fungicide k and insecticide use equal)	124 kg	9,412,840	2
Herbicides	2 kg	199,820	2
Transportation	3,236 kg	831,652	
Total		**24,834,509**	
	Output		
Peach yield[b]	29,500 kg	11,121,000[c]	
Protein yield	177 kg		
kcal output/kcal input		0.45	
kcal output/labor hour		40,469	

[a] Assumes solid-set sprinkler system.
[b] Assumes 13 ton/acre average yield.
[c] Assumes 380 kcal/kg peach flesh.

Table 2
EASTERN U.S. PEACH CULTURE

Item	Quantity/ha	kcal/ha	Ref.
	Input		
Labor	277 hr	Variable	
Machinery/tools	3200 kg	1,100,000	
Gasoline	16.7 ℓ	T168,820	3
Diesel	28 ℓ	319,592	3
Electricity	695 kWh	1,989,785	3
Nitrogen	55 kg	770,000	4
Phosphorus	55 kg	165,000	4
Potassium	55 kg	88,000	4
Lime	1515 kg	477,907	4
Irrigation	10 cm	4,350,000[a]	
Pesticides	164 kg	12,449,240	4
Herbicides	4 kg	251,080	4
Transportation	3,236 kg	831,652	
Total		**22,961,076**	
	Output		
Yield[b]	22,700 kg	8,626,000[c]	
Protein yield	136 kg		
kcal output/kcal input		0.38	
kcal output/labor hour		32,249	

[a] Assumes solid-set sprinkler system.
[b] Assumes 10 ton/acre average yield.
[c] Assumes 380 kcal/kg peach flesh.

no standard peach cultural situation. Tree longevity itself can vary within the extremes of 6 to 20 years. Nevertheless, it is hoped that the figures are representative.

REFERENCES

1. **Childers, N. F., Ed.,** *The Peach,* Horticultural Publications, New Brunswick, N. J., 1975.
2. Spray Guide for Tree Fruits in Eastern Washington, Extension Bulletin 419, Cooperative Extension Service, Washington State University, Pullman, 1973.
3. A Guide to Energy Savings for the Orchard Grower, U.S. Department of Agriculture, Washington, D.C., 1977.
4. Christ, E. G., Ed., Commercial Tree Fruit Production Recommendations, New Jersey Extension Service Bulletin 407-D, Rutgers University, New Brunswick, 1977.

ENERGY USE IN PEAR PRODUCTION

R. L. Stebbins

Adout 80% of the U.S. pear crop is produced in the three Pacific Coast states. In 1975, these states produced about 720,000 tons of pears. Varieties mainly of high quality and European origin are grown. The pear areas are characterized by dry summers with abundant sunshine. Pears cannot be grown commercially in most other states because winters are too cold or weather in summer is too conducive to the deadly fire blight disease.

The energy cost of producing pears varies greatly, both between districts and between orchards within districts. By far the biggest variable is the need for frost protection in spring. This varies from zero in a few locations to huge amounts of energy in others. It also varies tremendously from year to year.

Pear production is and will remain highly labor intensive. Mechanical harvesting has not proven to be economically feasible. Each tree must be pruned each year by hand.

Finally, it should be mentioned that pears are sold because they are enjoyable to eat and not because they have calories or protein. Additionally, the health value of fresh fruit, including pears, is well recognized.

Table 1
AN EXAMPLE OF TYPICAL ENERGY INPUTS PER HECTARE PER YEAR FOR A MATURE PEAR ORCHARD IN THE 1970s, WITH HEATING FOR FROST PROTECTION

Item	Quantity/ha	kcal/ha	Ref.
	Input		
Labor	210 hr	—	2
Machinery	42 kg[a]	1,084,554	
Gasoline (except harvest)	350 ℓ	2,818,616	2
Orchard heating oil	2,525 ℓ	23,530,475	1
Electricity	10 kWh[b]	26,060	
Nitrogen	224 kg	3,439,744	2
Lime	900 kg	1,710,500	
Irrigation	61 cm	11,989,578	2
Insecticides[c]	26 cm	1,939,387	2
Transportation [machinery (kg) + fuel (kg)]	2,829.5 kg	727,182 (257 kcal/kg transported)	
Transportation by truck	749ℓ	6,031,838	2
Move-a-load bins	81 ℓ	652,308	2
Buildings, labor, housing	0.57 m[2d]	79,230	3
Service buildings	1.45 m[2d]	55,100	3
Total/ha/year		**54,084,072**	
	Output		
Pear yield/ha[f]	44,833 kg/g	27,348,029	
Protein yield	314 kg		
kcal output/kcal input		0.51	
kcal output/labor hours		130,229	

Note: Based on data from Oregon, California, and Washington.

[a] Estimates of equipment weights listed in source bl; see Table 2. Some of these are based on manufacturer's brochures which give weights for typical equipment.

[b] Author's best estimate.

[c] Used source b and a price of $3 per pound for insecticide.

[d] Labor housing calculated at 100 ft² per worker, 20 workers, 10-year depreciation, 80 acres. Service buildings calculated at 5,000 ft² for 80 acres, 20-year depreciation.

[f] Pear yield is assumed to be 44.8 metric tons/ha (20 tons/acre) which is a typical yield for pears in the most productive region in the U.S.

Table 2
EQUIPMENT FOR A 32-ha PEAR ORCHARD, SONOMA COUNTY, CALIFORNIA AND ESTIMATES IN WEIGHTS PER KILOGRAM

Item	Weight estimate	Life (years)	Weight/ year	Weight/ year/ha
Tractor, wheel D 45 hp	2000	20	100	3.125
Tractor, crawler 40 hp (used)	4000	10	400	12.500
Disc	300	12	25	0.781
Rototiller, 5 ft	272	10	27	0.844
Pruning tower	1000	10	100	3.125
Sprayer, 400 gal	2000	10	200	6.250
Sprayer, weeds, Pto	82	10	8	0.250
Forklift attachment	80	12	7	0.219
Truck (used)	2500	7	357	11.156
Pickup truck	800	10	80	2.500
Ladders (20)	90	10	9	0.281
Picking bags (30)	30	5	6	0.188
Buckrake (front end)	100	10	10	0.313
Total/ha/year				**41.532**

Table 3
AN EXAMPLE OF TYPICAL ENERGY INPUTS PER HECTARE PER YEAR FOR A MATURE PEAR ORCHARD IN THE 1970s, WITHOUT HEATING FOR FROST PROTECTION

Input		
Total kcal/ha from Table 1		54,084,072
Less kcal/ha for heating		23,530,475
Total/ha/year		**30,553,597**
Output		
Pear yield/ha	44,833 kg	27,348,029
kcal output/kcal input		0.90
kcal output/labor hour (from Table 1)		130,229

REFERENCES

1. **Rackham, R. L.,**Pears Sample Production Costs Rogue River Valley 1974, Mimeo, Oregon State University Agricultural Extension Service, Jackson County, 1974.
2. **Parsons, P. H.,** Sonoma County Sample Costs to Produce Pears, 1975 Fresh Shipping and Processing, Mimeo, University of California, Davis and John J. Smith Farm Advisors, Davis, 1975.
3. **Hannon, B., Stein, R., Segal, B., and Serber, D.,** Energy Use in Building Construction, C.A.C. Document 228, Center for Advanced Computation, University of Illinois, Urbana, 1977.
4. **Krummel, J. and Chick, S.,** Food Energy and Protein Content of Various Foods, Cornell University, Ithaca, N.Y.

ENERGY USE FOR THE PRODUCTION OF PLUMS AND PRUNES

David W. Ramming*

The two principal types of plums grown commercially are the European plum, *Prunus domestica* L., and the Japanese plum, *P. salicina* Lindl. The Japanese plums originated in China, but after being brought to the U.S., they were hybridized by Luther Burbank with *P. cerasifera* Ehrh., *P. simonii* Carr., *P. americana* Marsh, and other species. Most of the commercial cultivars of Japanese plums are a mixture of these species. The term "plum" usually refers to the Japanese and a few European plum cultivars that are used for fresh consumption and processing. Prunes are European-type plums and are used primarily for drying, although some are sold fresh or processed. They can be dried without fermentation because they have a high enough sugar content. More energy is required to produce dried prunes and processed plums than to produce the same plums or prunes for fresh consumption.

The U.S. produced about 517,200 metric tons (fresh weight)[1] of plums and prunes in 1975. The crop was valued at $89 million. The major part of the plum crop is produced in California, Oregon, Washington, Michigan, and Idaho. In 1975, California produced over 70% of the total crop, 340,200 tons of prunes and 114,300 tons of plums, valued at $79 million. Oregon, Washington, Michigan, and Idaho produced 24,900 tons, 18,300 tons, 16,300 tons, and 3,200 tons, respectively, of which the majority were prunes. California and Oregon are the only states reporting prunes sold as dried fruit; prunes grown in Washington, Michigan, and Idaho are utilized only fresh and processed.

Environmental conditions and cultural practices that prevail during production of plums and prunes affect the amount of energy used and the total amount of final product produced. Certain prune-producing areas, such as Michigan and Oregon, usually receive enough rainfall annually for crop production, so in these areas irrigation generally is not needed. The relatively high rainfall, however, increases disease problems and thus the need for additional pesticide sprays. The energy needed under these conditions (Tables 1 and 2) is compared with the energy needed in California (Tables 3 and 4) where irrigation is required for crop production but fewer sprays are required for disease control.

The occurrence of spring frosts at certain sites where plums and prunes are grown necessitates the use of frost protection, which leads to increased use of energy. Without this expenditure of energy, the crop might be lost; even with it, the crop may be reduced if the frost is severe enough. Japanese plums bloom earlier than European plums[2] and thus are more susceptible to frost injury. More energy, therefore, is needed to protect Japanese plums than to protect European plums grown at the same site where frost hazards occur.

The increasing mechanization of pruning, thinning, and harvesting will reduce the amount of labor needed and increase the input of energy, but will not necessarily affect the size of the crop. At present, most plums for fresh consumption are harvested by hand, whereas prunes for drying and processing are harvested mechanically. Individual farmers also have their own practices that affect energy consumption. For example, either clean cultivation, with tillage equipment used to keep the orchard free of weeds or strip-non-tillage, with herbicides used in the row and mowing used in the row cen-

* Research Horticulturist, Horticultural Crops Production, Federal Research, Science and Education Administration, U.S. Department of Agriculture

ters, may be practiced. The energy consumpion of these practices is compared for prune production in California (Tables 3 and 4).

The time of maturity, as well as the higher requirements for nitrogen fertilizer and the larger crop production of later maturing cultivars, affects the amount of energy expended to produce a crop. Energy use in production of an early-maturing plum and a midseason plum is shown in Tables 5 and 6, respectively.

Finally, the establishment of an orchard requires considerable energy before any production is obtained. Usually plum and prune orchards start to produce a crop the third growing season after planting and reach full production about the fifth growing season. The energy input for establishing a plum orchard in California is shown in Table 7.

Table 1
AN EXAMPLE OF ENERGY INPUTS PER HECTARE FOR PRODUCTION OF PRUNES IN MICHIGAN (WITH NO IRRIGATION OR CULTIVATION)

Item	Quantity/ha	kcal/ha[a]	Ref.
	Input		
Labor	51 hr	—	3
Machinery	47.2 kg[b]	630,612	
Gasoline	350.8 ℓ[c]	3,546,231	4
Diesel	369.5 ℓ[c]	4,217,473	4
Nitrogen	111.0 kg	1,631,700	3
Herbicide	3.4 kg	260,168	3
Insecticide	60.8 kg	5,057,779	3
Fungicide	49.0 kg	1,254,075	3
Transportation	815.5 kg[d]	209,584	
Orchard establishment	—	1,776,061[e]	
Total		**18,583,683**	
	Output		
Yield of prunes	6277 kg	4,425,990	5
kcal output/input		0.238	
Protein yield	47 kg		6
Carbohydrate yield	1162 kg		6

[a] Data on kcal input per unit are from standard conversions used in this volume.
[b] Assumes that 19.1 metric tons of machinery (tractors, trucks, harvester, etc.) were used for production, that this machinery serves an average of 40 hectares, and that it is depreciated over 10 years.
[c] Gasoline and diesel used for operation of tractors, trucks, and harvester.
[d] Total weight of machinery, fuel, fertilizer, and pesticides transported to crop a hectare.
[e] Assumes that three years are required to establish an orchard and the life of the orchard is 15 years.

Table 2
AN EXAMPLE OF ENERGY INPUTS PER HECTARE FOR PRODUCTION OF PRUNES IN THE WILLAMETTE VALLEY, OREGON (WITH NO IRRIGATION, BUT WITH CULTIVATION)

Item	Quantity/ha	kcal/ha[a]	Ref.
	Input		
Labor	58.7 hr	—	7
Machinery	45.5 kg[b]	769,640	
Gasoline	425.6 ℓ[c]	4,302,390	4
Diesel	313.4 ℓ[c]	3,577,147	4
Nitrogen	145.7 kg	2,141,790	8
Potash (K_2O)	363.2 kg	581,120	8
Insecticide	74.0 kg	6,372,158	9
Fungicide	26.7 kg	741,459	9
Herbicide	1.0 kg[d]	109,520	
Transportation	1,211 kg[e]	311,227	
Orchard establishment	—	1,776,061[f]	
Total		**20,682,512**	
	Output		
Yield of prunes	14,493 kg	10,217,250	7
kcal output/kcal input		0.494	
Protein yield	109 kg		6
Carbohydrate yield	2,684 kg		6

[a] Data on kcal input per unit are from standard conversions used in CRC handbook.
[b] Assumes that 18.4 metric tons of machinery (tractors, trucks, harvester, etc.) were used for production, that this machinery serves an average of 40 hectares, and that it is depreciated over 10 years.
[c] Gasoline and diesel used for operation of tractors, trucks, and harvester.
[d] Determined from average and recommended application rates.
[e] Total weight of machinery, fuel, fertilizer, and pesticides transported to crop a hectare.
[f] Assumes that three years are required to establish an orchard and the life of the orchard is 15 years.

Table 3
AN EXAMPLE OF ENERGY INPUTS PER HECTARE FOR PRODUCTION OF PRUNES IN CALIFORNIA (WITH IRRIGATION AND CULTIVATION)

Item	Quantity/ha	kcal/ha[a]	Ref.
	Input		
Labor	93.5 hr	—	10
Machinery	46.1 kg[b]	788,909	
Gasoline	488.7 ℓ[c]	4,940,268	4
Diesel	532.0 ℓ[c]	6,072,248	4
Nitrogen	112.1 kg	1,647,870	10
Herbicide	1.0 kg[d]	109,520	
Insecticide	58.3 kg[d]	4,952,373	
Irrigation	493 cm	608,127	10
Transportation	989 kg[e]	254,173	
Orchard establishment	—	1,852,128[f]	
Total		**21,225,616**	
	Output		
Yield of prunes	22,417 kg	15,803,985	10
kcal output/input		0.744	
Protein yield	169 kg		6
Carbohydrate yield	4,151 kg		6

[a] Data on kcal input per unit are from standard conversions used in CRC handbook.
[b] Assumes that 18.7 metric tons of machinery (tractors, trucks, harvester, etc.) were used for production, that this machinery serves an average of 40 hectares, and that it is depreciated over 10 years.
[c] Gasoline and diesel used for operation of tractors, trucks, and harvester.
[d] Determined from average and recommended application rates.
[e] Total weight of machinery, fuel, fertilizer, and pesticides transported to crop a hectare.
[f] Assumes that three years are required to establish an orchard and the life of the orchard is 15 years.

Table 4
AN EXAMPLE OF ENERGY INPUTS PER HECTARE FOR PRODUCTION OF PRUNES IN CALIFORNIA (WITH IRRIGATION, BUT NOT CULTIVATION)

Item	Quantity/ha	kcal/ha[a]	Ref.
	Input		
Labor	84.3 hr	—	10
Machinery	42.2 kg[b]	713,012	
Gasoline	298.4 ℓ[c]	3,016,526	4
Diesel	532.0 ℓ[c]	6,072,248	4
Nitrogen	201.8 kg	2,966,460	10
Herbicide	2.0 kg[d]	219,040	
Insecticide	58.3 kg[d]	4,952,373	
Irrigation	49.3 cm	608,127	10
Transportation	936 kg[e]	240,552	
Orchard establishment	—	1,852,128[f]	
Total		**20,640,466**	
	Output		
Yield of prunes	22,417 kg	15,803,985	10
kcal output/input		0.765	
Protein yield	169 kg		6
Carbohydrate yield	4,151 kg		6

[a] Data on kcal input per unit are from standard conversions used in CRC handbook.
[b] Assumes that 17.1 metric tons of machinery (tractors, trucks, harvester, etc.) were used for production, that this machinery serves an average of 40 hectares, and that it is depreciated over 10 years.
[c] Gasoline and diesel used for operation of tractors, trucks, and harvester.
[d] Determined from average and recommended application rates.
[e] Total weight of machinery, fuel, fertilizer, and pesticides transported to crop a hectare.
[f] Assumes that three years are required to establish an orchard and the life of the orchard is 15 years.

Table 5
AN EXAMPLE OF ENERGY INPUTS PER HECTARE FOR PRODUCTION OF AN EARLY-MATURING PLUM (SANTA ROSA) IN CALIFORNIA (WITH IRRIGATION AND CULTIVATION)

Item	Quantity/ha	kcal/ha[a]	Ref.
	Input		
Labor	250 hr	—	11
Machinery	33.4 kg[b]	604,589	
Gasoline	280.6 ℓ[c]	2,836,585	4
Nitrogen	112.1 kg	1,647,870	11
Zinc	10.1 kg[d]	20,200	
Herbicide	1.0 kg[d]	109,520	
Insecticide	58.3 kg[d]	4,952,373	
Irrigation	43.2 cm	532,882	11
Transportation	420 kg[e]	107,940	
Orchard establishment	—	1,717,272[f]	
Total		**12,529,231**	
	Output		
Yield of plums	17,933 kg	8,091,369	11
kcal output/kcal input		0.645	
Protein yield	84 kg		6
Carbohydrate yield	2,073 kg		6

[a] Data on kcal input per unit are from standard conversions used in CRC hndbook.
[b] Assumes that 13.5 metric tons of machinery (tractors, trucks, etc.) were used for production, that this machinery serves an average of 40 hectares and that it is depreciated over 10 years.
[c] Gasoline used for operation of tractors and trucks.
[d] Determined from average and recommended application rates.
[e] Total weight of machinery, fuel, fertilizer, and pesticides transported to crop a hectare.
[f] Assumes that three years are required to establish an orchard and the life of the orchard is 15 years.

Table 6
AN EXAMPLE OF ENERGY INPUTS PER HECTARE FOR PRODUCTION OF A MIDSEASON PLUM (LARODA) IN CALIFORNIA (WITH IRRIGATION AND CULTIVATION)

Item	Quantity/ha	kcal/ha[a]	Ref.
	Input		
Labor	354 hr	—	12
Machinery	33.4 kg[b]	604,589	
Gasoline	301.7 ℓ[c]	3,049,885	4
Nitrogen	140.1 kg	2,059,470	12
Zinc	10.1 kg[d]	20,200	
Herbicide	1.0 kg[d]	109,520	
Insecticide	58.3 kg[d]	4,952,373	
Irrigation	49.3 cm	608,127	12
Transportation	464 kg[e]	119,248	
Orchard establishment	—	1,649,840[f]	
Total		**13,173,252**	
	Output		
Yield of plums	22,900 kg	12,137,280	12
kcal output/input		0.921	
Protein yield	126 kg		6
Carbohydrate yield	3.110 kg		6

[a] Data on kcal input per unit are from standard conversions used in CRC handbook.

[b] Assumes that 13.5 metric tons of machinery (tractors, trucks, etc.) were used for production, that this machinery serves an average of 40 hectares and that it is depreciated over 10 years.

[c] Gasoine used for operation of tractors and trucks.

[d] Determined from average and recommended application rates.

[e] Total weight of machinery, fuel, fertilizer, and pesticides transported to crop a hectare.

[f] Assumes that three years are required to establish an orchard, and the life of an orchard is 15 years.

Table 7
AN EXAMPLE OF ENERGY INPUTS PER HECTARE FOR ESTABLISHMENT OF A PLUM ORCHARD IN CALIFORNIA (WITH IRRIGATION AND CULTIVATION)

Item	Quantity/ha			kcal/ha[a]			Ref.
	1st year	2nd year	3rd year	1st year	2nd year	3rd year	
Input							
Labor	110	59	97 hr	—	—	—	13
Machinery	27.5	19.3	27.4 kg[b]	437,668	412,610	486,256	
Gasoline	264.2	131.0	179.5 ℓ[c]	2,670,798	1,324,279	1,814,566	4
Diesel	28.1	—	— ℓ[c]	320,733	—	—	4
Nitrogen	42.6	49.3	72.9 kg	626,220	724,710	1,071,630	13
Zinc	2.2	4.5	6.7 kg[d]	4,400	9,000	13,400	
Insecticide	18.9	37.8	56.7 kg[d]	1,713,607	3,427,214	5,140,821	
Fumigation	654.8	—	— ℓ[d]	5,095,435	—	—	
Irrigation	24.7	30.8	37.0 cm[e]	304,680	379,925	456,404	
Trees	267	5	—[f]	1,068,000	20,000	—	
Transportation	506	209	295 kg[g]	130,042	53,713	75,815	
Total				**12,371,583**	**6,351,451**	**9,058,892**	
Output							
Yield of plums	0	0	5,977 kg[h]	0	0	2,696,822	
kcal output/input				0	0	0.297	
Protein yield	0	0	28 kg				6
Carbohydrate yield	0	0	691 kg[a]				

[a] Data on kcal input per unit are from standard conversions used in CRC handbook.
[b] Assumes that 11.0, 7.7, and 11.0 metric tons of machinery (tractors, trucks, etc.) were used for the 1st, 2nd, and 3rd year, respectively, that this machinery serves an average of 40 hectares, and that it is depreciated over 10 years.
[c] Gasoline and diesel were used for operation of tractors and pickup.
[d] Determined from average and recommended application rates.
[e] Assumes that 4/5 of the water was applied from an irrigation canal (10-meter pumping lift) and 1/5 was applied from an irrigation well (100-meter pumping lift); furrow irrigation.
[f] Number of trees per hectare based on 6- × 6-meter planting.
[g] Total weight of machinery, fuel, fertilizer, pesticides, and trees transported to establish and crop a hectare.
[h] Assumes the crop is about 30% of that produced by mature trees.

REFERENCES

1. Agricultural Statistics 1976, U.S. Department of Agriculture, Washington, D.C., 1976, 238.
2. Fruits and Tree Nuts, Bloom, Harvesting, and Marketing Dates and Producing Counties, Agric. Handbook 186, Crops Reporting Board, U.S. Department of Agriculture Statistical Reporting Service, Washington, D.C., 1968.
3. Kelsey, M. and Van Derbech, S., Michigan State University Agricultural Facts Ext. Bull. E-1116, 1977.
4. Nebraska Tractor Test, Implement and Tractor Red Book, 92(3), Intertech, Kansas City, 1977, A168-169.
5. Michigan Agricultural Statistics, CRSB-76-03, Michigan Crop Reporting Service, Lansing, Mich., 1976.
6. **Watt, B. K. and Merrill, A. L.,** Handbook No. 8, U.S. Department of Agriculture, Washington, D.C., 1975.
7. Italian Prunes, Willamette Valley, Oregon State University Extension Service, Corvallis, 1974.
8. **Stebbins, R. L., Gardner, E. H., and Chaplin, M. H.,** Prunes (Oregon), Circ. FG24, Oregon State University Extension Service, Corvallis, 1977.
9. **Fisher, G. C. and MacSwan, L. C.,** Ext. Circ. 675, Oregon State University Extension Service, Corvallis, 1977.
10. **Reed, A. D. and Horel, L. A.,** Leaflet 2897, Division of Agricultural Sciences, University of California, Berkeley, 1976.
11. **Brown, L. C., Gerdts, M. H., Hench, K. W., Horel, L. A., LaRue, J. H., Leavitt, G. M., Obenauf, G. L., and Yeary, E. A.,**Special Publication 3076, Division of Agricultural Science, University of California, Berkeley, 1976.
12. **Brown, L. C., Gerdts, M. H., Hench, K. W., Horel, L. A., LaRue, J. H., Leavitt, G. M., Obenauf, G. L., and Yeary, E. A.,**Special Publication 3075, Division of Agricultural Sciences, University of California, Berkeley, 1976.
13. **Brown, L. C., Hench, K. W., Horel, L. A., LaRue, J. H., Leavitt, G. M., Obenauf, G. L., and Yeary, E. A.,** Special Publication 3088, Division of Agricultural Sciences, University of California, Berkeley, 1976.

ENERGY REQUIREMENTS FOR GRAPE PRODUCTION IN THE U.S.

Robert M. Pool

Grapes and grape products have historically formed a significant portion of the diet of persons living in the temperate regions of the northern hemisphere. This is because pretechnologic societies were able to preserve grapes as wine or dried raisins. Even though wine and raisin production are now highly sophisticated industries, grapes remain an important crop. On a world-wide basis, there are more grapes grown than any other fruit crop.[1] In 1976, grape growing occupied 10,343,000 ha of land and 57,147,800 metric tons of grapes were produced.[2] Eighty-three percent of these grapes were crushed and used to produce 323,015,000 hℓ of wine. Fresh sales took 12%, and 5% of the production was dried to raisins.[2]

In the U.S. in 1976, 297,000 ha of land was planted to grapes,[2] and 3,811,400 metric tons of grapes worth $620,787,400 were grown.[3] Ninety percent of the grape production in the U.S. was from one state, California, although ten other states, New York, Washington, Michigan, Pennsylvania, Ohio, Arizona, Arkansas, North Carolina, Missouri, and New Jersey, in order of production, produced enough grapes to have separate production statistics recorded.

California and Arizona are the only states that predominately grow the European grape *Vitis vinifera*. In the rest of the U.S., grape production is unique in the world in that, in the bulk of the vineyards, the cultivated varieties are grape species native to the U.S. or hybrids between native species and *V. vinifera*. This is primarily because they are more resistant to diseases and winter cold. Much of these grapes are used to produce fresh juice, although significant amounts are used to make wine.[4]

The tables in the grape section were chosen to be representative of the areas and methods of grape production in the U.S. The Central Valley area of California is where most of the grapes in the U.S. are produced. Production is characterized by heavy crops of high quality table and raisin grapes and standard quality wine grapes.[4] One variety, Thompson Seedless, is the primary variety used for all three outlets, so the three examples given for the Central Valley area are all of Thompson Seedless (Tables 2 to 5). Many older vineyards of the coastal areas of California were planted so that only minimal inputs are required (no irrigation, no trellis, etc.). Tables 6 and 7 show the inputs for this area. Newer vineyards in this area reflect more recent trends in production and, thus, would have inputs more like those used in the Central Valley regions. Because of a cooler, shorter growing season, production in these coastal areas is inherently lower than that in the Central Valley region,[4] but the high quality of the wines produced justifies the decreased return of grapes per energy input.

Two examples of Concord grape production are reported, Washington State and the Great Lakes region. The potential yield in Washington (Tables 9 and 10) is greater than in the Great Lakes (Tables 11 and 12) because of more hours of sunshine, but the Great Lakes is considered to produce grapes of very high quality for fresh Concord juice production, and, too often, the high potential yield of Washington vineyards has not been realized. Because of the arid summers and the lack of indigenous grape pests, the pest and disease control requirements are very low in Washington State for powdery mildew-resistant Concord grapevines. The final example given is wine grape production in the Great Lakes region (Tables 13 and 14). This represents an area of production that has increased in recent years. Most of this wine is made from varieties that are interspecific hybrids between native American species and *V. vinifera*.

Table 1
SOME EQUIPMENT USED IN VINEYARDS

Implement gross weight (kg)		kg/acre/year	kg/ha/year
Brush chopper	476	0.476	1.176
Disc	205	0.205	0.506
Fertilizer spreader	114	0.114	0.282
Grape harvester	6818	1.363	3.367
Grape hoe	137	0.137	0.338
Post driver	137	0.137	0.338
Trailer	53	0.053	0.131
Sprayer	1136	1.136	2.806
Vineyard tractor	1909	1.909	4.715
Weed sprayer	176	0.176	0.434
Total		**5.706**	**14.093**

Note: 5.7 kg × 2.47 = 14.1 kg/ha/year = 253,800 kcal = 18,000 kcal/kg.

Table 2
INPUTS TO ESTABLISH A THOMPSON SEEDLESS VINEYARD IN THE CENTRAL VALLEY OF CALIFORNIA TO PRODUCE TABLE GRAPES — YEAR 1[5]

Item	Quantity/ha	kcal/ha
	Input	
Labor	138 hr	—
Machinery	14.1 kg	253,800
Diesel	958 ℓ	10,934,612
Irrigation	40 cm	634,000
Insecticide	0.6 kg	52,146
Fungicide	44 kg	1,172,090
Transportation	9204.4 kg	2,365,531
Wire	958 kg	246,206
Stakes	7418 kg	34,124,000
Total		**49,782,385**
	Output	
Grape yield	0	

From Peacock, D., Swanson, F., Christensen, P., Kasimatis, A. N., Jensen, F., Yeary, E. A., and Horel, L. A., University of California, Agr. Sci. Publ. 3093, 1976.

Table 3
INPUTS TO ESTABLISH A THOMPSON SEEDLESS VINEYARD IN THE CENTRAL VALLEY OF CALIFORNIA TO PRODUCE GRAPES — YEAR 2[5]

Item	Quantity/ha	kcal/ha
	Input	
Labor	103 hr	—
Machinery	14.1 kg	253,800
Diesel	296 ℓ	3,374,207
Nitrogen	80 kg	1,176,000
Irrigation	60 cm	952,954
Insecticide	1.2 kg	125,150
Fungicide	18 kg	1,000,800
Transportation	796.7 kg	204,752
Wire	531 kg	11,190,294
Total		**18,277,957**
	Output	
Grape yield	0	

Table 4
INPUTS TO ESTABLISH A THOMPSON SEEDLESS VINEYARD IN THE CENTRAL VALLEY OF CALIFORNIA TO PRODUCE GRAPES — YEAR 3[5]

Item	Quantity/ha	kcal/ha
Input		
Labor	103 hr	—
Machinery	14.1 kg	253,800
Diesel	334 ℓ	3,806,797
Nitrogen	80 kg	1,176,000
Irrigation	75 cm	1,191,194
Insecticide	3.4 kg	295,494
Fungicide	57.2 kg	3,180,320
Transportation	298 kg	76,586
Total		**9,980,191**
Total establishment inputs		**78,040,533**
Output		
Grape yield	6,740 kg	4,583,200
Protein yield	40 kg	
Net establishment inputs		**73,457,333**
kcal output/kcal input (Year 3 only)		0.459
kcal output/labor hour (Year 3 only)		44,497
kcal output/labor hour (Total Establishment)		13,323
kcal output/kcal input (Total Establishment Input)		0.059

Table 5
ANNUAL INPUTS PER HECTARE TO PRODUCE AND PACK THOMPSON SEEDLESS TABLE GRAPES IN THE CENTRAL VALLEY OF CALIFORNIA[6]

Item	Quantity/ha	kcal/ha
Input		
Labor	603 hr	—
Establishment[a]	—	2,938,293
Machinery	14.1 kg	253,800
Diesel	296 ℓ	3,374,207
Nitrogen	160 kg	2,352,000
Irrigation	75 cm	1,035,821
Insecticide	3.4 kg	295,494
Fungicide	119 kg	6,616,400
Growth hormone	0.3 kg	29,700
Wire	29 kg	611,146
Stakes	300 kg	627,273
Transportation	594.7 kg	152,838
Total		**18,286,972**
Output		
Grape yield	19,094 kg	12,983,920
Protein yield	115 kg	
kcal output/kcal input		0.710
kcal output/labor hour		30,327

[a] One twenty-fifth of net establishment costs.

Table 6
INPUTS TO ESTABLISH A THOMPSON SEEDLESS VINEYARD IN THE CENTRAL VALLEY OF CALIFORNIA FOR WINE OR RAISIN PRODUCTION — YEAR 1[7]

Item	Quantity/ha	kcal/ha
Input		
Labor	133 hr	—
Machinery	14.1 kg	253,800
Diesel	1,487 ℓ	16,957,552
Irrigation	30 cm	325,440
Insecticide	0.6 kg	52,146
Fungicide	9 kg	500,400
Transportation	1462 kg	375,747
Stakes	184 kg	846,818
Total		**19,311,903**
Output		
Grape yield	0	

Table 7
INPUTS TO ESTABLISH A THOMPSON SEEDLESS VINEYARD IN THE CENTRAL VALLEY OF CALIFORNIA FOR WINE OR RAISIN PRODUCTION — YEAR 2[7]

Item	Quantity/ha	kcal/ha
Input		
Labor	178 hr	—
Machinery	14.1 kg	253,800
Diesel	1,349 ℓ	15,400,225
Nitrogen	80 kg	1,176,000
Irrigation	40 cm	635,303
Insecticide	1.2 kg	104,292
Fungicide	18 kg	1,000,800
Wire	276 kg	581,642
Transportation	1,436.75 kg	369,245
Total		**19,521,307**
Output		
Grape yield	0	

Table 8
INPUTS TO ESTABLISH A THOMPSON SEEDLESS VINEYARD IN THE CENTRAL VALLEY OF CALIFORNIA FOR WINE OR RAISIN PRODUCTION — YEAR 3[7]

Item	Quantity/ha	kcal/ha
Input		
Labor	144 hr	—
Machinery	14.2 kg	253,800
Diesel	159 ℓ	1,816,881
Nitrogen	128 kg	1,881,660
Irrigation	50 cm	690,547
Insecticide	1.7 kg	147,747
Fungicide	34.9 kg	1,940,440
Transportation	149.3 kg	38,370
Total		**6,769,343**
Total establishment		**45,602,533**
Output		
Grape yield	6,740 kg	4,583,200
Protein yield	40 kg	
kcal output/kcal input (Year 3 only)		0.677
kcal output/kcal input (Total Establishment Input)		0.101
Net establishment		**41,019,353**
kcal output/labor hour (Year 3 only)		31,828
kcal output/labor hour (All 3 years)		10,073

Table 9
ANNUAL INPUTS PER HECTARE TO PRODUCE AND DRY THOMPSON SEEDLESS SUN-DRIED RAISINS IN THE CENTRAL VALLEY OF CALIFORNIA[8]

Item	Quantity/ha	kcal/ha
Input		
Labor	161 hr	—
Establishment[a]		1,640,774
Machinery	14.1 kg	253,800
Diesel	349 ℓ	3,979,834
Nitrogen	160 kg	2,352,000
Irrigation	50 cm	794,129
Insecticide	1.7 kg	147,747
Fungicide	7.7 kg	428,120
Transportation	3,534.65 kg	908,405
Wire	28 kg	590,072
Stakes	360 kg	1,656,000
Paper trays	2850 kg	2,057,700
Total		**14,808,581**
Output		
Raisin output	5054 kg	14,61,006
Protein yield	121 kg	
kcal output/kcal input		0.986
kcal output/labor hour		90,689

[a] One twenty-fifth of net establishment inputs.

Table 10
ANNUAL INPUTS PER HECTARE TO PRODUCE THOMPSON SEEDLESS GRAPES FOR UTILIZATION AS WINE IN THE CENTRAL VALLEY OF CALIFORNIA[8]

Item	Quantity/ha	kcal/ha
Input		
Labor	108 hr	—
Establishment[a]	—	1,640,774
Machinery	14.1 kg	253,800
Diesel	326 ℓ	3,720,279
Nitrogen	161 kg	2,366,700
Irrigation	50 cm	794,129
Insecticide	1.7 kg	147,747
Fungicide	68.6 kg	3,814,160
Transportation	696.2 kg	178,923
Wire	30 kg	632,220
Stakes	375 kg	1,725,000
Total		**15,273,732**
Output		
Grape yield	22,464	15,275,520
Protein yield	135	
kcal output/kcal input		1.000
kcal output/labor hour		141,440

[a]One twenty-fifth net establishment inputs.

Table 11
INPUTS TO ESTABLISH A NONIRRIGATED VINEYARD IN NORTH-COASTAL VITICULTURAL AREAS OF CALIFORNIA — YEAR 1[9]

Item	Quantity/ha	kcal/ha
Input		
Labor	123 hr	—
Machinery	14.1 kg	253,800
Diesel	318 ℓ	3,633,761
Insecticide	0.6 kg	52,146
Fungicide	11.2 kg	622,720
Transportation	15,584.4	4,005,191
Stakes	15,300 kg	7,380,000
Total		**15,947,618**
Output		
Grape yield	0	

Table 12
INPUTS TO ESTABLISH A NONIRRIGATED VINEYARD IN NORTH-COASTAL VITICULTURAL AREAS OF CALIFORNIA — YEAR 2[9]

Item	Quantity/ha	kcal/ha
	Input	
Labor	109 hr	—
Machinery	14.1 kg	253,800
Diesel	152 ℓ	1,730,362
Insecticide	1.2 kg	104,292
Fungicide	22.4 kg	1,245,440
Transportation	143.3 kg	36,828
Total		**3,370,722**
	Output	
Grape yield	0	

Table 13
INPUTS TO ESTABLISH A NONIRRIGATED VINEYARD IN NORTH COASTAL VITICULTURAL AREAS OF CALIFORNIA — YEAR 3[9]

Item	Quantity/ha	kcal/ha
	Input	
Labor	93 hr	—
Machinery	14.1 kg	253,800
Diesel	318 ℓ	3,633,761
Insecticide	1.7 kg	147,747
Fungicide	22.4 kg	1,245,440
Transportation	284.4 kg	73,091
Total		**5,353,839**
Total Establishment		**24,672,179**
	Output	
Grape yield	3,370 kg	1,536,720
Protein yield	29 kg	
kcal output/kcal input (Year 3 inputs only)		0.287
Net establishment inputs		**23,135,459**
kcal output/kcal input (Total establishment input)		0.062
kcal output/labor hour (Year 3 only)		16,524
kcal output/labor hour (Total establishment)		4,728

Table 14
ANNUAL INPUTS PER HECTARE TO PRODUCE WINE GRAPES IN OLDER NONIRRIGATED VINEYARDS IN NORTH-COASTAL CALIFORNIA[10]

Item	Quantity/ha	kcal/ha
	Input	
Labor	142 hr	—
Establishment[a]		925,418
Machinery	14.1 kg	253,800
Diesel	508 ℓ	5,796,714
Nitrogen	160 kg	2,352,000
Insecticide	1.7 kg	147,747
Fungicide	57.2 kg	3,180,320
Transportation	445.9 kg	114,596
Total		**12,770,595**
	Output	
Grape yield	11,232 kg	5,121,792
Protein yield	97 kg	
kcal output/kcal input		0.401
kcal output/labor hour		36,065

[a]One twenty-fifth of net establishment inputs.

Table 15
INPUTS TO ESTABLISH 1 ha OF IRRIGATED CONCORD GRAPES IN WASHINGTON — YEAR 1[11]

Item	Quantity/ha	kcal/ha
	Input	
Labor	100 hr	—
Machinery	14.1 kg	253,800
Diesel	781 ℓ	8,911,366
Irrigation	15 cm	238,239
Transportation	678 kg	174,246
Total		**9,577,651**
	Output	
Grape yield	0	

Table 16
INPUTS TO ESTABLISH 1 ha OF IRRIATED CONCORD GRAPES IN WASHINGTON — YEAR 2[11]

Item	Quantity/ha	kcal/ha
Input		
Labor	245 hr	—
Machinery	14.1 kg	253,800
Diesel	470 ℓ	5,364,123
Irrigation	25 cm	397,065
Transportation	1,491 kg	383,187
Wire	327 kg	6,891,198
Stakes	750 kg	3,450,000
Total		**16,739,373**
Output		
Grape yield	0	

Table 17
INPUTS TO ESTABLISH 1 ha OF IRRIGATED CONCORD GRAPES IN WASHINGTON — YEAR 3[11]

Item	Quantity/ha	kcal/ha
Input		
Labor	178 hr	—
Machinery	14.1 kg	253,800
Diesel	477 ℓ	5,104,569
Vetch seed	7.9 kg	102,700
Rye seed	31.4 kg	125,600
Irrigation	35 cm	511,918
Insecticide oil	28 ℓ	319,592
Insecticide	2.2 kg	191,202
Transportation	464.85 kg	119,467
Wire	6 kg	126,444
Total		**6,855,292**
Total establishment		**33,172,316**
Output		
Grape yield	6,739	3,072,984
Protein yield	58 kg	
kcal output/kcal input (Year 3 input only)		0.488
Net establishment		**30,099,332**
kcal output/kcal input (Total establishment)		0.093
kcal output/labor hour (Year 3 only)		17,264
kcal output/labor hour (Total establishment)		5,876

Table 18
ANNUAL INPUTS PER HECTARE TO GROW CONCORD GRAPES IN WASHINGTON STATE[12]

Item	Quantity/ha	kcal/ha
Input		
Labor	122 hr	—
Establishment[a]	—	1,203,973
Machinery	14.1 kg	253,800
Diesel	364 ℓ	4,152,869
Nitrogen	192 kg	2,822,400
Irrigation	40 cm	635,303
Insecticide oil	56 ℓ	639,184
Insecticide	2.2 kg	191,202
Transportation	800.5 kg	205,728
Wire	27 kg	568,998
Stakes	450 kg	2,074,000
Total		**12,747,547**
Output		
Grape yield	17,971 kg	8,194,776
Protein	155 kg	
kcal output/kcal input		0.643
kcal output/labor hour		67,170

[a]One twenty-fifth net establishment inputs.

Table 19
INPUTS TO ESTABLISH 1 ha OF CONCORD GRAPES IN THE GREAT LAKES VITICULTURAL AREAS OF NORTH AMERICA — YEAR 1[13]

Item	Quantity/ha	kcal/ha
Input		
Labor	160 hr	—
Machinery	14.1 kg	253,800
Diesel	2,039 ℓ	23,273,374
Nitrogen	112 kg	1,646,400
Rye seed	62.3 kg	249,200
Insecticide	1.1 kg	95,601
Fungicide	2.0 kg	129,820
Transportation	5,307.55 kg	1,364,040
Stakes	3,498 kg	16,089,545
Total		**43,101,780**
Output		
Grape yield	0	

Table 20
INPUTS TO ESTABLISH 1 ha OF CONCORD GRAPES IN THE GREAT LAKES VITICULTURAL AREAS OF NORTH AMERICA — YEAR 2[13]

Item	Quantity/ha	kcal/ha
Input		
Labor	96 hr	—
Machinery	14.1 kg	253,800
Diesel	394 ℓ	4,490,942
Nitrogen	112 kg	1,646,400
Rye seed	62.3 kg	249,200
Insecticide	3.7 kg	321,567
Fungicide	3.0 kg	194,730
Herbicide	2.7 kg	269,757
Transportation	863.3 kg	221,868
Wire	452 kg	9,525,448
Total		17,173,712
Output		
Grape yield	0	

Table 21
INPUTS TO ESTABLISH 1 ha OF CONCORD GRAPES IN THE GREAT LAKES VITICULTURAL AREAS OF NORTH AMERICA — YEAR 3[13]

Item	Quantity/ha	kcal/ha
Input		
Labor	173 hr	—
machinery	14.1 kg	253,800
Diesel	417 ℓ	4,758,497
Nitrogen	224 kg	3,292,800
Rye seed	62.3 kg	249,200
Insecticide	4.4 kg	382,404
Fungicide	10.0 kg	649,100
Herbicide	2.7 kg	269,757
Transportation	435.85 kg	112,013
Wire	5 kg	105,370
Total		**10,072,941**
Total establishment		**70,348,433**
Output		
Grape yield	3,370 kg	1,536,720
Protein yield	29 kg	
kcal output/kcal input (Year 3 input only)		0.153
Net establishment inputs		68,811,713
kcal output/kcal input (Total establishment)		0.022
kcal output/labor hour (Year 3 only)		8,883
kcal output/labor hour (Total establishment)		3,582

Table 22
ANNUAL INPUTS PER HECTARE TO PRODUCE CONCORD GRAPES IN THE GREAT LAKES VITICULTURAL AREAS OF NORTH AMERICA[14]

Item	Quantity/ha	kcal/ha
Input		
Labor	176 hr	—
Establishment[a]	—	2,752,469
Machinery	14.1 kg	253,800
Diesel	2319 ℓ	26,474,545
Nitrogen	235 kg	3,454,500
Potassium	450 kg	720,000
Rye seed	56.2 kg	224,800
Insecticide	9.0 kg	782,190
Fungicide	10.2 kg	662,082
Herbicide	5.4 kg	539,514
Transportation	2170.45 kg	557,806
Wire	27 kg	568,998
Stakes	102 kg	470,455
Total		**37,461,159**
Output		
Grape yield	10,109 kg	4,609,704
Protein yield	87 kg	
kcal output/kcal input		0.123
kcal output/labor hour		26,192

[a]One twenty-fifth net establishment inputs.

Table 23
INPUTS TO ESTABLISH 1 ha OF WINE GRAPES IN THE GREAT LAKES VITICULTURAL AREAS OF NORTH AMERICA — YEAR 1[13]

Item	Quantity/ha	kcal/ha
Input		
Labor	179 hr	—
Machinery	14.1 kg	253,800
Diesel	887 ℓ	10,122,620
Nitrogen	112.3 kg	1,650,810
Rye seed	56 kg	224,000
Insecticide	1.1 kg	95,601
Fungicide	3.3 kg	183,480
Transportation	4,785 kg	1,229,745
Wire	450 kg	9,483,300
Stakes	3,511 kg	16,152,272
Total		**39,395,628**
Output		
Grape yield	0	

Table 24
INPUT TO ESTABLISH 1 ha OF WINE GRAPES IN THE GREAT LAKES VITICULTURAL AREAS OF NORTH AMERICA — YEAR 2[13]

Item	Quantity/ha	kcal/ha
Input		
Labor	82 hr	—
Machinery	14.1 kg	253,800
Diesel	152 ℓ	1,730,362
Nitrogen	112.3 kg	1,650,810
Rye seed	56 kg	224,000
Insecticide	4.4 kg	347,640
Fungicide	13.4 kg	745,040
Herbicide	0.9 kg	89,919
Transportation	199.3 kg	51,220
Total		**5,092,791**
Output		
Grape yield	0	

Table 25
INPUT TO ESTABLISH 1 ha OF WINE GRAPES IN THE GREAT LAKES VITICULTURAL AREAS OF NORTH AMERICA — YEAR 3[13]

Item	Quantity/ha	kcal/ha
Input		
Labor	183 hr	—
Machinery	14.1 kg	253,800
Diesel	250 ℓ	2,855,098
Nitrogen	224.6 kg	3,301,620
Rye seed	56 kg	224,000
Insecticide	4.4 kg	347,640
Fungicide	22.8 kg	1,479,971
Herbicide	1.8 kg	179,838
Transportation	288.6 kg	74,170
Wire	6 kg	126,444
Total		**8,842,581**
Total establishment		**53,331,000**
Output		
Grape yield	3,370 kg	1,536,720
Protein yield	29 kg	
kcal output/kcal input (Year 3 input only)		0.174
Net establishment		**51,794,280**
kcal output/kcal input (Total establishment)		0.029
kcal output/labor hour (Year 3 input only)		8,397
kcal output/labor hour (Total establishment)		3,461

Table 26
ANNUAL INPUTS PER HECTARE TO PRODUCE WINE GRAPES IN THE GREAT LAKES VITICULTURAL AREAS OF NORTH AMERICA[14]

Item	Quantity/ha	kcal/ha
	Input	
Labor	199 hr	—
Establishment[a]		2,071,771
Machinery	14.1 kg	253,800
Diesel	591 ℓ	6,748,413
Nitrogen	235 kg	3,454,500
Potassium	450 kg	720,000
Rye seed	56 kg	724,000
Insecticide	9.0 kg	782,190
Fungicide	18.6 kg	1,207,326
Herbicide	5.4 kg	539,514
Transportation	681.5 kg	175,146
Wire	7 kg	147,518
Stakes	102 kg	470,455
Total		**17,294,658**
	Output	
Grape yield	11,232 kg	5,121,792
Protein yield	97 kg	
kcal output/kcal input		0.296
kcal output/labor hour		25,738

[a] One twenty-fifth net establishment inputs.

REFERENCES

1. **Harlan, J. R.,** *Sci. Am.*, 235, 89, 1976.
2. **Mamon, P.,** *Bull. OIV*,50, 739, 1977.
3. **Anon.,** Agricultural Statistics 1977, U.S. Department of Agriculture, Government Printing Office, Washington, D.C., 1977, 231—232.
4. **Winkler, A. J., Cook, J. A., Kliewer, W. M., and Lider, L. A.,** *General Viticulture*, 2nd ed., University of California Press, Berkeley, 1974.
5. **Peacock, D., Swanson, I., Christensen, P., Kasimatis, A. N., Jensen, F., Yeary, E. A., and Horel, L. A.,** University of California, Agricultural Science Publication, 3093, 1976.
6. **Luvisi, D., Swanson, I., Peacock, B., Christensen, P., Kasimatis, A. N., Jensen, F., Yeary, E. A., and Horel, L. A.,** University of California, Agricultural Science Publication 3094, 1976.
7. **Swanson, F., Christensen, P., Leavitt, G., Peacock, B., Luvisi, D., Jensen, F., Kasimatis, A. N., Yeary, E. A., and Horel, L. A.,** University of California Agric. Sci. Spec. Public. 3091, 1976.
8. **Swanson, F., Luvisi, D., Christensen, P., Leavitt, G., Peacock, B., Kasimatis, A. N., Jensen, F., Yearey, E. A., and Horel, L. A.,** University of California, Agricultural Science Special Publication 3092, 1976.
9. **Bowers, K. W., Sissin, R. L., Bearden, B. E., Kasimatis, A. N., Horel, L. A., and Reed, A. D.,** University of California, Agricultural Science Special Publication 3086, 1976.

10. **Bowers, K. W., Sisson, R. L., Bearden, B. E., Kasimatis, A. N., and Reed, A. D.,** University of California Agricultural Extension Bulletin, ACT-158, 1974.
11. **Doran, S. M. and Anderson, F. A.,** Washington State University, Agricultural Extension Service, EM 2958, Pullman, 1975.
12. **Doran, S. M. and Anderson, F. A.,** Washington State University, Cooperative Extension Bulletin, EM 2259, Pullman, 1975.
13. **Good, D. and Jordan, T. D.,** Cornell University Agricultural Economics Extension Bulletin, 76-31, Ithaca, 1976.
14. **Good, D. L. and Jordan, T. D.,** Cornell University Agricultural Economics Extension Bulletin, 75-18, Ithaca, 1975.

ENERGY USE IN U.S. CITRUS PRODUCTION AND HARVESTING

Herman J. Reitz

U.S. citrus production in 1976—77 was 13,855,000 metric tons.[1] This amount exceeded the total of all other tree fruits produced in the U.S. Production was centered in Florida, with 73% of the crop, followed by California, Texas, and Arizona. California and Arizona dominated the lemon industry.

Florida processors utilized 95% of Florida orange production and 86% of the total Florida citrus crop. In the other states, processors utilized between one-third and one-half of the crop. The remainder is utilized in the fresh market.

Production practices are widely divergent between the humid southeastern U.S. and the semi-arid southwestern U.S. In the southeast, less irrigation and cold protection are required, but higher rainfall and humidity necessitate more fertilizer, more spraying activity to control fungus diseases, and more problems with weed control. Requirements for various inputs vary widely — for example, from total dependence on irrigation to complete independence.

Because of its dominant position in citrus production, examples have been selected from Florida circumstances. Examples from California would show higher expenditures for irrigation and cold protection and lower expenditures for fertilization and weed control. These examples apply to operation of an established mature bearing grove and do not include developmental costs incurred prior to productive years.

Florida oranges produced for the processor constituted 55% of the total U.S. citrus crop. Energy requirements for production and harvesting of irrigated bearing orange groves in Florida are estimated in Table 1. Labor requirements consist of 63 man-hours per hectare per year for production and 147 man-hours for harvesting.[2,3] Assuming a 200-ha production unit, equipment required includes two 60- to 65-hp tractors, one air-carrier sprayer, one pickup truck, three flatbed trucks, one specialized hydraulic fruit loader with picking containers, two 10-ft disks, two mowers, one fertilizer distributor, and one hèrbicide sprayer. Machinery input is an estimate of manufacturing only. Transportation input includes allowances only for bringing fuel, chemicals, and machinery to the grove site. Gasoline and diesel fuel requirements are taken from Fuel Use Estimates for Florida Agricultural Production[4] and from Batty and Keller.[5] Machinery and fuel requirements for irrigation are based on a traveling volume gun, putting on a total of 25 cm of supplemental irrigation water per hectare per year.[6] Fertilizer requirements are based on Florida Agricultural Experiment Station Bulletin 536C,[7] and unpublished data by Pimentel.[8] Pesticide use is based on two applications of spray material per year at rates listed in Extension Circular 393C.[9]

Caloric output[10,11] per caloric input is 1.68. Caloric output was calculated to include not only the food portion of the fruit but also the peel and other residue which is dried down to a feed product for cattle.

Citrus can successfully be produced in many non-irrigated sites (Table 2). In this case, yield would be approximately 25% less, with corresponding reduction in labor, machinery, fuel, fertilizer, and transportation of materials to the grove. The decrease in intensity of energy utilization results in an increase in ratio of energy output to input (2.03).

Grapefruit constitutes approximately one-fifth of the U.S. total citrus crop, and approximately 60% of all grapefruit is processed. Table 3 gives estimates of the annual energy input for production and harvesting of non-irrigated bearing grapefruit groves managed for production of fruit for processing. The energy input in this case is essentially the same as for non-irrigated bearing orange groves, but grapefruit groves char-

Table 1
ANNUAL ENERGY INPUT FOR PRODUCTION AND HARVESTING OF IRRIGATED BEARING ORANGE GROVES MANAGED FOR PRODUCTION OF FRUIT FOR PROCESSING

Item	Quantity/ha	kcal/ha/year
Input		
Labor	210 hr	
Machinery	24 kg	432,000
Gasoline for trucks	90 ℓ	910,000
Diesel fuel		
Excluding irrigation	96 ℓ	1,096,000
For irrigation	305 ℓ	3,516,000
Fertilizers		
Nitrogen as N	225 kg	3,308,000
Phosphorus (as triple super)	113 kg	339,000
Potassium as K_2O	225 kg	360,000
Lime	1,288 kg	406,000
Pesticides		
Synthetic organics active ingredient	5.6 kg	487,000
Pesticidal oil emulsion	66 ℓ	753,000
Herbicides (on 10% of the area) active ingredient	0.5 kg	31,000
Transportation of materials to the grove	488.5 kg	125,545
Total		**11,763,545**
Output		
Orange yield (988 boxes/ha)	40,370 kg	19,781,000
Protein yield (1%)	404 kg	
kcal output/kcal input		1.68
kcal output /labor hour		94,195

Table 2
ANNUAL ENERGY INPUT FOR PRODUCTION AND HARVESTING OF NON-IRRIGATED BEARING ORANGE GROVES MANAGED FOR PRODUCTION OF FRUIT FOR PROCESSING

Item	Quantity/ha	kcal/ha/year
Input		
Labor	168 hr	
Machinery	16 kg	288,000
Gasoline for trucks	75 ℓ	758,000
Diesel fuel for tractors	96 ℓ	1,096,000
Fertilizers		
Nitrogen as N	196 kg	2,881,000
Phosphorus (as triple super)	98 kg	294,000
Potassium as K_2O	196 kg	314,000
Lime	1,120 kg	353,000
Pesticides		
Synthetic organics active ingredient	5.6 kg	487,000
Pesticidal oil emulsion	65 ℓ	742,000
Herbicides (on 10% of the area) active ingredient	0.5 kg	31,000
Transportation of materials to the grove	209.1 kg	53,739
Total		**7,297,739**
Output		
Orange yield (741 boxes/ha)	30,277 kg/ha	14,836,000
Protein yield (1%)	303 kg	
kcal output/kcal input		2.03
kcal output/labor hour		88,310

Table 3
ANNUAL ENERGY INPUT FOR PRODUCTION AND HARVESTING OF NON-IRRIGATED BEARING GRAPEFRUIT GROVES MANAGED FOR PRODUCTION OF FRUIT FOR PROCESSING

Item	Quantity/ha	kcal/ha/year
Input		
Labor	183 hr	
Machinery	16 kg	288,000
Gasoline for trucks	100 ℓ	1,011,000
Diesel fuel for tractors and sprayers	96 ℓ	1,096,000
Fertilizers		
Nitrogen as N	196 kg	2,881,000
Phosphorus (as triple super)	98 kg	294,000
Potassium as K_2O	196 kg	314,000
Lime	1,120 kg	353,000
Pesticides		
Synthetic organics active ingredient	5.6 kg	487,000
Pesticidal oil emulsion	65 ℓ	742,000
Herbicides (on 10% of the area) active ingredient	0.5 kg	31,000
Transportation of materials to the grove	227.9 kg	58,570
Total		**7,555,570**
Output		
Grapefruit yield (1235 boxes/ha)	47,655 kg/ha	22,300,000
Protein yield (1%)	477 kg	
kcal output/kcal input		2.96
kcal output/labor hour		121,858

acteristically produce a greater tonnage of fruit than do orange groves. Consequently, additional man-hours of labor and additional fuel for harvesting the crop are required. The characteristically greater grapefruit yield results in a higher ratio of caloric output compared with input (2.96) than in any other case.

While the major outlet for citrus produced in the U.S. is through processing channels, a substantial amount of citrus reaches the consumer in fresh form. Table 4 gives estimates of energy input for production and harvesting of irrigated bearing grapefruit groves managed for production of fruit for the fresh market. The production program in this case differs from the production program for processing fruit primarily in that one additional application of spray material would ordinarily be required to maintain a good rind appearance. A comparison of Table 1 and Table 4 reflects primarily an increase in the amount of energy input for pesticides.

The production inputs for oranges and tangerines destined for the fresh market would be similar to the inputs shown in Table 4. Grapefruit was specified in Table 4 in order to calculate the input-output ratio based on greater grapefruit yield. For oranges for the fresh market, the orange output in Table 1 (11,763,545 kcal) may be compared with the fresh grapefruit input specified in Table 4 (12,687,292 kcal) for a ratio of 0.93.

Orchard heating for protection against cold weather has not been included in any of the above calculations, although Jesse[11] emphasized this energy input. According

Table 4
ANNUAL ENERGY INPUT FOR PRODUCTION AND HARVESTING OF IRRIGATED BEARING GRAPEFRUIT GROVES MANAGED FOR PRODUCTION OF FRUIT FOR THE FRESH MARKET

Item	Quantity/ha	kcal/ha/year
Input		
Labor	225 hr	
Machinery	24 kg	432,000
Gasoline for trucks	112 ℓ	1,132,000
Diesel fuel		
Excluding irrigation	103 ℓ	1,176,000
Irrigation only	305 ℓ	3,516,000
Fertilizers		
Nitrogen as N	225 kg	3,308,000
Phosphorus (as triple super)	113 kg	339,000
Potassium as K_2O	225 kg	360,000
Lime	1,288 kg	406,000
Pesticides		
Synthetic organics active ingredient (including fungicides)	8.4 kg	730,000
Pesticidal oil emulsion	98 ℓ	1,119,000
Herbicides (on 10% of the area) active ingredient	0.5 kg	31,000
Transportation of materials to grove	538.1 kg	138,292
Total		**12,687,292**
Output		
Grapefruit yield (160 boxes/ha)	61,976 kg	29,129,000
Protein yield (1%)	620 kg	
kcal output/kcal input		2.30
kcal output/labor hour		129,462

to Gerber,[13] adequate protection against cold on severe nights in the present citrus growing districts would require the release of up to 5 million Btu per acre per hour (3,107,680 kilocalories per hectare per hour). On this basis, energy utilization for orchard heating on one cold night could exceed energy utilization for all other production and harvesting operations for an entire year combined. Orchard heating has never been a general practice in Florida or Texas and has been general only for protection of the fruit in parts of California. Advances in price for heating oil since 1973 have discouraged orchard heating, and an even smaller proportion of citrus groves will be heated in future years. Consequently, heating was not included in any of the tabular material provided. Unavailability of inexpensive heating oil will gradually lead to abandonment of the colder locations where citrus is now grown.

Energy utilization in citrus production is similar to that for production of other crops, with irrigation and nitrogen fertilization being the two most energy-consumptive items. Less intensive production practices will nearly always lead to higher ratios of caloric output to caloric input, but with coincident reduction in yield. The most distinctive feature of fruit production inputs is the high requirement for harvesting labor. The "best" solution is dependent upon subjective evaluations of the importance of energy use compared with the need for food production.

REFERENCES

1. **Small, R. P.**, Florida Agricultural Statistics, Citrus Summary 1977, Florida Crop and Livestock Report Service, Orlando, Fla., 1977.
2. **Fairchild, G. F.**, Socioeconomic Dimensions of Florida Citrus Harvesting Labor, ERD Rep. 75-2, Economic Research Department, Florida Department of Citrus, University of Florida, Gainesville, 1975.
3. **Dow, J. K.**, Impact of Mechanical Harvesting on the Demand for Labor in the Florida Citrus Industry, Agricultural Economics Report 10, Department of Agricultural Economics, Institute of Food and Agricultural Sciences, University of Florida, Gainesville, 1970.
4. **Smerdon, E. T., Baird, D., Adams, D., Beman, C., Brasher, J., Clay, R., Farber, E., Gerber, J., Griffith, J., Harrison, M., Langhan, M., McCloud, D., Rose, J., Schirrard, W., Waters, W., Whiten, M., and Bayley, F.**, Fuel Use Estimates for Florida Agricultural Production, Institute of Food and Agricultural Sciences, University of Florida, Gainesville, 1975.
5. **Batty, J. C., Hamad, S. N., and Keller, J.**, Energy Inputs to Irrigation, *J. Irrigation Drainage Div.*, ASCE 101(IR4) American Society of Civil Engineers, 293, 1975.
6. **Harrison, D. S.**, Irrigation systems for crop production in Florida, WRC-8, *Univ. Fla., Inst. Food Agric. Sci. Publ.*, Water Resources Council, 1977.
7. **Reitz, H. J., Leonard, C. D., Stewart, I., Koo, R. C. J., Anderson, C. A., Reese, R. L., Calvert, D. V., and Smith, P. F.**, Recommended fertilizers and nutritional sprays for citrus, *Fla. Agric. Exp. Stn. Bull.*, 536C, 1972.
8. **Pimentel, D.**, Energy Inputs for the Production, Formulation, Packaging, and Transport of Various Pesticides, unpublished data, 1977.
9. **Anon.**, Florida citrus spray and dust schedule, Circular 393C, Univ. of Fla. Coop. Ext. Serv., 1978.
10. **Adams, C. F.**, Nutritive Value of American Foods, Agric. Handbook 456, Agricultural Research Service, U.S. Department of Agriculture, 1975.
11. **White, A., Handler, P., and Smith, E. L.**, General metabolism, in *Principles of Biochemistry*, McGraw-Hill, New York, 1964, 282.
12. **Jesse, E. V.**, Energy Inputs in U.S. Fruit and Vegetable Production: Relative Costs and Efficiencies, paper presented at 9th Int. Congr. NORCOFEL, Dijon, France, Nov. 3 to 5, 1976.
13. **Gerber, J. F. and Martsolf, J. D.**, Protecting Citrus from Cold Damage, Circular 287, University of Florida Agricultural Extension Service, 1966.

ENERGY REQUIREMENTS FOR BANANA PRODUCTION IN SELECTED AREAS

F. W. Liu

Banana (*Musa* spp.) is one of the most important fruits in the world, second only to grape and citrus in productions.[1] Some 38,504,000 t of bananas were produced in 1976. Compared to some other fruits, bananas are easy to grow and are widely distributed in tropical and subtropical countries. About 36% of the bananas in the world are produced in South America, 31% in Asia, 18% in Central America, and 11% in Africa.

Although 96% of bananas are grown in developing countries, modern technology has been extensively adopted for banana growing and handling. Most large scale plantations that produce bananas for international trade are either directly operated by, or under certain control of, large companies such as United Fruit Company and Castle and Cooke. Their management and technical data are often kept as company secrets and, therefore, are not readily available to the public. The data presented in this article are limited to those from Hawaii, New South Wales, and Taiwan. The author is deeply grateful to Mr. W. Yee, Fruit Specialist from the University of Hawaii, Mr. G. R. Gregory, Chief of the Division of Horticulture, New South Wales Department of Agriculture, and Mr. C. K. Chu, Chief of the Department of Horticulture, Chiayi Agricultural Experiment Station, who kindly supplied invaluable data of their own areas.

Hawaii is the only state of the U.S. that has sizable commercial banana plantations. The acreage is rather small and has been declining in the last decade. Some 3290 t of bananas were produced on 436 ha in 1965, and 2820 t were produced on 276 ha in 1975.[6] Australia had a total production of 119,000 t in 1976[1] mainly produced in two areas, northern Queensland and the subtropical coastal strip in New South Wales. Taiwan was once an important banana exporter in the world, but its production has declined sharply in recent years. It had a total production of 654,000 t in 1967 and only 213,000 t in 1976.[3,4] The cultural practices in Hawaii and Australia are highly mechanized. Compared to these two areas, considerably more manual labor is used in growing bananas in Taiwan.

Most banana plantations in Hawaii and in southern Taiwan are on low-altitude, flat land. Energy input for irrigation water pumping lift in these areas is relatively small compared to that in Coffs Harbour, Australia where most of the plantations are on hilly land. On the other hand, although many plantations in central Taiwan are on slope land, irrigation water is usually introduced into the plantations from even higher altitude water sources so that pumping lift is often not necessary.

Some 1870 and 2400 kg/ha of bamboo sticks are used in central and southern Taiwan, respectively, for propping banana stems. Some wooden props are also used in Hawaii and Australia, but there are no quantitative data available. Since the propping materials are crude forest materials whose production requires only solar energy and soil resources, they are not included in the energy input data for banana production. Excluding propping materials, there are four major energy inputs for banana production: (1) gasoline, (2) pumping lift for irrigation water, (3) nitrogen fertilizer, and (4) machinery and equipment. The total input is considerably lower in Taiwan than in the other two areas because the former uses lower inputs on items 1 and 2.

Banana flesh has an average of 1040 kcal/kg energy and 1.2% protein content.[2] The ratio of flesh to peel is about 2 at eating ripe state.[5] These figures and factors were the basis of calculation of energy output and protein yield.

Table 1
ANNUAL INPUTS PER HECTARE TO PRODUCE BANANAS IN HAWAII

Item	Quantity/ha	kcal/ha	Ref.
	Input		
Labor	400 hr	—	6, 7
Machinery and equipment	83 kg	1,494,000	7
Gasoline	551 ℓ	5,570,059	7
Diesel	133 ℓ	1,518,062	7
Pumping lift for irrigation water	20-m lift for 80 cm water[a]	2,015,552	6
Nitrogen	270 kg	3,969,000	6
Phosphate (P_2O_5)	135 kg	175,500	6
Potash (K_2O)	540 kg	864,000	6
Lime	2,550 kg[b]	804,398	6
Insecticides	3 kg[c]	184,410	
Fungicides	34 kg	944,180	6
Herbicides	6.2 kg	389,174	6, 7
Farm building	1.16 m^2	489,778	7
Transportation	609.30kg	156,590	
Total		**18,575,103**	
	Output		
Banana yield	22,000 kg	15,253,000	6, 7
Protein yield	176 kg		
kcal output/kcal input		0.821	
kcal output/labor hour		38,133	

[a] On the supposition of 20-m pump lift for irrigation water.
[b] Lime is used only in a fraction of plantations.
[c] On the supposition that similar amount is used in Hawaii as in Australia.

Table 2
ANNUAL INPUTS PER HECTARE TO PRODUCE BANANAS IN COFFS HARBOUR, NEW SOUTH WALES, AUSTRALIA

Item	Quantity/ha	kcal/ha
Input		
Labor	600 hr	—
Machinery and equipment	53 kg	954,000
Gasoline	540 ℓ	5,458,860
Electricity	520 kWh	1,488,760
Pumping lift for irrigation water	90-m lift for 45 cm water	5,101,866
Polyethylene film for bunch cover	100 kg	2,600,000
Nitrogen	250 kg	3,675,000
Phosphate (P_2O_5)	60 kg	78,000
Potash (K_2O)	550 kg	880,000
Lime	1,500 kg[a]	473,175
Insecticides	3 kg[b]	184,410
Fungicides	4.83 kg	134,129
Oil for disease control	80 ℓ	913,120
Herbicides	1.5 kg	94,155
Packing house	1.2 m^2	506,666
Transportation	626.0 kg	160,882
Total		**19,395,523**
Output		
Banana yield	18,000 kg	12,480,000
Protein yield	144 kg	
kcal output/kcal input		0.643
kcal output/labor hour		20,800

[a] 50% growers use no lime.
[b] 10% growers use additional 120 kg of Mocap® for nematode control.

Quantitative data of inputs and banana yield supplied by Gregory, G. R., Chief of the Division of Horticulture, Department of Agriculture, New South Wales, 1978.

Table 3
ANNUAL INPUTS PER HECTARE TO PRODUCE BANANAS IN CENTRAL TAIWAN

Item	Quantity/ha	kcal/ha
Input		
Labor	2,527 hr	—
Machinery and equipment	99 kg	1,782,000
Gasoline	30 ℓ	303,270
Diesel	100 ℓ	1,141,400
Electricity	48 kWh	137,424
Polyethylene film for bunch cover	47 kg	1,222,000
Nitrogen	352 kg	5,174,400
Phosphate (P_2O_5)	176 kg	228,800
Potash (K_2O)	704 kg	1,126,400
Insecticides	12 kg	737,640
Fungicides	24 kg	666,480
Oil for disease control	48 ℓ	547,872
Herbicides	2 kg	125,540
Packing house	1.66 m^2	700,889
Transportation	294.3 kg	75,635
Total		**13,969,750**
Output		
Banana yield	19,000 kg	13,173,333
Protein yield	152 kg	
kcal output/kcal input		0.943
kcal output/labor hour		5,213

Unpublished surveyed data supplied by Chu, C. K., Chief of the Department of Horticulture, Chiayi Agricultural Experiment Station, 1978.

Table 4
ANNUAL INPUTS PER HECTARE TO PRODUCE BANANAS IN SOUTHERN TAIWAN

Item	Quantity/ha	kcal/ha
Input		
Labor	3,100 hr	—
Machinery and equipment	99 kg	1,782,000
Gasoline	40 ℓ	404,360
Diesel	120 ℓ	1,369,680
Electricity	52 kWh	148,876
Pumping lift for irrigation water	5-m lift for 50 cm water	314,930
Polyethylene film for bunch cover	57 kg	1,482,000
Nitrogen	495 kg	7,276,500
Phosphate (P_2O_5)	275 kg	357,500
Potash (K_2O)	900 kg	1,440,000
Insecticides	2 kg	122,940
Fungicides	20 kg	555,400
Oil for disease control	80 ℓ	913,120
Herbicides	2 kg	125,540
Packing house	0.7 m^2	295,555
Transportation	299 kg	76,843
Total		**16,665,244**
Output		
Banana yield	33,000 kg	22,888,000
Protein yield	264 kg	
kcal output/kcal input		1.373
kcal output/labor hour		7,383

Unpublished surveyed data supplied by Chu, C. K., Chief of the Department of Horticulture, Chiayi Agricultural Experiment Station, 1978.

REFERENCES

1. **Production Yearbook** 1976, Food and Agriculture Organization, Rome, 1977.
2. **Simmonds, N. W.,** *Bananas,* 2nd ed., Longmans, London, 1966.
3. Taiwan Agricultural Yearbook, Taiwan Provincial Department of Agriculture and Forestry, 1968.
4. Taiwan Agricultural Yearbook, Taiwan Provincial Department of Agriculture and Forestry, 1977.
5. **Von Loesecke, H. W.,** *Bananas,* Interscience, New York, 1949.
6. **Yee, W., Hamilton, R. A., Kumabe, B., and Martinez, A. P.,** Banana Production in Hawaii, undated.
7. **Anon.,** Bananas — 4 Year Cost Estimate, Hawaii, undated.

REPRESENTATIVE UNITED STATES STRAWBERRY ENERGY BUDGETS

G. J. Galletta* and R. C. Funt

Strawberries are the small fruit crop that has almost universal appeal. In the U.S. over 6.6 million kilograms of strawberries were produced for commercial harvesting in 1977 on more than 10,500 ha. The value of the crop was almost $188 million. This included only the winter strawberry production in Florida under black plastic mulch, the phenomenally high California production under clear polyethylene mulch, and the production of Michigan, Washington, and Oregon for the processing trade.

California and Florida produce practically all of the fresh strawberries for shipment, with some small shipping areas still persisting in Louisiana, Arkansas, and Maryland. California, Oregon, Washington, and Michigan produce practically all of the strawberries for processing in the U.S. As California captured most of the U.S. wholesale strawberry market, the majority of the states east of the Mississippi River started producing strawberries for direct customer harvest (U-Pick) and retail sale at the farm. U-Pick has become a very promising business. Conservative estimates are that there are 5000 ha producing an average of 13,478 kg/ha. The berries are selling for an average price of 50c/lb (0.45 kg) which equals $14,976 gross per hectare. Strawberry production for commercial trade and for customer harvesting are both on the increase.

Strawberry yields have increased dramatically in the last 15 years as new, disease-resistant, and well-adapted regional cultivars have replaced the older cultivars. The new cultivars bear larger and more fruit under a wider array of climate and soil conditions than previous cultivars did.

Assuming proper soil and site selection and a suitable cultivar, weed and pest control, water supply, and labor supply have always been critical elements of strawberry production. Improvements in strawberry culture have also contributed to strawberry yield increase. These improvements have generally been aimed at reducing labor needs and plant and yield loss due to weed or pest competition or insufficient moisture at critical times. However, the improvements all tend to require high energy input. For example, raised-bed culture necessitates more and heavier machinery with increased fuel consumption. Broad-spectrum soil fumigation requires additional chemicals, polyethylene covers, and special machinery for application. Increased use of fertilizer and pesticides also increases the energy input. Supplemental irrigation for the several uses indicated later necessitates the use of expensive electrical energy.

One could estimate that if a grower were to plant 2 ha of a modern strawberry cultivar without preplanting soil fumigation, without irrigation, with a minimum of fertilization and pest control, effect weed control with hand hoeing and tractor and cultivator, sell the fruit directly to the consumer on the farm, and do all the other chores with family labor, his energy budget might look like that in Table 1.

Even with the modest projected yield in this predominatly hand-labor situation, the kilocalorie output per kilocalorie input ratio is much better than in all but the very highest yield in California. Naturally, improved yields at any energy level situation will improve the ratio for this high energy-requiring crop.

Table 2 estimates the energy requirements for a typical Florida winter productionn shipping situation using a good deal of harvest labor, high water supply, preplant soil fumigation, plastic mulch culture, heavy fungicide applications, and single fruiting season (energy output to input ratio = 0.18).

* Science and Education Administration, U.S. Department of Agriculture, Beltsville, Md.

Table 1
STRAWBERRY PRODUCTION BY SMALL-SCALE GROWER DOING MOST OF THE LABOR HIMSELF

Item	Quantity/ha	kcal/ha
	Input	
Labor	1,369 hr	—
Machinery	40 kg	720,000
Gasoline	112 ℓ	1,132,208
Nitrogen	190 kg	2,793,000
Phosphorus	45 kg	135,000
Potassium	45 kg	72,000
Plants set	13,454 plants	1,062,866
Insecticides	9 kg	553,230
Fungicides	9 kg	249,930
Transportation	512 kg	131,584
Total		**6,849,818**
	Output	
Yield	15,163 kg	5,610,310
Yield protein	106 kg	
kcal output/kcal input		0.82
kcal output/labor hour		4,098

Assumptions: 3.2 metric tons of machinery depreciated over 10 yr on 2 ha of strawberries and 8 ha total farm; 3 fruiting years with 6739 kg yield the first year, 5054 kg the second year, and 3370 kg the third year; the P and K added preplant, the N added 78 kg during the establishment year, and 56 kg in each of the next two seasons.

Table 3 estimates the energy requirements for a Maryland three-harvest season fresh-market shipping scheme. With a smaller number of plants set per hectare than in Florida, it takes three seasons to recover the yield. Fuel, lime, and herbicide higher energy requirements than Florida help to offset the plastic and fumigation requirements of the latter. (Maryland output to input ratio = 0.21).

Table 4 estimates the energy required to produce a strawberry plant (79 kcal/ha) for field setting by a modern nursery operating in a state plant certification program. This figure would be applicable to the entire U.S. except California, where improved plant yields lower the energy requirements per plant, even though total energy requirements are higher than in Maryland.

Table 5 estimates the energy required to produce strawberries in the traditional strawberry production region of California. This area is one of the principal suppliers of fresh strawberries for the U.S. In this example, yield is maximized by employing summer setting, a double-hill planting system, irrigation, soil fumigation, raised beds, and polyethylene bed mulch. The potential improvement in yield using the newest varieties is indicated (Central Coast output to input ratio = 0.55).

Table 6 estimates the energy required to produce early spring strawberries for shipping and processing in southern California. In this example, grower expenses are reduced by having the plants in the ground a shorter period of time and by increasing the number of plants set per hectare. Fuel, electricity, and fumigation remain high energy requirements under this system (output to input ratio = 0.48).

Table 2
ESTIMATED ENERGY USE FOR WINTER STRAWBERRY PRODUCTION IN CENTRAL FLORIDA

Item	Quantity/ha	kcal/ha
	Input	
Labor	6,445.6 hr	—
Machinery[a]	38 kg	684,000
Gasoline	786 ℓ	7,945,674
Electricity[b]	3,088.8 kWh	8,843,234
Nitrogen	224 kg	3,292,800
Phosphorus	56 kg	168,000
Potassium	185 kg	346,320
Calcium	140 kg	296,000
Fungicides	259 kg	7,192,430
Insecticides and miticides	34 kg	2,089,980
Herbicides	4 kg	251,080
Soil fumigant (MC 33)	224 kg	5,280,352
Black plastic mulch	215 kg	6,450,000
Cover crop seed (Sorghum)	56 kg	784,000
Plants set[c]	59,304 plants	4,685,016
Transportation[d]	2,621 kg	673,597
Total		**48,982,483**
	Output	
Strawberry yield[d]	23,538 kg	8,709,060
Protein yield	164.8 kg	
kcal output/kcal input		0.18
kcal output/labor hour		1,351

Note: Plants were set in fall, used double-hill raised-bed system with black polyethylene mulch, were harvested the following January to April, and were cultured as an annual — fruit largely shipped fresh. (Data kindly furnished by Dr. E. E. Albregts, University of Florida, IFAS)

[a] Assumes 7.6 metric tons of machinery used per year, that this machinery serves 20 ha of strawberries, and is depreciated over 10 years (includes chicken manure applied to cover crop prior to planting).

[b] This quantity of electricity is used to put on 5653 to 9662 m^3/ha of water per year for moisture, plant establishment, and frost control; manufacturing and installing energy for irrigation pipe not included.

[c] Each plant required 79 kcal for production (see Table 4).

[d] Transportation figure is the sum of all items transported to the farm. Gasoline used for hauling.

Table 3
ESTIMATED ENERGY USE FOR 4-YEAR CYCLE (THREE PRODUCING SEASONS) OF STRAWBERRY PRODUCTION IN MARYLAND

Item	Quantity/ha	kcal/ha
Input		
Labor — first year	184 hr	
Labor — second year	2,165 hr	
Labor — third year	1,698 hr	
Labor — fourth year	1,402 hr	
Total	5,449 hr	—
Machinery at 61 kg/year[a]	244 kg	4,392,000
Gasoline exclusive of irrigation	802 ℓ	8,107,418
Gasoline for irrigation[b]	685 ℓ	6,924,665
Diesel	359 ℓ	4,097,626
Electricity at 20 kWh/year	80 kWh	229,040
Nitrogen	182 kg	2,675,400
Phosphorus	92 kg	276,000
Potassium	92 kg	147,200
Lime	3,408 kg	8,207,998
Plants	12,301 plants	971,779
Insecticides	28 kg	1,721,160
Fungicides	11 kg	305,470
Herbicides	74 kg	4,644,980
Transportation	5,766 kg	1,481,862
Total		**44,182,598**
Output		
Strawberry yield second year	10,227 kg	
Strawberry yield third year	7,954 kg	
Strawberry yield fourth year	6,818 kg	
Total	**24,999 kg**	**9,249,630**
Protein yield	175 kg	
kcal output/kcal input		0.21
kcal output/labor hour		1,697

Note: Plants set in spring and permitted to form a narrow matted row. (Data kindly furnished by Dr. R. C. Funt, Department of Horticulture University of Maryland, largely from Maryland Agricultural Experiment Station Bulletin 922).

[a] Assumes 12.1 metric tons of machinery which serve 20 ha of fruit and 20 ha of other crops and depreciated over an average of 10 years.

[b] Amount of fuel to apply 2032 m^3 of water per hectare per season for irrigation and frost protection.

Table 4
ESTIMATED ENERGY USE FOR STRAWBERRY NURSERY PLANT PRODUCTION IN THE EASTERN UNITED STATES FOR PLANTS SET IN SPRING AND DUG, STORED, CLEANED, PACKAGED, AND SHIPPED DURING THE DORMANT PERIOD FOLLOWING THE SINGLE GROWING SEASON

Item	Quantity/ha	kcal/ha
Input		
Labor	311 hr	—
Machinery[a]	24 kg	432,000
Gasoline	918 ℓ	9,280,062
L.P. gas (irrigation)[b]	280 ℓ	2,157,400
Electricity[c]	550 kWh	1,574,650
Nitrogen	232 kg	3,410,400
Phosphorus	103 kg	309,000
Potassium	99 kg	158,400
Magnesium	22 kg	46,376
Insecticides and fumigants	116 kg	7,130,520
Fungicides	8 kg	222,160
Herbicides	13 kg	816,010
Transportation	1605 kg	412,485
Total		**25,949,463**
Output[d]		
Yield	328,643 plants	—

Note: Data provided by Curt Massey, Raynor Brothers Nursery, Salisbury, Md.

[a] Based on 29 metric tons of machinery used per year; serves 40 ha of strawberries and 81 ha of other crops and is depreciated over 10 years.

[b] Manufacturing and installing energy for irrigation pipe not included.

[c] Includes electricity for cold storage, cleaning, sorting, and packing plants.

[d] A total of 17,297 plants/ha are set to receive an average return of 328,643 plants/ha.

Table 5
ESTIMATED ENERGY USE FOR CALIFORNIA CENTRAL COAST STRAWBERRY PRODUCTION

Item	Quantity/ha	kcal/ha
Input		
Labor first year	3,941 hr	—
Labor second year	3,188 hr	—
Labor total	7,129 hr	—
Machinery at 44 k/year[a]	88 kg	1,584,000
Gasoline first year	215 ℓ	—
Gasoline second year	364 ℓ	—
Gasoline total	579 ℓ	5,853,111
Electricity at 1677 kWh/year[b]	3,354 kWh	9,602,502
Nitrogen (at 180 kg first year, 90 kg second year)	270 kg	3,969,000
Phosphorus	135 kg	405,000
Potassium	135 kg	216,000
Lime	4,493 kg	10,821,166
Plants set[c]	61,503 plants	2,244,860
Insecticides and miticides (17 kg/year)	34 kg	2,089,980
Fungicides (17 kg/year)	34 kg	944,180
Herbicides	20.2 kg	1,267,954
Soil fumigant	280 kg	6,600,440
Clear plastic mulch		
For fumigation	224.5 kg	6,750,000
For plant mulch	258 kg	9,675,000
Transportation	7,008 kg	1,800,979
Total		**63,824,172**
Output		
Strawberry yield first year[d]	47,693 kg	
Strawberry yield second year[d]	47,693 kg	
Strawberry yield total	95,386 kg	35,292,820
Protein yield	688 kg	
kcal output/kcal input		0.55
kcal output/labor hour		4,951

Note: Plants were set in summer, used double-hill raised-bed system with clear polyethylene mulch, were surface irrigated, and were harvested for two seasons (April to September; approximately two thirds of crop shipped fresh and one third processed). Information kindly supplied by Victor Voth of University of California, largely from California Agricultural Extension publications.)

[a] Assumes 8.8 metric tons of machinery which serve 20 ha of strawberries per grower per year and are depreciated over an average of 10 years, not including a yearly airplane custom oil spray.

[b] This amount of electricity needed to pump 75.9 ha-cm per year; manufacturing and installing energy for irrigation pipe not included.

[c] Plants set on 106.7-cm bed (row center to row center) with plants spaced at 30.5 cm in the row and 36.5 kcals per plant produced at low elevation sites with 6 months refrigerated storage.

[d] Based on 1977 California average of 8751 12-pint flats (5.45 kg) per hectare. Using the newest varieties and cultural methods, certain California growers in this area have realized yields as high as 22,651 flats/ha, which would yield 246,896 kg/ha over 2 years and 91,351,483 kcal/ha or have a ratio of kcal output/input of 1.4401.

Table 6
ESTIMATED ENERGY USE FOR SOUTHERN CALIFORNIA STRAWBERRY PRODUCTION

Item	Quantity/ha	kcal/ha
Input		
Labor	6,214.5 hr	—
Machinery[a]	86.5 kg	1,557,000
Gasoline	539 ℓ	5,448,751
Electricity[b]	1,419 kWh	4,062,597
Nitrogen[c]	45 kg	661,500
Phosphorus[c]	22 kg	66,000
Potassium[c]	22 kg	35,200
Plants set[d]	114,812 plants	5,763,562
Insecticides and miticides	17 kg	1,044,990
Fungicides	17 kg	472,090
Soil fumigant[e]	281 kg	6,624,013
Clear plastic		
For fumigation (1 mil)	225 kg	6,750,000
For plant mulch (1.25 mil)	225 kg	8,437,500
Transportation	2,426.5 kg	623,611
Total		**41,546,814**
Output		
Strawberry yield[f]	53,891 kg	19,939,670
Protein yield	377.2 kg	
kcal output/kcal input		0.48
kcal output/labor hour		3,209

Note: Plants were set in winter, used four-hill raised-wide-bed system with clear polyethylene mulch, were surface and trickle irrigated, and were harvested March to July in one season only (approximately four fifths were shipped fresh and one fifth processed). Production information kindly supplied by Victor Voth of University of California.

[a] Assumes 17.3 metric tons of machinery which serve 20 ha of strawberries per grower per year and are depreciated over an average of 10 years.

[b] Amount of electricity needed to surface and trickle irrigate 64.2 ha-cm water/per year (manufacturing and installing energy of irrigation pipe not included).

[c] Slot placement preplant IBDU mix (slow release) at 20-10-10.

[d] Plants set on 152.4-cm bed (row center to row center measure), 4 rows per bed with the plants spaced 22.9 cm apart in the row, and each plant requiring 50.2 kcals to produce at high elevation nurseries with 1 month of cold storage.

[e] A mixture of 2 parts methyl bromide to 1 part chloropicrin.

[f] Yield is 9884 12-pint flats/ha.

Table 7
ESTIMATED ENERGY USE FOR WESTERN OREGON STRAWBERRY PRODUCTION

Item	Quantity/ha	kcal/ha
Input		
Labor — establishment year	35 hr	—
Labor — 3 producing years @ 276 hr	828 hr	—
Labor — total 4 years	863 hr	—
Machinery[a]	26 kg	468,000
Gasoline	498 ℓ	5,034,282
Electricity[b]	860 kWh	2,462,180
Nitrogen	180 kg	2,646,000
Phosphorus[c]	40 kg	120,000
Potassium[c]	84 kg	134,400
Lime[c]	4,493 kg	10,819,144
Plants set	32,370 plants	2,557,230
Insecticides and miticides	18 kg	1,106,460
Fungicides	20 kg	555,400
Herbicides	46 kg	2,887,470
Transportation	5612 kg	1,442,284
Total		**30,120,678**
Output		
Strawberry yield[d]	26,955 kg	8,275,185
Protein yield	189 kg	
kcal output/kcal input		0.27
kcal output/labor hour		9,589

Note: Plants were set in spring and used single-hill culture; field was maintained for 4 years — three fruiting seasons (crop almost entirely processed). Production information kindly supplied by Dr. F. J. Lawrence of U.S. Department of Agriculture and Oregon State University.

[a] Assumes 15.8 metric tons of machinery which serve 8.09 ha of strawberries on a 60.7-ha farm and which are depreciated over an average of 10 years.

[b] Amount of electricity needed to pump 8001 m^3/ha of water over the 4-year period. Manufacturing and installing energy for irrigation pipe not included.

[c] Assuming that soil test showed low to medium range need for these elements.

[d] Average yield of 8985 kg/ha/year (4T/A/year).

Table 7 estimates the energy required to produce strawberries in western Oregon for processing. The energy requirements are very modest, but modest yields keep the output to input ratio (0.27) similar to that of Maryland.

Table 8 estimates the energy required to produce strawberries in Indiana for direct customer sales. In this very modern and technical production scheme, yields and energy

Table 8
ESTIMATED ENERGY USE FOR A CUSTOMER HARVEST STRAWBERRY FARM IN INDIANA

Item	Quantity/ha	kcal/ha
	Input	
Labor — establishment year	238 hr	—
Labor — producing years @ 94 hr/year	283 hr	—
Labor — harvest supervisory @ 900 hr/year[a]	2,700 hr	—
Machinery[b]	71.5 kg	1,287,000
Gasoline	140 ℓ	1,415,260
Diesel	163 ℓ	1,860,482
Electricity (258 kWh/year)[c]	1,032 kWh	2,954,616
Nitrogen (203 kg first year, 124 kg other years)[d]	451 kg	6,629,700
Phosphorus (186 kg first year, 94 kg other years)[d]	374 kg	1,220,000
Potassium (186 kg first year, 94 kg other years)[d]	374 kg	598,400
Plants set	15,382 plants	1,215,178
Cover crop (Sudex)	79 kg	948,000
Straw mulch (4492 kg/year)	13,476 kg	2,695,200
Insecticides and miticides (10 kg/year)	40 kg	2,458,800
Fungicides (3.36 kg/year)	13 kg	361,010
Herbicides (22.46 kg/year)	90 kg	5,649,300
Soil fumigation (methylbromide)	359 kg	8,921,150
Plastic cover for fumigation	224.5 kg	6,735,000
Parking lot seeding (rye and bluegrass)	79 kg	1,841,490
Transportation	16,043 kg	4,123,051
Total		**50,815,637**
	Output	
Strawberry yield[e]	45,666 kg	16,896,542
Protein yield	320 kg	
kcal output/kcal input		0.33
kcal output/labor hour		5,262

Note: Plants were set in spring, used matted-row culture, and were maintained for three fruiting seasons; crop was harvested and purchased by customers driving to the farm. Production information kindly supplied by Philip C. Ahrens of Huntingburg, Ind.

[a] Assumes four field supervisors and one person at check-out point for 18 days/year and 10 hr/day.

[b] Assumes 14.3 metric tons of machinery which serve 10 ha of fruting strawberries and 10 ha of strawberries for plant production and which are depreciated over an average of 10 years.

[c] Electrical power necessary to irrigate, frost protect, and field cool, with an average of 11.7 ha-centimeters of water per year, energy for irrigation equipment manufacture and installation not included.

[d] Includes cover crop fertilizer prior to planting.

[e] First and second year yields of 17,168 kg/ha and third year yield of 11,331 kg/ha.

Table 9
STRAWBERRY PRODUCTION ENERGY REQUIREMENT SUMMARY

Production area	Purpose of production[a]	Number of cropping seasons	Yield (kcal/ha)	Ratio kcal yield/ input	Ratio kcal, yield/ labor hour
Florida	Wholesale (F)	1	8,709,060	0.18	1351
Maryland	Wholesale (F)	3	9,249,630	0.21	1697
California (central)	Wholesale (F,P)	2	35,292,820	0.55	4951
California (southern)	Wholesale (F,P)	1	19,939,670	0.48	3209
Oregon	Wholesale (P)	3	8,275,185	0.27	9589
Indiana	Retail (D)	3	16,896,542	0.33	5261

[a] F, for fresh market; P, for processing market; D, direct farm to consumer sales (pick-your-own).

requirements are both high (output to input ratio = 0.33), so that the result is similar to Oregon.

The tabular examples are summarized in Table 9.

Thus, we see that there are a number of strawberry production energy levels that are highly profitable as long as strawberry demand remains high and there is ample energy available to manufacture the products needed to produce strawberries.

ENERGY USE IN AGRICULTURE; RED RASPBERRY PRODUCTION

Donald K. Ourecky

The red raspberry, once a common fresh fruit in the grocery store, has become a scarce and expensive delicacy. In 1900, the U.S. census reported the red raspberry acreage to be equivalent to 24,662 ha. By 1940, the area of production stood at 23,906 ha. From 1940 to 1960, the area of production decreased 50%. In 1900, the New York State area of production was 5010 ha, with a gradual decline to 3412 ha by 1940; but by 1960 the area of production had decreased 75%. The major reasons for this trend were (1) the lack of good quality virus-indexed planting stock, (2) a change in labor laws restricting the age of youth working in the field, and (3) movement away from a crop with high energy inputs or competition with crops that are more readily mechanized.

The 1969 Census of Agriculture[1] reports the total U.S. red raspberry acreage to be equivalent to 3391 ha. The five leading states, Washington, Oregon, Michigan, New York, and California, report a total of 3165 ha. Production figures on 2607 ha were over 7.5 million kg. The Oregon data[2] for a similar period show that the combined production of Oregon and Washington alone was over 12 million kg.

Without accurate yield data and estimates on area of production, the total market value of red raspberries grown in the U.S. is difficult to compute. The Oregon and Washington data[2] indicated that 95.5% of the total 1969 production, or 12 million kg, was processed. The farm price for processed fruit was 66.7¢/k or $3,240,000 for the two states. A small portion of the total U.S. production is sold fresh. Fresh market prices in the eastern U.S. ranged from $1.48/k on a "U-pick" basis to over $3.00/k retail. The price per kilo output varies greatly, influencing to some extent the type of cultural inputs.

Estimates of various inputs or expenditures of energy[4,5] in the production of red raspberries are presented on a hectare unit. Table 1 is a brief outline of the energy input components required for establishment of a raspberry planting. Raspberries belong to the genus *Rubus,* which also includes such fruits as blackberries, dewberries, loganberries, and boysenberries. Energy inputs are similar for these fruit plants, but, for the purposes of this study, the discussion will be limited to the red raspberry, *Rubus ideaus* L.

Raspberries are a long-term crop compared to strawberries, but short-term compared to tree fruit crops. If a planting is well maintained and kept free from weed, insect, and disease problems, it may remain productive for 10 to 15 years. Therefore, the establishment cost may be prorated for the life of the planting. During the establishment year (Table 1) a herbicide input may be necessary (or optional, depending on the region of production, growing conditions, and length of growing season). A trellis or support system is used in areas where the plants are extremely vigorous and production is high. In areas with less vigor, they are grown free-standing without support.

The fertilizer requirement varies with the region, available moisture, soil, and site. The length of the growing season also determines the number of cultivation inputs. The irrigation input is optional in many plantings, as it depends upon its availability and need. Weed control depends upon the length of the growing season and type of weed problems present. Insect and disease control inputs are highly dependent variables. The red raspberry aphid is the major vector of mosaic viruses, requiring several sprays during the growing season. Cane and foliage disease may also require inputs during the growing season. Prior to and during fruiting, some plantings require sprays for the control of fruit rots and fruit worm. Late season sprays may be required for

Table 1
INPUTS REQUIRED TO ESTABLISH A HECTARE OF RED RASPBERRIES

Item	Quantity/ha	kcal/ha
Input		
Labor	84 hr	—
Machinery	14.1 kg[a]	253,800
Diesel fuel	322 ℓ	3,675,308
Insecticide	1.1 kg	67,617
Fungicide	1.1 kg	30,547
Herbicide	13 kg	1,298,830
Transportation	8713 kg	2,239,241
Nitrogen	18 kg	257,400
Phosphorus	54 kg	112,000
Potassium	18 kg	28,800
Wire	958 kg[b]	246,206
Stakes	7418 kg[b]	34,124,000
Total for one year		**40,094,508**

[a] Energy input for machinery assumed to be similar to grape production in California and Oregon.

[b] Wire and stakes are assumed to be similar to grape production.

control of cane and crown borers. In summary, the length of growing season and availability of moisture determine to a large extent the total number and kinds of energy inputs required each year (Table 2). The annual inputs per hectare to produce red raspberries is presented in Table 2. The output is dependent upon the rate of establishment. For prime areas with irrigation and a long growing season, 2 to 3 years are required to bring a planting into full production. In marginal regions with limiting factors, 3 to 4 years may be required.

Pruning of summer-fruiting raspberries is one of the most time consuming and expensive inputs required in raspberry culture. All canes that have fruited the previous season must be removed. This may be done in the late fall or during the winter months. The current season's canes are topped in early spring for a balanced pruning effect. If supports are used, all canes must be placed within the supports.

The development of fall-bearing raspberries has presented the industry with a new concept in culture and fruiting. Instead of topping, cane removal, and trellising, all fruiting canes are brush-hogged to ground level in late fall, with fruiting occurring the following summer on current season's growth. For control of over-wintering pests, it is recommended that all canes be destroyed in late fall. This system would eliminate the hand labor required for pruning.

The West Coast growing areas are rapidly advancing toward completely mechanized harvest. A mechanized operation may require six people to operate a machine that can harvest 1 ha in 2.5 hr. A total of five harvestings may be required to maximize yield potential. This involves 75 man-hr plus the operation of a mechanical harvester.

In the eastern U.S., the "U-pick" or "pick-your-own"system is extremely popular. In a number of small plantings or roadside marketing operations, hand harvesting of fruit is required. Fruit may be selectively sorted at the time of picking, requiring no postharvest sorting for retailing. A specialty market makes use of stored fruit for shipment to distant markets. With a highly perishable product, this requires special atten-

Table 2
ANNUAL INPUTS PER HECTARE TO PRODUCE RED RASPBERRIES IN FULL PRODUCTION

Item	Quantity/ha	kcal/ha
	Input	
Labor	60 hr	—
Establishment[a]	3,341,209 kcal	3,341,209
Machinery	14.1 kg[b]	253,800
Diesel	332 ℓ	3,785,168
Nitrogen	18 kg	257,400
Phosphorus	54 kg	112,000
Potassium	18 kg	28,800
Insecticide	1.1 kg	67,617
Fungicide	1.1 kg	30,547
Herbicide	3 kg	299,730
Transportation	347 kg	89,179
Total		**8,265,450**
	Output	
Raspberry yield	4940 kg	2,815,800
Protein yield	59.5	
kcal output/kcal input		0.34
kcal output/labor hour		46,930

[a] One twelfth of net establishment costs, Table 1.
[b] Energy input for machinery assumed to be similar to grape production in California and Oregon.

For irrigation
5 cm/month for 4 months = 20 cm
317,652 kcal/ha

tion. For a hand-harvesting operation, approximately 14 people per hectare are required, working 15 days for a total of 210 man-hr.

The output from a raspberry planting varies greatly. West Coast yields are much higher than Eastern yields because of the longer growing season, freedom from high summer temperatures during fruiting, and reduced winter and spring low-temperature injury problems. Raspberry output averages 2360 kg/ha in Washington and Oregon, while prime sites have realized nearly twice this average. Eastern U.S. outputs may vary from 1300 to 2500 kg/ha, depending on limiting factors of site, soil, and water potential.

Regional energy inputs vary in four major areas: (1) most eastern plantings are not trellised or supported — they are grown in a hedgerow self-supporting system, (2) because of the longer growing season in the West, more cultivations may be required, as well as (3) additional applications of insecticides and fungicides, and (4) the methods of harvesting are shifting toward mechanization in the West, while Eastern fruit is mostly hand harvested.

Marginal regions that grow red raspberries are generally located in areas with climatic stresses. With greatly reduced output, the inputs are likewise reduced. The prime areas of production are not found in regions of stress. Prime areas generally have a long growing season, adequate moisture supply throughout the entire growing season, and moderate temperatures for good plant growth and fruit development. Under these conditions, output may vary as much as five- to sixfold.

The energy input for red raspberry production in a dryland vs. an irrigated region are nearly identical to those of a marginal vs. prime growing area. In a dryland operation, an additional year or two may be required for a planting to reach maturity or full production. An irrigated planting should be in full production by the 3rd year. A dryland or irrigated operation is likely to be located in areas that have other climatic stresses, such as low or fluctuating winter temperatures. Output is more or less a reflection of stress.

The energy inputs into red raspberry production under a mechanized vs. an intensive hand-labor system present the widest range in types of input. Most all of the cultural requirements can be done by hand or by the manipulation of hand equipment. An operation requiring hand labor may not be able to provide sufficient fertilizer requirements. Weed, insect, and disease control can be done by making use of a hand-operated sprayer. Output under these systems may vary at least threefold.

SUMMARY

The energy inputs in the production of red raspberries are many and varied. Planting establishment is of major importance, as a decrease in plant vigor or lack of good plant survival may delay full production as much as 2 years. Marginal plantings generally have reduced inputs and outputs. Moisture supply is frequently one of the most limiting factors in raspberry production, as raspberries require a good level of soil moisture during the entire growing season. High and low temperature stresses are also important variables. A prime growing region has a good distribution of soil moisture throughout a long growing season, adequate chilling period, and lack of extreme temperature stress. Southern regions lack an adequate chilling period and are subject to high and fluctuating temperatures. Eastern regions have a shorter growing season, low or fluctuating winter temperatures with high temperatures during the fruiting season, compared to many of the prime growing areas on the West Coast.

Advances are being made in the development of mechanical harvesters in the U.S. and Scotland. Most blackberries and a high percentage of black raspberries are harvested mechanically. Each year there is a significant increase in the volume of red raspberries harvested mechanically. The development of new fruiting types can potentially shift this fruit crop, now requiring expensive hand labor, into a completely mechanized enterprise.

REFERENCES

1. Census of Agriculture, Fruits, Nuts, Berries, Vol. 5 (Part 6), U.S. Department of Agriculture, 1969.
2. Oregon Commodity Data Sheet, Oregon State University Cooperative Extension Service and Crop and Livestock Reporting Service, June 22, 1970.
3. Agricultural Statistics, U.S. Department of Agriculture, 1975.
4. **Funt, R. C.,** An Economic Comparison of Several Small Fruit Production and Harvest Systems in Maryland, 1975, MP 922, Agricultural Experiment Station, University of Maryland, College Park, 1977.
5. **Funt, R. C. and Wildman, S. L.,** Commercial raspberry and blackberry culture in Maryland, HE 40-75, 1976.

BLUEBERRIES

Max E. Austin

Blueberry fruit is produced commercially in many parts of the country, but production is dependent on climatic and edaphic conditions. Several species are native to North America. The most important of these species are lowbush (*Vaccinium augustifolium*, formerly *V. lamarckii)*, Canada *(V. myrtilloides)*, highbush *(V. australe* and *V. corymbosum)*, rabbiteye *(V. ashei)*, dryland *(V. alto-montanum* and *V. pallidum)*, evergreen *(V. ovatum)*, and mountain *(V. membranaceum)*.[1]

The lowbush blueberry, the most important commercial wild species, is harvested from Maine to Minnesota and parts of Canada and southward in the Alleghenies to West Virginia. Commercial plantings of the highbush blueberry are found in southeastern North Carolina, northward to southern Maine and Nova Scotia, west to Michigan, western Oregon, Washington, and British Columbia.[2] The rabbiteye blueberry is grown commercially in northern Florida, northward to Alabama and Georgia.[3]

The lowbush blueberry fruit ripens during July and August and may extend into September in the most northern areas. The fruit is gathered for the canneries with hand-held rakes. For the fresh market, the fruit is harvested by hand as well as with rakes. Lowbush blueberries come naturally into forest areas that have been cut and/or burned. In eastern Maine the fields are reburned every second or third year using oil or liquefied petroleum gas. This kills weeds and underbrush and prunes the blueberry plants.[2] However, the burnings do not get all the weeds and other techniques must be used in attempting to control them. These techniques include hand-pulling, mowing, and applying herbicides.

The early highbush or rabbiteye cultivars are harvested around April 25 in Forida, May 15 in North Carolina, June 15 in Georgia and New Jersey, and July 10 in southern Michigan. Generally, there are three to five harvests in 5- to 7-day intervals with hand harvesting.

A mechanical aid used in harvesting blueberries is a hand-held vibrator that is powered by two 12-V batteries and a catching assembly. Machine harvesting consists of a self-propelled over-row harvester that detaches fruit by striking sides of the plant with steel slapper bars.

Fruit harvested by these two methods must be cleaned and sorted before being sold. Carefully graded fruit harvested with the hand-held vibrator can be marketed fresh as well as for processing. Machine-harvested fruit should not go on the fresh market.

Table 1
ENERGY INPUTS PER HECTARE FOR RABBITEYE BLUEBERRIES AT ALAPAHA, GEORGIA[4-6]

Item	Quantity/ha	kcal/ha
Input		
Labor	363 hr[a]	—
Machinery	84 kg[b]	1,512,000
Gasoline	102.7 ℓ	1,038,194
Electricity	195 kWh[c]	558,285
Nitrogen	101 kg	1,384,490
Phosphorus	15 kg	45,000
Potassium	56 kg	89,600
Plants	1,481[d]	—
Peat moss	70 hℓ[e]	1,400,000
Irrigation (water applied)	109 cm[f]	1,227,923
Herbicides	7.2 kg	623,520
Transportation[g]	221 kg[h]	56,797
Ryeseed	60 kg	720,000
Total		**8,655,809**
Output		
Blueberry yield	10.1 metric tons	6,262,000
Protein yield	293 kg	
kcal output/kcal input		0.72
kcal output/labor hour		17,251

[a] Labor used to harvest, grade, and transport blueberry fruit, apply pesticides, cultivate land, prune plants, and transplant and remove plants.
[b] Trucks, tractors, cultivators, mowers, grader, harvesters.
[c] Electricity estimated to run grader, charge four 12-V batteries for harvesters, fan for cooling grading room, and air conditioner to cool storage room.
[d] Plants set 1.8 m in rows 3.6 m apart.
[e] Energy in production of peat moss is calculated to be 200 kcal/ℓ.
[f] Sprinkle irrigation, pumping water from 10 m head.
[g] Transportation = [machinery (kg) + fuel (kg) + seed (rye) (kg)] × 257 kcal/kg transportation.
[h] Total weight of goods included in transport calculation.

Table 2
ENERGY INPUTS FOR BLUEBERRY PRODUCTION IN NORTH CAROLINA[5,9,10]

Item	Established blueberry quantity/ha	kcal/ha
	Input	
Labor	272 hr	—
Machinery	434 kg	7,812,000
Gasoline	400 ℓ	4,093,600
Diesel	100 ℓ	1,141,400
Nitrogen	45 kg	643,500
Phosphorus	25 kg	75,000
Potassium	37 kg	59,200
Insecticides	56 kg	2,787,120
Fungicides	3.3 kg	214,203
Transportation[a]	819 kg	210,483
Herbicides	2.2 kg	219,802
Total		**17,256,308**
	Output	
Blueberry yield	7,002 kg	4,337,827
Protein yield	49 kg	
kcal output/kcal input		0.25
kcal output/labor hour		15,948

[a]Transportation = [Machinery (kg) + fuel (kg) + seed (if any) (kg)] × 257 kcal/kg.

Table 3
ENERGY INPUTS IN BLUEBERRY PRODUCTION IN NOVA SCOTIA ON A 2-YEAR ROTATION[6]

Item	Quantity/ha[a]	kcal/ha
	Input	
Labor	800 hr	—
Machinery	10 kg	180,000
Diesel	8 ℓ	91,312
Nitrogen	56 kg	800,800
Burning (#2 oil)	702 ℓ	8,012,628
Transportation	614 kg	157,798
Total		**9,242,538**
	Output	
Blueberry yield	1,120 kg	693,854
Protein yield	8 kg	
kcal output/kcal input		0.075
kcal output/labor hour		867

[a]Inputs estimated from data supplied by J. T. MacAulay of Nova Scotia Agricultural College.

REFERENCES

1. **Eck, P. and Childers, N. F.,** *Blueberry Culture,* Rutgers University Press, New Brunswick, New Jersey, 1966.
2. **Scott, D. H., Draper, A. D., and Darrow, G. M.,** Commercial blueberry growing, *U.S. Dep. Agric. Farmers Bull.,* 2254, (revised), 1973.
3. **Brightwell, W. T.,** Rabbiteye blueberries, *Univ. Ga. Agric. Exp. Sta. Bull.,* 100, 1971.
4. **Nelson, J.,** Michigan Blueberry Growers Association, Grand Junction, personal communication.
5. **Mainland, C. M.,** Blueberry Specialist, Horticulture Crop Research Station, Castle Hayne, N.C., personal communication.
6. **MacAulay, J. T.,** Department of Agricultural Engineering, Nova Scotia Agricultural College, Truro, personal communication.
7. **Crocker, T. E.,** Department of Fruit Crops, University of Florida, Gainesville, personal communication.
8. **Moore, J. N.,** Department of Horticulture, University of Arkansas, Fayetteville, personal communication.
9. Proceedings 9th Annual Open House, Southeastern Blueberry Council, Marmora, N.J., Jan. 22, 1976.
10. Planning for Profit, North Carolina Agricultural Extension Service, Circular 518, North Carolina State University, Raleigh.

ENERGY INPUTS IN CRANBERRY PRODUCTION

Alvan R. Brick, Jr.

The cranberry, in many ways the most American of our fruits, was first introduced to western civilization by the pilgrims of Plymouth colony. For untold centuries it had been used by the Amerindian tribes inhabiting the northern reaches of the Atlantic coastal plain as a staple ingredient of pemmican, a mix of dried venison, fat, and cranberries. These early New Englanders adopted the Indian uses of this tart, deep-red berry just as they adopted the uses of maize, both of which were quite unlike anything found in Bristol or Coventry. Its tart flavor, color, and remarkable keeping qualities made the cranberry well worth the trouble of gathering from its scattered bog habitat. A colonial prize, it was sealed in casks of swamp water and sent to England where it was transfered to small water-filled glass bottles and sold for astoundingly high prices.

Yet it wasn't until 1816 that the observant Bay State man, Henry Hall, interested himself in the cultivation of the wild land crop. Thanks to his astute observations, diligence, and interest, we have the cultivated crop which today is extensively grown in five states; which has stimulated an exciting blend of scientific study and grower innovation; and which has just produced a record crop of over 2 million barrels (1 bbl = 100 lb = 45.5 kg).

Careful grower selection has produced a group of highly successful cultivars from the original *Vaccinium macrocarpon* stocks. These out-produce the original vine populations, and are known to have distinctive cultural and market qualities which provide the guides for a promising hybridization program.

The cranberry is an unusual fruit requiring an unusual, highly specialized, and remarkably advanced horticulture. This trailing swamp vine presents unique opportunities to and imposes unique restraints on the grower. It demands an acid soil mix of sand and peat (pH 4 to 5), and grows in natural forest pockets. It is easily damaged by trampling or improper drainage. It has a specialized pest complex which is well adapted to normal conditions of cranberry habitat. These factors make cranberry growing a difficult and high risk occupation. High capital investment, the ever-present risk of complete crop loss, the necessity for a highly sophisticated scheme of water management, coupled with the current depressed condition of the cranberry market, have driven all but the most dedicated and capable growers from this risky enterprise.

Only five states, within the U.S., and Nova Scotia account for the world's entire commercial cranberry crop. The early and sustained efforts in Massachusetts have made it the leading producer and the model of the industry with a record 1976 crop of 935,000 bbl grown on 4413 ha. New Jersey followed by quickly expanding its growing area after the Civil War and for many years occupied second place, but has since fallen to third despite a dramatic increase in yield per acre. In 1976 the Jersey crop was 276,000 bbl and was grown on 1255 ha. Wisconsin is the strapping young giant in the cranberry industry. Heady with its own youth and benefiting from its own major innovation "harvesting on the flood", it has grown since World War II to become the second largest producer. In 1976 the Wisconsin growers produced 1,004,000 bbl on 2834 ha and for the first time surpassed Massachusetts. On the West Coast, Washington and Oregon fill out the roster of cranberry states with 1976 crops of 103,000 and 89,300 bbl respectively, grown on 445 and 364 ha. West Coast growers benefiting from mild climate, natural peat bogs, intensive management, and several outstanding varieties (most notably the Crowley) have consistently led the nation in yield per hectare.

Interestingly, this region boasts the smallest commercial bogs in existence. Some are as small as 1/8 of an acre (0.05 ha).

Unlike most other crop systems, cranberry growing is a form of swamp agriculture. Only blueberries, mint, and rice can claim this distinction. In fact, only the colonial rice production of 18th century South Carolina provides a comparable agricultural model in terms of scale, water management structures and regimes, pest control practices, and, until very recently, labor intensity. Successful water management is the *sine qua non* of cranberry culture and the cultural center around which all else revolves. Bog (Wisconsin marsh) sites are chosen with respect to air drainage, water drainage, and water availability. The entire complex establishment of reservoirs, dikes, ditches, trunks (sluiceways), and bogs proper has only one aim: the precise control of water distribution. Cranberry properties are most usefully distinguished and classified by their water management systems. Their long term success is most accurately predicted by the successes of these systems.

The major growing regions differ profoundly in primary water control patterns. Massachusetts and New Jersey being most traditional are distinguished by gravity dependent flowage systems where water is diverted directly from a surface water course as needed, or more commonly stored in a system of reservoirs until needed and flowed by gravity through an intricate system of canals and ditches. Control is achieved by a series of gates or flumes and trunks, which are most often made of wood. A property with 81 acres set to vines may employ as many as 100 such devices. A number of growers have increased their control over flow rates by introducing high capacity pumps (up to 94,750 lpm). Pumping has been common practice in Wisconsin for years due to the greater danger of frost in this region's continental climate. There is actually no absolutely frost-free period of the year in Wisconsin. As many as 30 frost nights have been reported during the summer. The availability of cheap electricity and the ever-present threat of frost stimulated Wisconsin growers to depend on pumping systems; thus opening additional sites to cranberry culture. The pratice of re-using flood waters by draining them back into low-lying, pump-equipped reservoirs provided flexibility and conservation. Currently the West Coast system of bog protection has spread throughout the cranberry industry.

The mild, nearly winterless climate of coastal Washington and Oregon made the traditional winter submergence of entire bogs (the winter flood) unnecessary. These growers needed protection from occasional frosty nights, but by no means needed the extensive reservoirs and complex water control systems of the East. Solid-set irrigation met the need. Currently this system is being rapidly installed in the East, particularly in Massachusetts. It provides effective, nearly instant, frost protection and irrigation. It conserves water and is an effective tool for the distribution of soluble fertilizers and pesticides. Only a few of the larger New Jersey growers have felt the investment in such a system to be justified. Wisconsin growers, fearing that freeze damage to pvc and polyethylene components would result from poor drainage and harsh Wisconsin winters, prefer moveable sprinkler systems.

Since the early part of this century, and accelerating after World War II, two major areas of innovation have caused the increased productivity, challenged the ingenuity of cranberry growers, and altered the energy use patterns in cranberry culture. The first of these was the successive improvement in the control of crop pests. Beginning with the now almost incredible use of up to 6064 l/acre of water-white kerosene for control of grasses and sedges, the use of arsenicals for insect control and Bordeaux mixture for fruit rots, pest control progressed through increasingly effective and benign measures to the currently used Casoron, Malathion, and Ferbam, and the development of integrated pest management programs.

The improvement of chemical controls was accompanied by an improvement in means of distribution which was just one example from the second area of innovation — mechanization.

It is difficult to believe that in less than 40 years all aspects of cranberry culture have been altered by mechanization. For example, four major harvest systems have successively replaced one another in that time. Hand picking is no longer practiced. It was completely superceeded by the use of the Jersey scoop or rocky dump. Immediately following World War II, mechanical harvesters were developed. Operated by one man and resembling large lawn mowers, these machines raked or paddled the berries off the vines and lifted them by conveyor into a waiting container. One even pruned the vines as it traveled. Damage to vines and fewer dropped berries improved productivity. Machines developed in Wisconsin followed the tradition of the Wisconsin water rake (the water borne analog of the Jersey scoop) by harvesting on the flood. The astounding increase in yields (often double) accompanying this method interested the entire industry in water raking. Soon the West Coast pioneered a light weight, inexpensive, and effective means of water harvest. Known as the *egg beater* and aptly described as a "cross between a lawn mower and a baby carriage," this device is being used more each year, and with it the reliance on water harvest is increasing.

Interestingly, this brings us full circle. The introduction of the West Coast egg beater has demanded modification of the water management systems. Bogs must be level, or nearly so, for effective use of the new methods, but original eastern deep bottomed bogs often varied by 10 ft or more. The doubling of yield which it promises, together with the increased dependence on and opportunities to market process fruit, has led growers to radically rebuild water control structures and abandon the difficulties and expense of dry harvest for the fresh market. All of these changes influence the energy effectiveness of cranberry culture and have implications for energy flows.

Water management, pest control, and mechanization make energy use in cranberry culture a rich and complex subject. Consider just one example: the switch to water harvest has the following consequences: the retooling of the cranberry property, the redesign of basic water control patterns, the switch from fresh to process (thus high energy) production. Formerly all cranberries were air-dried and stored in cranberry houses or screen houses without temperature controls. Today, water-harvested berries must be immediately dried, sorted, and quick frozen. They then must be kept that way until used.

Curiously, the remarkable keeping quality, which made the cranberry so valuable to the Pequots, the Leni-Lenape, and the frugal pilgrims, and which allowed its introduction to European markets, has been completely factored out of current cranberry economics and energetics.

Has the doubling of yield justified the increased energy costs? No one yet knows. At current energy prices and under present market conditions, it seems to be.

Today cranberry growers can boast of crops exceeding 494 bbl/ha; with occasional areas reporting 741 bbl/ha. A Massachusetts crop of 98.8 bbl/ha was outstanding 30 years ago and a typical New Jersey crop was 49.4 bbl/ha. Today cranberry growing is a highly mechanized and efficient form of horticulture, requiring less than one third the man-hours per acre required just 20 years ago. Innovation has reduced drudgery, and increased yields and crop reliability. To be sure, this has involved an increased use of fossil fuels, but overall the scale of innovation has tended toward efficiency, rather than excess. Cranberry culture is still low-powered agriculture. Tractors are small and engines low horsepower.

Perhaps most impressive scientifically are the programs of the new Cranberry and Blueberry Research lab in Chatsworth, N.J. This carefully planned and executed facility centers on 20 replicate 1-acre bogs where all aspects of cranberry growing (including energy relations) can be experimentally manipulated. The current study of cranberry pest management being carried out there may be the most sophisticated of its kind for any crop in the U.S.

Perhaps more important than the facility is the philosophy which governs it, one which has pervaded the cranberry industry for many years: namely, that cranberry growing is an important, difficult, but enjoyable enterprise which can only be successful if dedicated growers and skilled scientists share their resources in an atmosphere of mutual respect, friendship, and good humor.

Table 1
ENERGY INPUTS PER HECTARE PER YEAR FOR A 16-ha CRANBERRY PROPERTY IN NEW JERSEY OR MASSACHUSETTS (WINTER FLOWAGE, DIESEL POWERED SOLID SET IRRIGATION, AERIAL APPLICATIONS, WATER HARVEST)

Item	Quantity/ha	kcal/ha	Ref.
	Input		
Labor	287.36 hr[a]	—	
Machinery (less irrigation)	60.64 kg[b]	1,091,520	
Irrigation equipment	18.65 kg	412,770	13, 14
Gasoline	36.0 ℓ[c]	363,924	
Aviation fuel	6.05 ℓ[d]	60,515	
Diesel fuel (pump only)	145.7 ℓ	1,662,837.2	14
Nitrogen	22.45	330,015	15
Phophorus	44.9	103,270	15
Potassium	22.45	40,410	15
Flooding and Irrigation (water)	67.1-173.7 cm[e]	—	
Insecticides	15.43 kg	562,114.9	16
Herbicides	7.8 kg	444,600	17
Fungicides	20.20 kg	308,050	16
Sand	30.52 m³	?	12
Pollination	2.47 hives	?	18
Transportation	371.38 kg[f]	95,434.4	
Total		**5,474,460.5**	
	Output		
Cranberry yield	9,363.5 kg (83.4 bbl)	4,307,230.9	
Protein yield	33.73 kg		
kcal output/kcal input		.787	

[a] Calculated from unpublished grower records and numerous published accounts in *Cranberries* Vols. 15—45, 1948—1978.

[b] Equipment weights listed in Tables 5-8. These data were assembled from manufacturers brochures, articles in *Cranberries,* Vols. 15—45, 1948—1978, the *National Farm Tractor and Implement Blue Book,* actual field measurements, and grower estimates.

[c] Calculated from unpublished grower records.

[d] Fuel use in aerial applications supplied by David Pimentel.

[e] Calculated from information on average flood depths, rates and quantities of water used by irrigation systems, the annual number of frost warnings and the methods of application typical of each geographical region. For example, in New Jersey a gravity powered system which is dry harvested may use the following quantities: winter flood of 15.24 cm, two spring and two fall floods of 15.24 cm each for a total of 76.22 ha cm. Deeper bogs may demand as much as 182.88 ha cm for the same number of floods. A Washington solid set irrigation system which is water harvested may use the following quantities: no winter flood, ten to twelve, 5 hr uses of 0.254 cm/hr both in the spring and the fall using in all 30.48 cm, and commonly nine weekly applications of 2.54 cm each for summer irrigation. An additional 15.25 cm is required for water harvest. The total annual water requirement of such a system is 68.58 ha cm.

[f] Total weight of machinery, fuel, fertilizers, seeds, and pesticides transported to crop hectare.

[g] Control measures suggested for Mass.

Table 2
ENERGY INPUTS PER HECTARE PER YEAR FOR A 16-HA CRANBERRY PROPERTY IN NEW JERSEY (ALL GRAVITY FLOWAGE, AERIAL APPLICATIONS, DRY HARVESTED)

Item	Quantity/ha	kcal/ha	Ref.
	Input		
Labor	247.1 hr[a]	—	
Machinery (less irrigation)	53.53 kg[b]	963,540	
Gasoline	46.97 ℓ[c]	474,838.43	
Aviation fuel	6.05 ℓ[d]	60,515	
Diesel	—	—	
Electricity	—	—	
Nitrogen	22.45 kg	330,015	12
Phosphorus	67.35 kg	154,905	12
Potassium	22.45 kg	40,410	12
Flooding and Irrigation (water)	76.2-182.88 cm[e]	—	
Insecticides	15.43 kg	562,114.9	19
Herbicides	7.8 kg	444,600	19
Fungicides	20.20 kg	308,050	19
Sand	—	—	
Pollination	2.47 hives	?	18
Transportation	244.27 kg[f]	62,777.39	
Total		**3,401,765.72**	
	Output		
Cranberry yield	5,613.64 kg (50 bbl)	2,582,274.4	
Protein yield	22.45 kg		
kcal output/kcal input		0.76	

[a] Calculated from unpublished grower records and numerous published accounts in *Cranberries* Vols. 15—45, 1948—1978.

[b] Equipment weights listed in Tables 5-8. These data were assembled from manufacturers brochures, articles in *Cranberries,* Vols. 15—45, 1948—1978, the *National Farm Tractor and Implement Blue Book,* actual field measurements, and grower estimates.

[c] Calculated from unpublished grower records.

[d] Fuel use in aerial applications supplied by David Pimentel.

[e] Calculated from information on average flood depths, rates and quantities of water used by irrigation systems, the annual number of frost warnings and the methods of application typical of each geographical region. For example, in New Jersey a gravity powered system which is dry harvested may use the following quantities: winter flood of 15.24 cm, two spring and two fall floods of 15.24 cm each for a total of 76.22 ha cm. Deeper bogs may demand as much as 182.88 ha cm. for the same number of floods. A Washington solid set irrigation system which is water harvested may use the following quantities: no winter flood, ten to twelve, 5 hr uses of 0.254 cm/hr both in sprng and the fall using in all 30.48 cm, and commonly nine weekly applications of 2.54 cm each for summer irrigation. An additional 15.25 cm is required for water harvest. The total annual water requirement of such a system is 68.58 ha cm.

[f] Total weight of machinery, fuel, fertilizers, seeds, and pesticides transported to crop hectare.

[g] Control measures suggested for Mass.

Table 3
ENERGY INPUTS PER HECTARE PER YEAR FOR A 16-ha CRANBERRY PROPERTY IN WISCONSIN (WINTER FLOWAGE, ELECTRICALLY POWERED, MOVEABLE IRRIGATION SYSTEM, AERIAL APPLICATION, WATER HARVEST)

Item	Quantity/ha	kcal/ha	Ref.
	Input		
Labor	263.35 hr[a]	—	
Machinery (less irrigation)	66.09 kg[b]	1,189,620	
Irrigation equipment	29.39 kg	422,100	13, 14
Gasoline	40.12 ℓ[c]	405,583.2	
Aviation fuel	6.05 ℓ[d]	60,515	
Electricity (only pump)	362.12 kWh	311,061.08	14
Nitrogen	44.9 kg	660,030	12, 20
Phosphorus	22.45 kg	51,625	12, 20
Potassium	22.45 kg	40,410	12, 20
Flooding and Irrigation (water)	67.06-97.54 cm[e]	—	
Insecticides	15.43 kg[f]	562,114.9	
Herbicides	7.8 kg[f]	444,600	
Fungicides	30.31 kg[f]	462,180.2	
Sand	30.52 m^3	?	12
Pollination	1.235 hives	?	21
Transportation	263.47 kg[g]	67,711.79	
Total		**4,677,561**	
	Output		
Cranberry yield	13,899.36 kg (123.8 bbl)	6,393,705.6	
Protein yield	55.60 kg		
kcal output/kcal input		1.37	

[a] Calculated from unpublished grower records and numerous published accounts in *Cranberries* Vols. 15—45, 1948—1978.

[b] Equipment weights listed in Tables 5-8. These data were assembled from manufacturers brochures, articles in *Cranberries,* Vols. 15—45, 1948—1978, the *National Farm Tractor and Implement Blue Bock,* actual field measurements, and grower estimates.

[c] Calculated from unpublished grower records.

[d] Fuel use in aerial applications supplied by David Pimentel.

[e] Calculated from information on average flood depths, rates and quantities of water used by irrigation systems, the annual number of frost warnings and the methods of application typical of each geographical region. For example, in New Jersey a gravity powered system which is dry harvested may use the following quantities: winter flood of 15.24 cm, two spring and two fall floods of 15.24 cm each for a total of 76.22 ha cm. Deeper bogs may demand as much as 182.88 ha cm. for the same number of floods. A Washington solid set irrigation system which is water harvested may use the following quantities: No winter flood, ten to twelve, 5 hr uses of 0.254 cm/hr both in the spring and the fall using in all 30.48 cm, and commonly nine weekly applications of 2.54 cm each for summer irrigation. An addtional 15.25 cm is required for water harvest. The total annual water requirement of such a system is 68.58 ha cm.

[f] Control measures suggested for Mass.

[g] Total weight of machinery, fuel, fertilizers, seeds, and pesticides transported to crop hectare.

Table 4
ENERGY INPUTS PER HECTARE PER YEAR FOR A 16-ha CRANBERRY PROPERTY IN WASHINGTON (FLOODED BY ELECTRICALLY POWERED PUMP, SOLID SET IRRIGATION SYSTEM, SPRINKLER APPLICATION, WATER HARVEST)

Item	Quantity/ha	kcal/ha	Ref.
	Input		
Labor	263.35 hr[a]	—	
Machinery (less irrigation)	58.65 kg[b]	1,055,700	
Irrigation equipment	18.65 kg	412,770	13, 14
Gasoline	36 ℓ[c]	363,623	
Electricity	301.8 kWh	259,205	14
Nitrogen	22.45 kg	330,015	15
Phosphorus	44.9 kg	103,270	15
Potassium	22.45 kg	40,410	15
Flooding and Irrigation (water)	68.58 cm[e]	—	
Insecticides	15.43 kg[f]	562,114.9	
Herbicides	7.8 kg[f]	444,600	
Fungicides	20.20 kg[f]	308,050	
Sand	30.52 m^3	?	12
Pollination	2.47 hives	?	18
Transportation	238.3 kg[g]	61,250.81	
Total		**3,193,598.7**	
	Output		
Cranberry yield	17,290 kg (154 bbl)	7,953,400	
Protein yield	69.16kg		
kcal output/kcal input		2.49	

[a] Calculated from unpublished grower records and numerous published accounts in *Cranberries* Vols. 15—45, 1948—1978.

[b] Equipment weights listed in Tables 5-8. These data were assembled from manufacturers brochures, articles in *Cranberries,* Vols. 15—45, 1948—1978, the *National Farm Tractor and Implement Blue Book,* actual field measurements, and grower estimates.

[c] Calculated from unpublished grower records.

[d] Fuel use in aerial applications supplied by David Pimentel.

[e] Calculated from information on average flood depths, rates and quantities of water used by irrigation systems, the annual number of frost warnings and the methods of application typical of each geographical region. For example, in New Jersey a gravity powered system which is dry harvested may use the following quantities: winter flood of 15.24 cm, two spring and two fall floods of 15.24 cm each for a total of 76.22 ha cm. Deeper bogs may demand as much as 182.88 ha cm. for the same number of floods. A Washington solid set irrigation system which is water harvested may use the following quantities: no winter flood, ten to twelve, 5 hr uses of 0.254 cm/hr both in the spring and the fall using in all 30.48 cm, and commonly nine weekly applications of 2.54 cm each for summer irrigation. An addtional 15.25 cm is required for water harvest. The total annual water requirement of such a system is 68.58 ha cm.

[f] Control measures suggested for Mass.

[g] Total weight of machinery, fuel, fertilizers, seeds, and pesticides transported to crop hectare.

Table 5
EQUIPMENT FOR A 16-ha CRANBERRY PROPERTY IN MASSACHUSETTS OR NEW JERSEY

Implements	Gross weight (kg)	No./16 ha	Life (years)	kg/ha/year
Pump (diesel)	245.5	2	12	2.58
Pipe				
Main (alum.)	5,188.8	—	20	16.03
Lateral (poly.)	68.0	—	10	0.17
Heads	278.12	340	7	2.44
Trucks				
1½ ton	3,182	2	30	13.09
Dump	3,636	1	30	7.48
Pickup	2,045	2	15	17.29
Jalopy	454.5	2	30	1.98
Field car	1,363.6	1	15	5.75
Tractor	1,420.0	1	30	2.91
Water reel	56.8	6	20	1.16
Bog conveyor	363.64	1	20	1.11
Clipper	56.8	3	20	0.52
Front-end loader (attach.)	80	1	7	0.74
Fork lift (attach.)	281.8	1	12	1.53
Brush "hog"	368.2	1	12	1.90
Disc	300	1	20	0.93
Hand tools	100	—	10	0.62
Pruner	363	1	20	1.12
Total				**79.29**

Table 6
EQUIPMENT FOR 16-ha MINIMUM INPUT, CRANBERRY PROPERTY IN NEW JERSEY (DRY HARVEST) (TABLE 2)

Implements	Gross weight (kg)	No./16 ha	Life (years)	kg/ha/year
Trucks				
1½ ton	3,182	2	30	13.09
Dump	3,636	1	30	7.48
Pickup	2,045	2	15	17.29
Jalopy	454.5	2	30	1.98
Field car	1,363.6	1	15	5.75
Tractor	1,420.0	1	30	2.91
Western picker	90.9	3	20	0.84
Front-end loader (attach.)	80	1	7	0.74
Brush "hog"	368.2	1	12	1.90
Disc	300	1	20	0.93
Hand tools	100	—	10	0.62
Total				**53.53**

Table 7
EQUIPMENT FOR A 16-ha CRANBERRY PROPERTY IN WISCONSIN (TABLE 3)

Implements	Gross weight (kg)	No./16 ha	Life (years)	kg/ha/year
Pump (electric)	200	2	15	1.65
Pipe				
Main (alum.)	5,188.8	—	20	16.03
Lateral (Alum.)	176.78	—	20	10.92
Heads	278.12	340	7	2.44
Trucks				
1½ ton	3,182	2	30	13.09
Dump	3,636	1	30	7.48
Pickup	2,045	2	15	17.29
Jalopy	454.5	2	30	1.98
Field car	1,363.6	1	15	5.75
Tractor	1,420.0	1	30	2.91
Dana Getsinger Picker	420.4	2	20	2.57
Picker Boats	18.2	6	30	0.22
Boat lift	3,362	1	30	6.91
Clipper	170.4	1	20	0.52
Front-end loader (attach.)	80	1	7	0.74
Fork lift (attach.)	281.8	1	12	1.53
Brush "hog"	368.2	1	12	1.90
Disc	300	1	20	0.93
Hand tools	100	—	10	0.62
Total				**95.48**

Table 8
EQUIPMENT FOR A 16-ha CRANBERRY PROPERTY IN WASHINGTON (TABLE 4)

Implements	Gross weight (kg)	No./16 ha	Life (years)	kg/ha/year
Pump (electric)	200	2	15	1.65
Pipe				
Main (alum.)	5,188.8	—	20	16.03
Lateral (poly.)	68.0	—	10	0.17
Heads	278.12	340	7	2.44
Trucks				
1½ ton	3,182	2	30	13.09
Dump	3,636	1	30	7.48
Pickup	2,045	2	15	17.29
Jalopy	454.5	2	30	1.98
Field car	1,363.6	1	15	5.75
Tractor	1,420.0	1	30	2.91
Water reel	56.8	6	20	1.16
Bog conveyor	363.64	1	20	1.11
Clipper	56.8	3	20	0.52
Front-end loader (attach.)	80	1	7	0.74
Fork lift (attach.)	281.8	1	12	1.53
Brush "hog"	368.2	1	12	1.90
Disc	300	1	20	0.93
Hand tools	100	—	10	0.62
Total				77.3

REFERENCES

1. **Beaton, G. T.,** The cranberry story (I,II,III,IV), *Cranberries,* 23(11,12), 16, 24(1,2), 10,14, 1959.
2. **Beckwith, C. S.,** Cranberry growing in New Jersey, *N.J. Agric. Exp. Stu. Cir. 246,* 1931.
3. **Chandler, F. B.,** Cranberries in North America, *Cranberries,* 23(5,7,8,10,12), 7,7,10,17,9; 24(2), 18, 1958, 1959.
4. *Cranberries,* 15—45, 1948—1978.
5. **Crowley, D. J.,** Cranberry growing in Washington, *Wash. Agric. Exp. Stu. Bull.,* 554, 1954.
6. **Dana, M. N. and Klingbeil, G. C.,** Cranberry growing in Wisconsin, *Univ. Wis. Coll. Agric. Cir., 654,* 1966.
7. **Franklin, H. J.,** Cranberry growing in Massachusetts, *Mass. Agric. Exp. Stu. Bull.,* 447, 1946.
8. **Hall, C. J.,** Our West Coast cranberry growers, *Cranberries,* 28(4), 9, 1963.
9. **Norton, J. S.,** Cultural practices and mechanization of cranberry and wild blueberry, *Cranberries,* 34(1,4), 14, 12, 1969.
10. **Shoemaker, J. S.,** *Small Fruit Culture,* 4th ed., AVI, Westport, Conn., 1975.
11. **Stevens, C. D., Piper, W. E., Franklin, H. J., and Chandler, F. B.,** The cranberry industry in Massachusetts, *Mass. Dep. Agric. Bull.,* 139, 1948.
12. **Norton, J. S.,** Cultural practices and mechanization of cranberry and wild blueberry, Part I, *Cranberries,* 33(12), 14, 1969.
13. **Norton, J. S.,** Sprinkler frost protection, II, *Cranberries,* 33(5), 8, 1968.
14. **Batty, J. C., Keller, J.,** Energy requirements for irrigation, in *Handbook of Energy Utilization in Agriculture,* Pimental, D., Ed., CRC Press, Boca Raton, 1980.
15. **Demoranville, I. E.,** Cranberry fertilizer chart, Extension Service, University of Massachusetts, Cranberry Experimental Station, East Wareham, 1976.
16. **Tomlinson, W. E., Jr., Zuckerman, B. M., and Demoranville, I. E.,** Cranberry insect and disease control chart, Extension Service University of Massachusetts, *Cranberries,* 40(11), 11, 1976.
17. **Demoranville, I. E. and Devlin, R. M.,** 1976 Cranberry weed control chart, Extension Service, University of Massachusetts; Published in *Cranberries,* 40(11) 10, 1976.
18. **Marucci, P.,** Cranberry Bulletin, Cooperative Extension Service, Rutgers University, New Brunswick, N.J., 1978.
19. **Marucci, P.,** Blueberry and Cranberry Insect, Disease and Weed Control, notes distributed by the Blueberry and Cranberry Research Laboratory, Chatsworth, New Jersey, 1978.
20. **Dana, M. N. and Klingbeil, G. C.,** Cranberry growing in Wisconsin, University of Wisconsin College Agriculture Circular 654, 1966.
21. **Sheomaker, J. S.,** *Small Fruit Culture,* 4th ed., A.V.I., Westport, Conn., 1975.

ENERGY REQUIREMENT FOR PECAN, *CARYA ILLINOENSIS* (WANG) K. KOCH

Darrell Sparks

Pecan production is limited primarily to the U.S., although plantings are being established in Brazil, Mexico, Australia, Israel, and South Africa. Generally, the export market has not been developed. The average annual production of pecan nuts in the U.S. is about 90 millon kg with a value of approximately $90 million.[1] Georgia leads the U.S. production with about 32 million kg/yr, followed by Texas with about 20 million kg. Alabama, Arkansas, Arizona, Florida, Louisiana, Mississippi, New Mexico, North Carolina, Oklahoma, and South Carolina produce the remainder of the production.[2] New plantings are expanding rapidly in Georgia, Texas, Arizona, and New Mexico. These plantings are expected to substantially increase total production within the next 10 to 15 years.

Historically, the pecan-producing region of the U.S. has been divided into the western (Texas and westward) and the eastern (Louisiana and eastward) belts. The division is based primarily on climatic and soil differences. The eastern climate is humid and the soils are residual, highly leached, and low in fertility. In contrast, the western climate is arid and the soils are often alluvial, young, and fertile. Generally, pecan production problems based on climate differences are encountered in the east more frequently than in the west. This results in more energy inputs per hectare. In the humid southeast, fruit and foliage diseases, especially scab (*Fusicladium effusum* Wint.), are a serious problem and frequent fungicide applications are a necessity, whereas in the arid west, diseases are a minor problem and less fungicide is required. In the east, nutritional deficiencies occur frequently, and heavy applications of fertilizer and lime are essential. In the west, lime is generally not needed, and supplemental nutrients are required in lesser amounts.[3] In much of Texas and most of New Mexico and Arizona, irrigation is a necessity, whereas in the east, irrigation is a supplement to rainfall and, in some seasons, irrigation effects are minimal.

Pecans are produced in either seedling or cultivar groves. Seedling groves are pecan stands or forests which have been thinned to remove trees with undesirable nut characteristics or to improve sunlight exposure. Often such groves receive little energy input, except for harvesting. This is because seedling groves are used primarily for livestock pasture and any nut production is considered "free goods". Yields from seedling groves are characteristically low. In addition, nut size is small with a low kernel yield of about 35% of the nut weight. Seedling groves are limited primarily to Texas, Louisiana, Mississippi, Oklahoma, and Arkansas, i.e., to their native range. In these states, 50% or more of the production comes from seedling groves.[1]

Cultivar groves are usually planted solely for nut production. The cultivars have been selected because they are precocious and prolific and have large nut size and high kernel yield (50% or more). Because such groves are being managed only for nut production, the cultural or energy inputs are more intense than for seedling groves.

The minimal economical production unit for pecan is about 160 ha. If good cultural practices are employed, a commercial yield may be expected during the sixth growing season. The current cost of getting a single hectare into commercial production is about $1300. Consequently, pecan production is a long-term and expensive investment.[2]

Energy output is less than energy input for pecan regardless of cultural system employed or geographic location (Tables 1—5). Energy output/input is greatest in seedling groves. Under dry land conditions, kcal output/kcal input is slightly greater in Texas than in Georgia (Table 2 vs. 4). The greater yield under dry land condition in

Table 1
ENERGY INPUTS AND OUTPUTS PER HECTARE FOR PECAN NUTS (IRRIGATED) IN SOUTH GEORGIA, CULTIVAR GROVE

Item	Quantity/ha	kcal/ha
Input		
Labor	49.4 hr	—
Machinery	35.9 kg[a]	646,200
Gasoline	93.5 ℓ[b]	945,191
Diesel	243.2 ℓ[b]	2,775,885
Electricity	98.9 kWh[c]	282,864
Nitrogen	336.2 kg[d]	4,034,400
Phosphorus	28.9 kg[d]	86,700
Potassium	279.1 kg[d]	437,130
Zinc	12.1 kg[d]	64,033
Lime	4483 kg[d]	1,414,162
Irrigation (water applied)	94.1 cm[e]	16,796,450
Insecticides	11.2 kg[d]	172,480
Herbicides	1.1 kg[d]	56,575
Fungicides	13.4 kg[d]	140,030
Transportation	312.7 kg[f]	80,364
Total		**27,932,464**
Output		
Pecan nut yield	784.5 kg[g]	5,389,515[h]
Protein yield	72.2kg	
kcal output/kcal input		0.19
kcal output/labor hour		109,099

[a] Assumes 45.4 metric tons of machinery (tractors, sprayers, mowers, trucks, etc.) were used for production and this machinery served an average of 240 hectares and was depreciated over a 5 year period.

[b] Gasoline and diesel use for operation of tractors, sprayers, trucks, harvester, sweepers, and tree shaker.

[c] Electricity for residence and service buildings.

[d] Based on data for Georgia.

[e] Average depth of water applied to irrigated pecan; 50 m pumping lift; solid set sprinkler.

[f] Total weight of machinery and fuel transported to crop hectare.

[g] Assumes a kernel yield of 50%.

[h] Assumes 6870 kcal/kg of kernel.[10]

Table 2
ENERGY INPUTS AND OUTPUTS PER HECTARE FOR PECAN NUTS (NON-IRRIGATED) IN SOUTH GEORGIA, CULTIVAR GROVE

Item	Quantity/ha	kcal/ha
Input		
Labor	49.4 hr	—
Machinery	35.9 kg[a]	646,200
Gasoline	93.5 ℓ[b]	945,191
Diesel	243.2 ℓ[b]	2,775,885
Electricity	98.8 kWh[c]	282,864
Nitrogen	224.2 kg[d]	2,690,400
Phosphorus	28.9 kg[d]	86,700
Potassium	186.1 kg[d]	291,420
Zinc	12.1 kg[d]	64,083
Lime	3362 kg[d]	1,060,543
Insecticides	11.2 kg[d]	172,480
Herbicides	1.1 kg[d]	56,575
Fungicides	13.4 kg[d]	140,030
Transportation	312.7 kg[e]	80,364
Total		**9,292,735**
Output		
Pecan nut yield	560.5 kg[f]	3,850,653[g]
Protein yield	51.6 kg	
kcal output/kcal input		0.41
kcal output/labor hour		77,948

[a] Assumes 45.4 metric tons of machinery (tractors, sprayers, mowers, trucks, etc.) were used for production and this machinery served an average of 240 hectares and was depreciated over a 5 year period.

[b] Gasoline and diesel use for operation of tractors, sprayers, trucks, harvester, sweepers, and tree shaker.

[c] Electricity for residence and service buildings.

[d] Based on data for Georgia.

[e] Total weight of machinery and fuel transported to crop hectare.

[f] Assumes a kernel yield of 50%.

[g] Assumes 6870 kcal/kg of kernel.[10]

Table 3
ENERGY INPUTS AND OUTPUTS PER HECTARE FOR PECAN NUTS (IRRIGATED) IN CENTRAL TEXAS, CULTIVAR GROVE

Item	Quantity/ha	kcal/ha
Input		
Labor	49.4 hr	—
Machinery	35.9 kg[a]	646,200
Gasoline	80.4 ℓ[b]	812,764
Diesel	159.2 ℓ[b]	1,817,109
Electricity	98.8 kWh[c]	282,864
Nitrogen	112.1 kg[d]	1,345,200
Zinc	8.1 kg[d]	42,865
Irrigation (water applied)	125.5 cm[e]	1,065,914
Insecticides	5.6 kg[d]	486,696
Herbicides	1.1 kg[d]	109,901
Fungicides	1.7 kg[d]	110,347
Transportation	231.5 kg[f]	59,496
Total		**6,779,356**
Output		
Pecan nut yield	506.5 kg[g]	3,850,653[h]
Protein yield	51.6 kg	
kcal output/kcal input		0.57
kcal output/labor hour		77,948

[a] Assumes 45.4 metric tons of machinery (tractors, sprayers, mowers, trucks, etc.) were used for production and this machinery serves an average of 240 hectares and was depreciated over a 5 year period.
[b] Gasoline and diesel use for operation of tractors, sprayers, trucks, harvester, sweepers, and tree shaker.
[c] Electricity for residence and service buildings.
[d] Based on data for Texas.
[e] Flood irrigation with concrete irrigation canals.
[f] Total weight of machinery and fuel transported to crop hectare.
[g] Assumes a kernel yield of 50%.
[h] Assumes 6870 kcal/kg of kernel.[10]

Table 4
ENERGY INPUTS AND OUTPUTS PER HECTARE FOR PECAN NUTS (NON-IRRIGATED) IN CENTRAL TEXAS, CULTIVAR GROVE

Item	Quantity/ha	kcal/ha
	Input	
Labor	49.4 hr	—
Machinery	35.9 kg[a]	646,200
Gasoline	80.4 ℓ[b]	812,764
Diesel	159.2 ℓ[b]	1,817,109
Electricity	98.8 kWh[c]	282,864
Nitrogen	112.1 kg[d]	1,345,200
Zinc	8.1 kg[d]	42,865
Insecticides	5.6 kg[d]	486,696
Herbicides	1.1 kg[d]	109,901
Fungicides	1.7 kg[d]	110,347
Transportation	231.5 kg[e]	59,496
Total		**5,713,442**
	Output	
Pecan nut yield	364.0 kg[f]	2,500,680[g]
Protein yield	33.5 kg	
kcal output/kcal input		0.44
kcal output/labor hour		50,621

[a] Assumes 45.4 metric tons of machinery (tractors, sprayers, mowers, trucks, etc.) were used for production and this machinery served an average of 240 hectares and was depreciated over a 5 year period.

[b] Gasoline and diesel use for operation of tractors, sprayers, trucks, harvester, sweepers, and tree shaker.

[c] Electricity for residence and service buildings.

[d] Based on data for Texas.

[e] Total weight of machinery and fuel transported to crop hectare.

[f] Assumes a kernel yield of 50%.

[g] Assumes 6870 kcal/kg of kernel.[10]

Table 5
ENERGY INPUTS AND OUTPUTS PER HECTARE FOR PECAN NUTS (NON-IRRIGATED) IN CENTRAL TEXAS, SEEDLING GROVE

Item	Quantity/ha	kcal/ha
	Input	
Labor	13.3 hr	—
Machiney	4.5 kg[a]	81,000
Gasoline	23.4 ℓ[b]	236,551
Diesel	60.8 ℓ[b]	693,971
Transportation	73.7 kg[c]	18,941
Total		**1,030,463**
	Output	
Pecan nut yield	98.0 kg[d]	637,260[e]
Protein yield	9.0 kg	
kcal output/kcal input		0.62
kcal output/labor hour		47,914

[a] Assumes 5.4 metric tons of machinery (harvesting equipment) were used for production and this machinery served an average of 240 hectares and was depreciated over a 5 year period.

[b] Gasoline and diesel use for operation of harvesting equipment.

[c] Total weight of machinery and fuel transported to crop hectare.

[d] Assumes a kernel yield of 35%.

[e] Assumes 6870 kcal/kg of kernel.[10]

Georgia is offset by larger energy inputs for fertilizers, pesticides, and the resulting larger energy input for transportation. In Georgia, energy output/input in irrigated cultivation is lower than in non-irrigated (Table 1 vs. 2). This is in contrast to Texas where irrigation increases energy output/input (Table 3 vs. 4). The difference is due to some extent to greater energy inputs for additional fertilizer and transportation in Georgia than in Texas, but the major factor is the substantially greater input for irrigation. Overall, the energy output/input is greater in the western than in the eastern belt (cf. Tables 1, 2 vs. 3, 4, 5).

The low energy output/input in pecan results predominantly from the low pecan yield/ha and the high energy inputs from fuel, fertilizer (especially nitrogen), irrigation, and transportation. An increase in yield from new cultivars and advanced cultural practices, which reduce inputs for fuel, fertilizer, and irrigation, has the potential to increase the energy output/input in the future.

Yield increase from new cultivars appear imminent. Cultivars that are more precocious and prolific than conventional cultivars have been released recently.[4] These cultivars promise to increase yields with the same or slightly higher energy inputs required for conventional ones. A better understanding of mineral nutrition and methods and time of fertilizer application may lessen the fertilizer input. Also, use of legumes to fix nitrogen may substantially reduce the nitrogen applied. Likewise, better information on the water requirement and methods and timing of irrigation may lower the energy input for irrigation. Integrated insect and mite control[5,6] would reduce the amounts and frequency of pesticide application and the fuel required for application and so would fungicide application based on environmental conditions[7-9] and on growth stages of the tree and fruit.[9]

Another potential method of increasing the energy output/input in pecan is utilization of the pecan shell. Currently, shells are primarily a waste product. The shell constitutes about 50% of the weight of the cultivar nuts and about 65% of seedling nuts. Thus, shells are a substantial and potential resource in pecan production.

In summary, under the present cultural systems for pecan, energy output is less than energy input. Although improvements of the systems may improve the energy output/input, the ratio for pecans will probably remain less than unity.

REFERENCES

1. Statistics of fruits, tree nuts, and horticultural specialties, Agricultural Statistics, U.S. Department of Agriculture, 1975, chap. 5.
2. **Seale, A. D., Jr.,** Cost and return for a pecan enterprise, *Proc. S.E. Pecan Growers Assoc.,* 59, 15, 1966.
3. **Sparks, D.,** Predicting the nutrient needs of pecan, West. Pecan Conf. Proc., 1979, in press.
4. **Madden, G. D.,** USDA develops 14 main varieties, *Pecan Quart.,* 8(2), 5, 1974.
5. **French, W. J. and Fluker, S. S.,** Commerical versus perfect control of pecan scab and insects, *Proc. S.E. Pecan Growers Assoc.,* 66, 141, 1973.
6. **Polles, S. G.,** Integrated pest management program for pecans: yellow pecan aphids, *Proc. S.E. Pecan Growers Assoc.,* 68, 97, 1975.
7. **Cole, J. R.,** Weather conditions are an important factor in pecan scab control, *Proc. S.E. Pecan Growers Assoc.,* 51, 77, 1958.
8. **Cole, J. R.,** Chemical control of pecan scab in areas of high rainfall, *Proc. S.E. Pecan Growers Assoc.,* 58, 60, 1965.
9. **Valli, V. J.,** Weather conditions as related to pecan scab infection, *Proc. S.E. Pecan Growers Assoc.,* 57, 85, 1964.
10. Composition of Foods, Agric. Handbook 8, Consumer and Food Economics Research Division, Agricultural Research Service, U.S. Department of Agriculture, Washington, D.C. 1963.

ENGLISH WALNUTS

George C. Martin

Presently, about 96% of all English walnut production occurs in California where the crop is valued at some $190 million. Soon the California production of walnut will exceed 273,000 t. The limited remaining production occurs in Oregon and Washington. Factors which increase the energy needs of this crop include pest control, cultivation, irrigation, harvest, and walnut drying. In areas where summer rain provides the moisture necessary for growth and where drying of the harvested nuts can take place in the sun, costs of energy input would be dramatically reduced. A site of this description does not presently exist in California. To compete with other crops in California, walnut must be grown intensively. The latter requires emphasis on yield, which is the direct result of sound cultivation, pest control, irrigation, and quality harvest. These items all require great inputs of energy.

Walnut is not acclimated to the cold winters of other locations in the U.S. Further, the requirement for a 200-day growing season precludes the greater part of the U.S. Finally, the best market price with the largest production is met in California.

Table 1
ENERGY INPUTS AND OUTPUTS PER HECTARE FOR WALNUT PRODUCTION IN THE IRRIGATED CENTRAL VALLEY OF CALIFORNIA[1]

Item	Quantity/ha	kcal/ha	Ref.
	Input		
Labor	105 hr	—	
Machinery	23 kg[a]	414,000	
Gasoline, diesel, and natural gas for drying nuts	560 *l*	6,552,000	2
Nitrogen $(NH_4)_2SO_4$	168.5 kg	2,476,950	3
Irrigation (water applied)	301.3 cm	15,065,000	4
Insecticides	1.5 kg	96,000	5
Transportation	240 kg[b]	61,680	6
Total		**24,665,630**	
	Output		
Walnut yield	1,136 kg	7,395,360	
Protein yield	170 kg		
kcal output/kcal input		0.30	
kcal output/labor hour		70,432	

[a] Assumes 18.08 t of machinery to operate 32.4 ha, and the machinery will be depreciated in 10 years.

[b] Accounts for entry of all equipment for the year per hectare.

Table 2
ENERGY INPUTS AND OUTPUTS PER HECTARE FOR WALNUT PRODUCTION IN THE DRYLAND REGION OF SAN LUIS OBISPO COUNTY, CALIFORNIA[1]

Item	Quantity/ha	kcal/ha	Ref.
	Input		
Labor	38.3 hr	—	
Machinery	18.7 kg[a]	336,600	
Gasoline, diesel, and natural gas for drying	148 ℓ	1,731,600	2
Nitrogen	22.5 kg	330,750	3
Transportation	55 kg[b]	14,135	6
Total		**2,413,085**	
	Output		
Walnut yield	454 kg	2,955,540	7
Protein yield	68 kg		
kcal output/kcal input		1.22	
kcal output/labor hour		77,168	

[a] Assumes 14.6 t of machinery to operate 32.4 ha, and the machinery will be depreciated in 10 years.

[b] Accounts for entry of all equipment for the year per hectare.

Table 3
ENERGY INPUTS AND OUTPUTS PER HECTARE FOR WALNUT PRODUCTION IN THE COOL COASTAL REGION OF SAN BENITO COUNTY, CALIFORNIA[1]

Item	Quantity/ha	kcal/ha	Ref.
	Input		
Labor	100 hr	—	
Machinery	23 kg[a]	414,000	
Gasoline, diesel, and natural gas for drying	540 ℓ	6,318,000	2
Nitrogen	168.5 kg	2,476,950	3
Irrigation	150 cm	7,500,000	4
Insecticides	1.4 kg	88,200	5
Transportation	240 kg[b]	61,680	6
Total		**16,858,830**	
	Output		
Walnut yield	1,136 kg	7,395,360	7
Protein yield	170 kg		
kcal output/kcal input		0.44	
kcal output/labor hour		73,954	

[a] Assumes 18.08 t of machinery to operate 32.4 ha, and the machinery will be depreciated in 10 years.

[b] Accounts for entry of all equipment for the year per hectare.

REFERENCES

1. **Reed, A. D.,** An anaylsis of walnut production costs in California, Agricultural Extension Service, University of California, AXT-333, Berkeley, March, 1970.
2. **Cervinka, V.,** Fuel and energy efficiency, in *Handbook of Energy Utilization in Agriculture,* Pimentel, D., Ed., CRC Press, Boca Raton, Fla., 1980.
3. **Lockeretz, W.,** Energy inputs for nitrogen, phosphorus, and potash fertilizers, in *Handbook of Energy Utilization in Agriculture,* Pimentel, D., Ed., CRC Press, Boca Raton, Fla., 1980.
4. **Batty, J. C. and Keller, J.,** Energy irrigation, in *Handbook of Energy Utilization in Agriculture,* Pimentel, Ed., CRC Press, Boca Raton, Fla., 1980.
5. **Pimentel, D.,** Energy inputs for the producton, formulation, packaging and transport of various pesticides, *in Handbook of Energy Utilization in Agriculture,* Pimentel, D., Ed., CRC Press, Boca Raton, Fla., 1980.
6. **Pimentel, D.,** Energy used for transporting supplies to the farm, in *Handbook of Energy Utilization in Agriculture,* Pimentel, D.,Ed., CRC Press, Boca Raton, Fla., 1980.
7. **Chick, S. and Krummel, J.,** personal communication.

REFERENCES

ALMOND PRODUCTION

Dale E. Kester and Lynn A. Horel

Because of its exacting environmental requirements, almond production in the U.S. is almost completely limited to California. By 1977, production involved 110,000 bearing hectares and around 136 million kg of edible kernels, with trends increasing in both categories. The 5-year average yield per acre was 1840 kg of kernels per hectare. On the other hand, the commercial yields per orchard can range from about 1120 kg/ha to about 3360 kg/ha, depending on orchard location and management.

Within California, almond production occurs from the north end of the Sacramento Valley (Red Bluff) to the south end of the San Joaquin Valley (Bakersfield).Since about 1965, there has been a shift in new planted areas from north to south

Most orchards are located in deep, fertile, well-drained soil areas associated with river systems. Under these conditions, yields are highest partly due to location and partly due to intensive management and high inputs of irrigation water, nitrogen, fertilizer, insecticides, fungicides, and herbicides. Some orchards are located on less productive marginal areas on terrace soils on edges of valleys where yields are on the lower end of range. A few orchards are located on hills in coastal areas (San Luis Obispo County) where no irrigation is practiced and there is low input of fertilizer and chemical controls. Although energy input may be low, the yield is also low (454 kg or less per acre).

The Sacramento Valley area has more natural rainfall and less requirement for irrigation (about 60 cm). Irrigation water comes from both wells and surface runoff, but much is exported to the more arid San Joaquin Valley. A management system involving nontillage, weed control involving mowing and chemicals, and sprinkling system is widespread. Frost may be a problem in this area and control, either by use of sprinkler system or by heaters, can be an important energy cost.

The eastern part of the San Joaquin Valley relies for irrigation water on a combination of wells and surface delivery from reservoirs in the western slope of Sierra Nevada mountains. Because of long establishment, the latter water supply is relatively inexpensive. Clean cultivation and flood or furrow irrigation is widely practiced, necessitating use of considerable field equipment. Larger amounts of irrigation water are required than in Sacramento Valley because of somewhat higher summer temperatures and a lower natural winter supply. Thus, the amounts required range from about 90 cm in the northern part of this valley to about 120 cm in the south.

The western side of the San Joaquin Valley has naturally poor-quality water and must rely on the long distance surface delivery of high-quality water from the Sacramento River. Thus, the energy use involved in the development and maintenance of such systems should be considered in the overall estimate.

In these areas, flood irrigation combined with nontillage is widely practiced. Trees are often planted on low ridges, making a permanent basin between rows for irrigation. Such a system has a low labor requirement, but the cost of permanent underground irrigation installation would have to be considered.

Almond production is highly mechanized, particularly for harvest. Under clean cultivation, the land surface must be well prepared by leveling and rolling and often by sprinkling. Nuts are knocked to the ground, where they dry over a period of 10 days to 2 weeks. The nuts are raked into rows, picked up by machine, and hauled to the huller, either on the ranch or at a central location, operated cooperatively or commercially. The energy output per input ratio for this almond production system is 0.27 kcal.

Table 1
ENERGY INPUTS AND OUTPUTS PER HECTARE OF ALMOND PRODUCTION IN THE IRRIGATED CENTRAL VALLEYS OF CALIFORNIA (UNDER CONDITIONS OF TILLAGE AND FLOOD IRRIGATION)[1]

Item	Quantity/ha	kcal/ha	Ref.
	Input		
Labor	198 hr	—	
Machinery[a]	30 kg	540,000	
Gasoline and diesel fuel	140 ℓ	1,275,000	2
Nitrogen	196 kg	2,725,000	3
Irrigation[b]	301 cm	4,472,000	
Insecticide	11.2 kg	492,800	4
Fungicide	11.2 kg	246,400	4
Herbicides	4.5 kg	255,400	4
Transportation	3361 kg	3,730,900	4
Total[c]		**13,737,500**	
	Output		
Almond yield			
Kernels	1792.63 kg	10,719,927	5
Shells	1232.43 kg		
Hulls	3585.25 kg		
kcal output/kcal input		0.78	

[a] Machinery required is calculated at about 30 kg/ha/year.
[b] Assuming installation of 6018 MJ/ha and assuming 1000 gal/min 50-hp pump, 100 ft head = 150 kWh/acre foot. Total electricity was 1482 kWh.
[c] In specific areas where frost control is needed, an extra expenditure of 25,903,300 kcal/ha is needed to supply 2805 ℓ oil.

With nontillage, the ground surface does not require preparation, except close mowing which provides an easy surface for picking up. The energy output per input ratio for this system is 0.23 kcal.

At the hulling plant, the hulls are separated from the nut (kernel and shell). The amount of hull produced varies by variety, but may be estimated at two times the kernel weight. The hulls are used for livestock feed. The nuts are delivered to a processing plant, where they are cracked to remove the shell. These shells are used for a variety of purposes, notably in the making of briquettes. The amount of shell varies by cultivar ranging from 30 to 150% of weight of kernels. Overall, approximately equal weight of shells may be produced.

Almonds are grown in other countries of the world, but primarily in Spain, Italy, Iran, Morocco, and Tunisia. Traditionally, almonds have been grown on hillside locations without irrigation. Although energy inputs may be low, the yield output is also low, ranging from about 560 kg/ha to perhaps 1680 kg of kernels per hectare. Most cultivars have low shelling percentage (kernel weight per in-shell weight) such that shell production may be 300% kernel production. Trends including shifts to better locations and higher inputs of cultivars are such that changes in energy use and production output can be expected.

Table 2
ENERGY INPUTS AND OUTPUTS PER HECTARE OF ALMOND PRODUCTION IN THE IRRIGATED CENTRAL VALLEYS OF CALIFORNIA UNDER CONDITIONS OF NONTILLAGE AND SPRINKLER IRRIGATION[1]

Item	Quantity/ha	kcal/ha	Ref.
	Input		
Labor	198 hr	—	
Machinery[a]	30 kg	540,000	
Gasoline and diesel fuel	155.2 ℓ	1,413,600	2
Nitrogen	224.1 kg	3,115,000	3
Irrigation[b]	301 cm	9,025,000	
Insecticide	11.2 kg	492,800	4
Fungicide	11.2 kg	246,400	4
Herbicide	4.5 kg	255,400	4
Propane	37 ℓ	2,331,500	2
Transportation	3361 kg	3,730,900	4
Total[c]		**21,150,000**	
	Output		
Almond yield			
Kernels	1792.63 kg	10,719,927	5
Shell	1232.43 kg		
Hulls	3585.25 kg		
kcal output/kcal input		0.51	

[a] Machinery required is calculated at about 30 kg/ha/year.

[b] Assuming 6018 MJ/ha for installation and assuming 1000 gal/min, 100-hp pump, 250 ft head = 380 kWh/acre foot. Total electricity was 3754 kWh.

[c] In specific areas where frost control is needed, an extra expenditure of 389,780 kcal/ha is required to supply 13 cm of irrigation water by sprinkler.

REFERENCES

1. **Reed, A. D., and Horel, L. A.**, Almond Production Cost in California, University of California Cooperative Extension, Service Leaflet 2231, Berkeley, 1975.
2. **Cervinka, V.**, personal communication, 1977.
3. **Lockeretz, W.**, personal communication, 1978.
4. **Pimentel, D.**, personal communication, 1978.
5. **Chick, S. and Krummel, J.**, personal communication, 1978.

MAPLE PRODUCTION IN VERMONT: A PROCESS ANALYSIS INVESTIGATION OF ENERGY UTILIZATION*

James R. Nolfi, George C. Burrill, and Michael Miltner

INTRODUCTION

The evolution of maple sugaring technology has been a gradual process. The following historical perspective should afford the reader a broader vantage point from which to view the present state of on-site technology in the Vermont maple syrup industry.

The Algonquin, Cree, Iroquois, and Abnakis were the first maple sugar makers. Their tapping practices usually involved slashing the maple trees with tomahawks and inserting reed spouts. Collection of the maple sap was done in birch bark vessels. Before European contact, there were no metallic vessels, and boiling was done in logs hollowed out like dug-out canoes.

The Indians' sugaring methods and technology were simple in comparison with present modes. Sugaring utensils were crafted from locally available materials. It was a very labor-intensive process, with young and old, male and female, joining in the annual making of sugar. Gallons per hour drawoff may have been relatively low, but with sugar making came spring festivities and the fusion of labor and religious custom.

The early settlers adopted the Indian methods and utensils, but soon introduced innovations into the process of maple sugaring. Changes were made in sugaring tools, including hand augers, spouts hand carved from birch limbs, and buckets fashioned from sawn ash or basswood trees.

However, tools and implements remained relatively simple and home crafted. Men, women, and children continued such tasks as sap gathering and transporting without animal or mechanical assistance, and the renewable fuel — wood — served as the sole heat source for evaporation.

With the coming of draft animals to the Vermont farmstead in the early 1800s, sugar making emerged from its "primitive" form. The draft animal elevated sugaring technology to an intermediate level as this new source of power became available to the sugar maker. Much of the strenuous labor of hauling sap to the boiling place could now be shared with oxen or horses with sled in tow. This method of gathering sap, in addition to the early experiments with wooden troughs and metal pipe systems, was predominant up until the time of widespread adoption of the petroleum-powered farm tractor.

In the mid-1800s there is evidence of the use of tin buckets, and around the 1870s stationary evaporator rigs with arch and flat-bottomed sheet steel pans were first employed. Use of the evaporator was gradually adopted among Vermont sugar makers. Like other metallic innovations in sugaring utensils, it brought increased dependence on costly materials that were not easily obtained and crafted on the farm or in the industry of rural communities. These new specialized sugaring apparata, along with the established stationary sugarhouse, significantly changed the sugar making operation. Many thought they had reached the limit of improvements in sugar making, but time and an industrialized economy yielded more innovations.

A major change occurred in maple technology in the 1940s when tractors began to replace draft animals as the means for hauling sap to the boiling place. This change marked the beginning of the sugar maker's direct dependence on fossil fuels for the

* Adapted from *Energy Utilization in Vermont Agriculture, Part I.* Center for Studies in Food Self Sufficiency, Burlington, Vermont, 1976.

production of his maple crop and the transformation of sugaring technology toward a highly industrial-dependent form.

The past 10 to 15 years have seen additional technological innovations in the process of producing maple syrup. Most significant were the introduction of plastic tubing collection systems and the replacement of the hand-fired evaporator with the oil or gas rig. Those Vermont syrup makers who have made the conversion from buckets to tubing and from wood to oil-fired evaporators have enjoyed both increased labor saving and higher syrup production. These innovations have also brought increased capital investment and vulnerability to future fossil energy shortages. The same can be said of other recent developments in maple utensils, such as the power tapper, vacuum pump, gas finishing rigs, and disposable sap bags.

The Vermont maple industry is no exception to the trend in agriculture toward dependency on finite energy sources, in spite of the maple industry's "failure to keep pace with the rest of agriculture in developing and adopting new technology". The recent energy crisis saw many sugar makers affected by oil, metal, and plastic shortages and by the escalated prices of these high-energy commodities. Vermont maple producers have become increasingly susceptible to future economic hardship by petroleum dependency, either directly — by the use of gas and diesel-powered sap-gathering vehicles, or by conversion to oil evaporators; or indirectly — by reliance on plastics, electrically processed metals, and other materials derived from petroleum feedstocks and produced by energy-intensive industry.

The maple economy is such that sugar makers can still exercise a degree of choice over the technology they utilize. The last 2 years have seen many skeptical maple producers resist pressures to convert to oil-fired evaporators as petroleum prices skyrocketed. Vermont sap boilers have probably felt more security with their axes, wedges, and local hardwood stands than with distant petroleum producers. Local evaporator manufacturers felt this skepticism and responded by discontinuing the fabrication of oil-burning evaporator units. The *National Maple Syrup Digest,* using information from Vermont Maple Industry Council's Energy Committee, published an article that contained illustrated plans for converting oil-burning units back to the renewable wood fuel. To accompany this, there has been somewhat of a revival of interest in the draft animal as a source of power in logging and maple sugaring operations.

How much, or to what degree, have syrup makers retained earlier methods and tools for producing maple? From an energy viewpoint, what is the current state of on-site maple producing technology in Vermont? The following maple study was undertaken to answer these questions.

METHODOLOGY

When the process analysis technique was used to determine energy inputs into maple sugaring, it was necessary to make arbitrary decisions to define the limits of energy accounting. Realizing that all possible inputs could not be accounted for within the project time frame and from available data, it was decided to attempt to carry the direct and indirect inputs through the making of all items used on the sugarbush. For example, the inputs for sap-gathering vehicles included (unless otherwise indicated), the making of the steel, rolling, casting, and assembling the machinery. In all probability, inputs beyond this point are insignificant in the total energy budget.

DATA RESOURCES

Existing reliable data wsre used as much as possible. From a conceptual analysis of

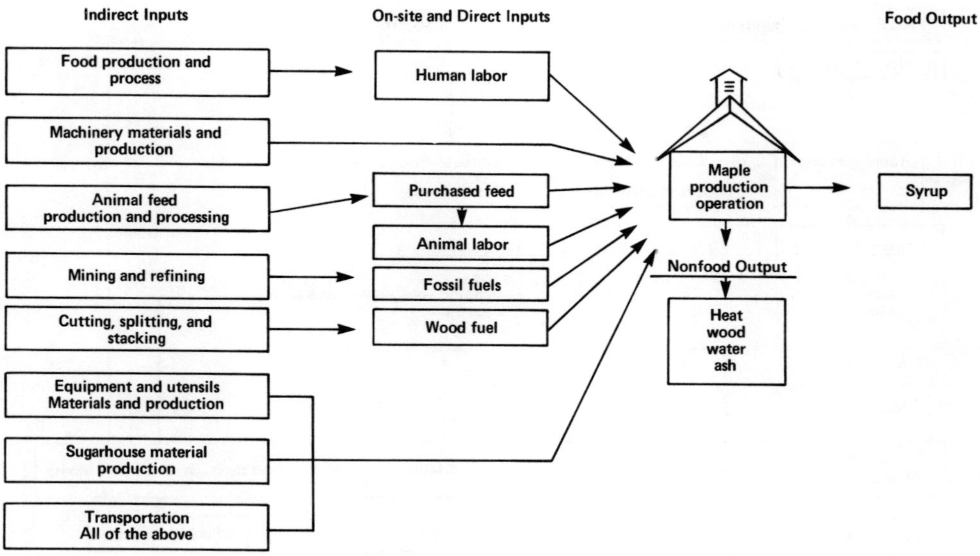

FIGURE 1. Energy flow for Vermont maple production.

the methods and equipment, the detailed process breakdown of sugar making described below was developed.

A recent in-depth study "Cost of Maple Sap Production for Two Types of Collection Systems" by Neil K. Huyler, Research Forester at the Northeastern Forest Experiment Station (1974) provided much of the necessary refined data needed to calculate energy input. It contains accurate field-gathered data on the major direct dollar and energy inputs for producing sap on 15 maple syrup operations in Vermont. These data, along with the raw data from Huyler's forthcoming "Cost Analysis of Processing Maple Sap to Syrup," was the basis for many of the on-site energy input calculations in the specific models of maple production. It also provided the basis for the energy input projections for the total maple crop in Vermont.

Additional data necesary to develop a technology profile of the Vermont maple industry was obtained from the Maple Survey conducted by the Center for Studies in Food Self-Sufficiency. The survey sampled a total of 134 maple producers in the state. These producers were chosen randomly from the 1973 listing in the Vermont Maple Survey conducted by the State Agricultural Department. The sample constituted approximately 12.3% of the total sugar maker population in Vermont for that year. The data from the survey and from Huyler's work provided the basis for projecting energy and equipment inputs for the state-wide maple production energy budget.

TYPES OF MAPLE PRODUCTION UNITS

Most individual maple production units follow one of three general models:

1. A bucket collection system using draft horse gathering and a wood evaporator (Figure 2).
2. A bucket collection system with tractor gathering and an oil evaporator (Figure 3).
3. A tubing collection system with an oil evaporator (Figure 3).

These flow charts point out at what points different forms of energy are utilized in

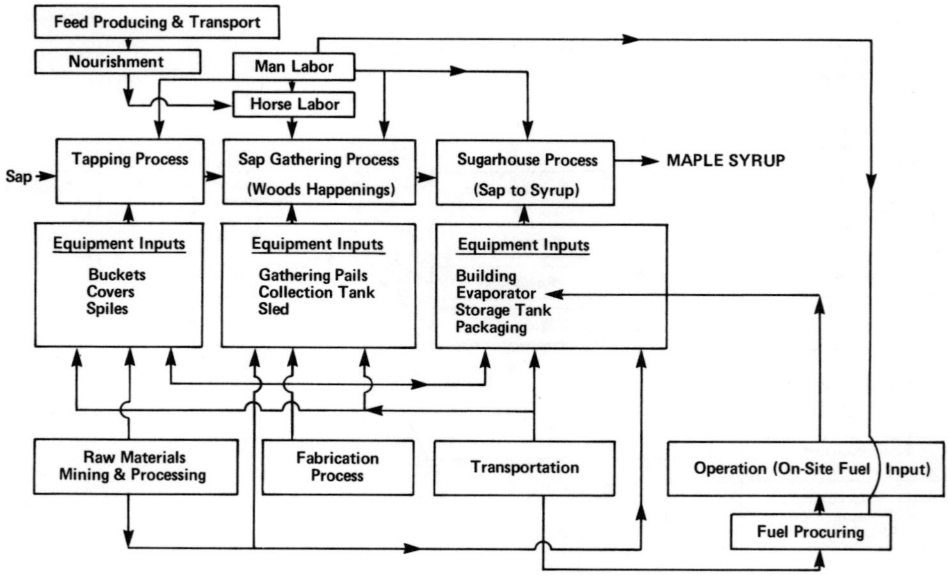

FIGURE 2. Energy flow for Model I: bucket collection system with draft horse gathering and wood-fired evaporator.

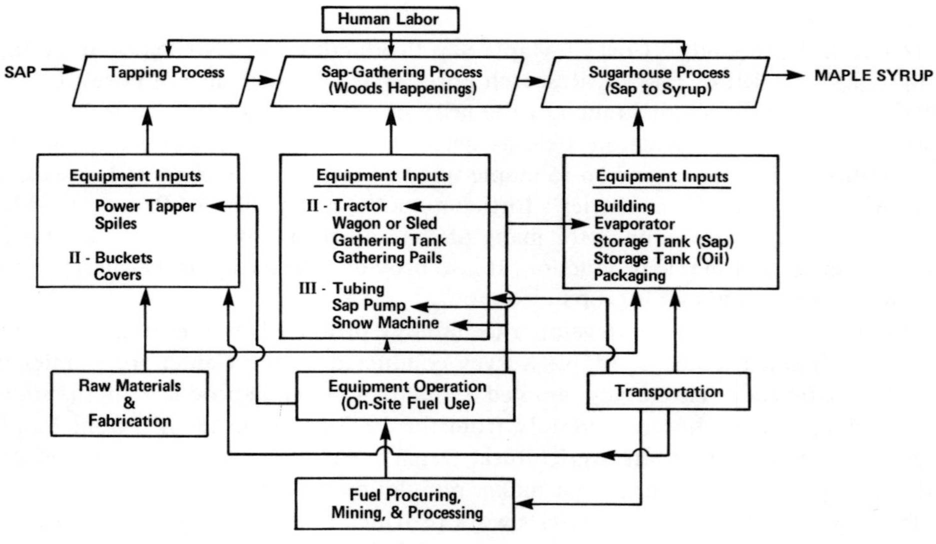

FIGURE 3. Energy flow for maple production units.

maple production. The following energy budgets for each model, Tables 1 through 3, show relative energy requirements on a 2000-tap operation of corresponding models.

In order to arrive at these energy budgets, process analysis and calculations were made following the pattern indicated in Figures 4 and 5.

ON-SITE DIRECT ENERGY INPUTS

The major direct energy inputs to the maple industry were found to be, in decreasing order of kilocalorie input: wood fuel consumption, fossil consumption (equipment), human labor, and horse labor. Their relative kilocalorie inputs are revealed on the State Energy Budget sheet. These values for process analysis were generally calculated

Table 1
ENERGY BUDGET FOR MODEL I: BUCKET COLLECT SYSTEM WITH DRAFT HORSE AND WOOD-FIRED EVAPORATOR

Inputs	**kcal**
Direct, on-site	
Human labor	—
Horse labor	1.0×10^6
Wood fuel	9.17×10^7
Indirect	
Sugarhouse	6.4×10^5
Wood procuring	2.27×10^5
Feeds production	9.4×10^4
Feeds processing	1.2×10^4
Feeds transporting	4.2×10^3
Equipment (all inputs)	2.6×10^7
Equipment (transporting)	1.5×10^4
Packaging (steel cans)	1.04×10^7
Packaging (transporting)	2.3×10^4
Total kcal input	**1.301×10^8 kcal/season**
Total kcal output	**7,125,000 kcal of food (syrup) (at 14,350 kcal/gal syrup)**
kcal in /kcal out	18.2 kcal in /kcal out (food)

Note: 2000 Tap operation at 4 taps/gal syrup, a 500 gal/season bush.

Table 2
ENERGY BUDGET FOR MODEL II: BUCKET COLLECTION SYSTEM WITH TRACTOR GATHERING AND OIL EVAPORATOR

Inputs	**kcal**
Direct, on-site	
Man labor	—
Tractor labor (fuel)	1.15×10^7
Fuels (oil)	7.56×10^7
Indirect	
Sugarhouse	6.4×10^5
Tractor fabrication	1.28×10^6
Tractor transport	1.16×10^4
Fuel mining, processing	2.15×10^6
Fuel transport	5.12×10^6
Evaporator pan, arch, stack, grates collection and storage tanks, buckets and covers	2.6×10^7
Equipment transportation	1.48×10^4
Packaging (cans)	1.04×10^7
Packaging transportation	2.3×10^4
Total kcal input	**1.34×10^8 kcal/season**
Total kcal output	**7.125×10^6 kcal of food (syrup) (at 14,350 kcal/gal syrup)**
kcal in /kcal out	18.8 kcal in /kcal out (food)

Note: 2000 tap operation at 4 taps/gal syrup, a 500 gal/season bush.

Table 3
ENERGY BUDGET FOR MODEL III: GRAVITY FLOW TUBING COLLECTION SYSTEM WITH OIL BURNING EVAPORATOR

Inputs	kcal
Direct, On-Site	
Man labor	—
Snow machine labor (fuel)	3.11×10^6
Evaporator fuel	7.56×10^7
Indirect	
Sugarhouse	6.4×10^5
Snow machine fabrication	5.66×10^5
Snow machine transport	1.54×10^4
Fuel production	1.91×10^6
Fuel transport	3.91×10^6
Tubing materials	1.32×10^6
Tubing transport	8.86×10^3
Evaporator materials	3.64×10^6
Storage tank materials	7.64×10^5
Evaporator and storage tank — transport	9.62×10^3
Packaging (cans)	1.04×10^7
Packaging (transporting)	2.3×10^4
Total kcal input	**1.02×10^8 kcal**
Total kcal output	**7.125×10^6 kcal**
kcal in /kcal out	14.3 kcal in /kcal out (food)

Note: 2000 tap operation at 4 taps/gal syrup, a 500 gal/season bush.

by utilizing the specific data gathered in Huyler's study and then projecting these figures on a state level from information generated in the Maple Survey conducted by the Center for Studies in Food Self-Sufficiency. A detailed description of these calculations follows.

Human Labor

Energy input for human labor was not included, but it can be calculated in two steps. First, a calculation for the total man-hours in sap production (woods activities) was made using the figure of 12.3 min per tap for bucket operations. This calculation multiplied by the number of bucket taps in the state yielded a total of 253,809 man-hr/year for bucket operations. Similarly, the figure of 9.6 min per tap on tubing, multiplied by the number of taps on tubing in Vermont, gives a total of 45,910 man-hr spent on tubing collections systems. These man-hour totals, when multiplied by the figure of 544 kcal/hr of farm labor, yielded the total human labor energy input for maple production. The second step involved calculating the total man-hours in sap processing: an average of 0.16 man-hr/gal of syrup in the sugarhouse was multiplied by the total gallons of syrup processed. When multiplied again by the number of kcals per hour of labor, this gives a figure for kcal input for syrup processing.

Horse Labor

The input for horse labor was derived first by multiplying the estimate for horse-gathered taps in Vermont by the average of 0.25 hr of horse labor per tap, giving

Table 4
ENERGY BUDGET FOR MAPLE PRODUCTION IN VERMONT — 1973 INPUTS

Inputs		kcal
Direct, On-Site		
Human labor:	sap production	
Horse labor:	syrup processing	1.66×10^8
	sap gathering	1.61×10^{10}
Fossil fuels:	evaporation	7.8×10^9
	equipment	
Wood fuels:	evaporation	3.95×10^{10}
Indirect		
Sugarhouse materials		6.91×10^8
Equipment and Machinery		
Tractor		
fabrication		5.4×10^9
materials		1.5×10^8
transportation		6.38×10^6
Crawler		
fabrication		2.83×10^8
materials		1.57×10^8
transportation		6.40×10^6
Snow machine		
fabrication		4.92×10^7
transportation		1.32×10^6
Evaporators		
10^9		
transportation		2.15×10^7
Collection and storage tanks		
materials		1.11×10^9
transportation		2.38×10^6
Buckets and covers		
fabrication		3.1×10^6
materials		2.89×10^9
transportation		6.66×10^6
Tubing		
materials		1.90×10^8
transportation		1.1×10^6
Packaging — cans		
material and fabrication		4.19×10^9
transportation		2.41×10^7
Packaging — steel drums		
material		1.9×10^9
transportation		1.3×10^6
Fuels		
fossil fuels — production		6.02×10^8
fossil fuels — transport		1.34×10^9
wood fuel — production		9.82×10^7
Feeds (draft animal)		
oats production		8.5×10^6
oats processing		2.0×10^6
oats transportation		3.1×10^5
hay production		3.1×10^6
hay transportation		3.9×10^5
Total kcal input		**8.74×10^{10} kcal**
Total kcal output		**4.6×10^9 kcal of food (at 14,350 kcal/gal syrup)**
kcal in /kcal out		19 kcal in /kcal out (food)

Output/input = 0.052 or 5.2%

[a] Arbitrarily set at 10% of materials production energy costs.

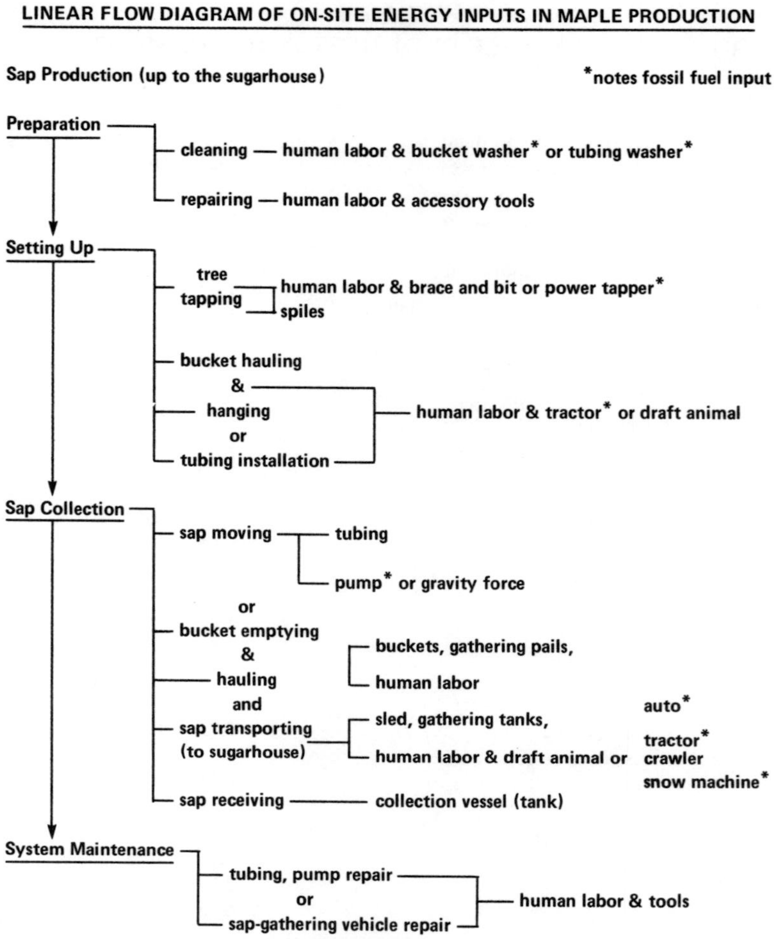

I. The Direct Inputs — maple sugaring tasks require cultural energy inputs that can be classified as: human labor, draft animal labor, mechanical labor, or fuel.

II. The Indirect or "Hidden" Inputs — each task in maple sugaring requires some piece of equipment, tool, or supply for its completion. Equipment, tools, and supplies — each piece of sugaring technology — has attached to it an energy cost of raw materials. It also has an energy cost of fabrication, which is the energy required to manufacture the item. Other indirect energy inputs include the energy required to produce the fuels consumed on the sugarbush and the energy cost of rendering the nourishment for the living beings involved in maple production.

Please observe that there are various alternative methods and technology used in producing maple syrup on the Vermont sugarbush today.

FIGURE 4. Process analysis of on-site energy inputs in maple production.

61,000 horse-hr per season in maple production. Using feed consumption figures of 0.264 kg of oats per hour, and 0.33 kg of hay per hour, totals of 16.1 t of oats and 20.3 tonnes of hay consumed per season by maple horses were derived. Using figures of 4669.3 kcal/kg of oats and 4500 kcal/kg of hay, the energy input per year attributed directly to horses in the maple industry was determined.

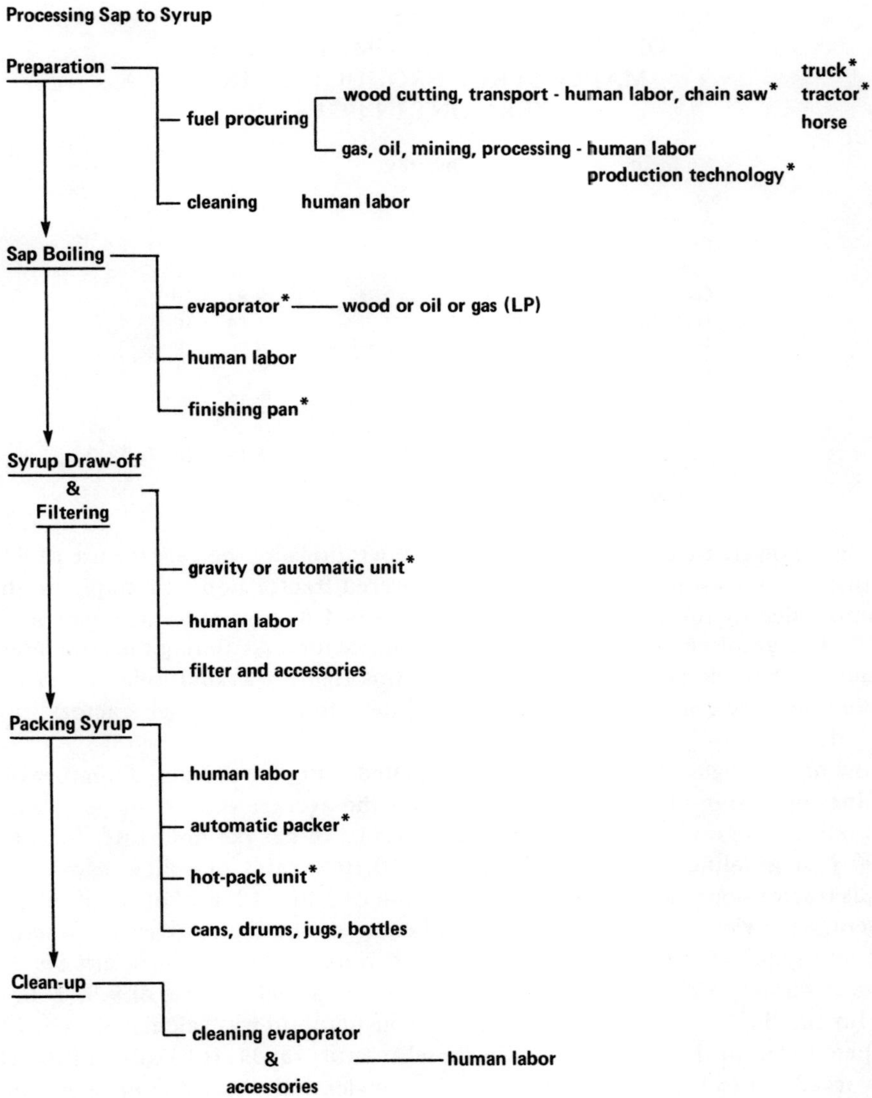

FIGURE 5. Process analysis of on-site energy inputs in maple production.

On-Site Fossil Fuel

On-site fossil fuel consumption was divided into two parts. First, the total consumption of No. 2 heating oil for evaporation was calculated by taking the average of 15.2 ℓ of oil per 3.79 ℓ of syrup times the total estimated gallons of syrup produced by oil. This estimate for gallons of oil-fired syrup was developed from the 1975 Cooperative Maple Survey figures of 33% oil-fired taps and 30% oil-fired evaporators in the state. Given these figures, it was assumed that 33% of the total syrup production, or 0.33 times 1,224,170 ℓ was produced by oil. This calculation produced a figure of 1,614,540 ℓ of oil used in sap to syrup evaporation in the state; when multiplied by 11,414 kcal/ℓ of heating oil, this gave the energy input for fossil use in evaporation.

The second fossil fuel input comes from the operation of sap-gathering vehicles in the production of maple. This figure includes fuels for tractors, crawlers, and snow machines, but neglects the input for the use of pickups for roadside collection in the state. Tractor fuel consumption was computed by multiplying the total estimated num-

Table 5
ON-SITE FUEL CONSUMPTION FOR MAPLE SYRUP PRODUCTION IN VERMONT IN 1973

Inputs	Quantity	kcal
LP gas[a]	—	—
Electricity[a]	—	—
Diesel	500,678 ℓ	5.71×10^{9}
Gasoline	479,238 ℓ	4.84×10^{9}
Heating oil	1,614,540 ℓ	1.84×10^{10}
Wood	17,773m³ (7845 cords)	3.95×10^{10}
Feed	16.1 t oats	
	20.3 t hay	8.32×10^{4}
	Total	**5.55×10^{10}**

[a] Not significant.

ber of gas-powered tractors in use on maple operations by the average use of 87.34 hr per tractor per season. This total for gas-powered tractor hours in maple production was multiplied by the average consumption of 159 ℓ of gas per hour, giving a total of 461,572 ℓ of gasoline attributed to gas-powered tractors. Assuming the remaining 50% of tractors were diesel, 87.34 hr per season, this figure was multiplied by 11.0 ℓ fuel per hour to derive a total of 265,971 ℓ of fuel (diesel) for diesel tractors in maple production.

Snow machine gas consumption was calculated using the estimated number of snow machines operating on tubing operation times the average of 49.7 hr per season use. Total hours were multiplied by an estimated 7.58 ℓ of gas per hour rate, for a total of 33,890 ℓ of gasoline. Taken together, using 10,109 kcal/ℓ, gasoline, snow machine, and gas tractor operation accounted for an input of 5.01×10^{9} kcal of fossil fuel.

Average crawler time usage was assumed equal to the figure given for tractors and when multiplied by an estimated 337 crawlers in use, total crawler hours were determined. Assuming 80% diesel crawlers and based on a fuel use rate of 9.97 ℓ diesel per hour for small (38 hp) crawlers, crawler fuel consumption was calculated to be 234,707 ℓ. When added to the total gallons for diesel tractors, at 11,414 kcal/ℓ of diesel fuel, this worked out to be 5.71×10^{9} kcal for the crawler and diesel tractor operation. The sum of these calculations is the total fossil fuel input for sap-gathering vehicles.

Wood

Energy in put in the form of wood fuel for evaporation was developed by using the figure of 2.9 ft³ of wood per gallon of syrup multiplied by an estimated 216,410 (67%) wood-fired gallons of syrup. This calculation gave a total of 17,773 m³ of wood, which at 2.27 m³ of solid wood per cord equaled 7845 cd. Using an average figure of 5.04×10^{6} kcal/cd for dried maple, birch, and elm, the total energy input for wood fuel was calculated.

INDIRECT INPUTS

Each piece of sugaring equipment, tool, or supply has attached to it an energy cost of raw materials, that is, the energy required to produce and transport its constituent materials. To calculate these inputs, existing literature was used to determine energy inputs per unit of material, and these were projected based on data from Huyler's study and the Cooperative Maple Survey. There also exists an energy cost of fabrica-

Table 6
ENERGY ECONOMIES OF SCALE — MODEL A

Inputs	kcal/season
DIRECT, ON-SITE	
Man labor	—
Tractor labor	5.21×10^6
Wood fuel	4.59×10^7
INDIRECT	
Sugarhouse materials	1.55×10^6
Tractor fabrication	1.28×10^6
Tractor transport	1.16×10^4
Fossil fuel mining and processing	1.52×10^5
Fossil fuel transport	3.62×10^5
Wood fuel procuring (labor and chain saw)	1.14×10^5
Equipment materials	
Buckets and covers	3.00×10^6
Storage and gatherings tanks	7.50×10^5
Evaporator	8.4×10^6
Transportation of materials	4.35×10^4
Packaging materials (and fabrication)	5.2×10^6
Packaging transport	1.15×10^4
Total kcal input	7.20×10^7
kcal output (at 250 gal syrup/season)	3.59×10^6
kcal in /kcal out	20.1 kcal in /kcal out

Note: Energy budget for a "typical" 1000 tap bucket system, tractor gathered, with wood-fired evaporator.

tion for equipment and supplies, which is essentially the energy input for the manufacturing of the item.

Calculating this energy proved to be much more difficult, as manufacturers were not responsive to our requests for accounting of factory energy consumption. Using available literature, and through phone conversations and letters, estimates were derived. Also significant are the energy costs of transportation for equipment and materials, which were developed from mileage approximations and data available on energy inputs per ton for modes of transport. Other indirect or "hidden" energy inputs included were the energy costs of producing (mining and refining) the fossil fuels and the transport of these fuels from faraway oil fields. Wood production (cutting, splitting, and stacking) was also included as an energy input, as well as the energy cost of producing and processing draft animal feed.

Sap-gathering equipment

The energy cost of tractor fabrication was determined by first calculating the estimated total tractor horsepower based on an average of 43.74 hp per tractor, multiplied by the total tractors in use. This figure, when taken times the estimated 2.65×10^6 kcal per tractor horsepower for fabrication, gives a total input of 6.43×10^{11} kcal for tractor fabrication. Depreciated over 20 years at 5%/year, and assuming that maple accounts for one sixth (2 months) of the tractor's yearly use, the yearly input for all tractor

Table 7
ENERGY ECONOMIES — MODEL B

Inputs	kcal/season
DIRECT, ON-SITE	
Man labor	—
Tractor labor	2.08×10^7
Wood fuel	1.84×10^8
INDIRECT	
Sugarhouse materials	1.96×10^6
Tractor fabrication	2.56×10^6
Tractor transport	2.32×10^4
Fossil fuel mining & processing	6.08×10^5
Fossil fuel transport	1.45×10^6
Wood fuel procuring (labor & chain saw)	4.55×10^5
Equipment materials:	
Buckets & covers	1.2×10^7
Storage & gathering tanks	2.82×10^6
Evaporator	1.21×10^7
Transport of materials	7.99×10^4
Packaging materials (and fabrication)	2.08×10^7
Packaging transport	4.6×10^4
Totals	
Total kcal input	2.59×10^8
Total kcal output (at 1000 gal syrup/season)	1.44×10^7
kcal in /kcal out	18.1 kcal in /kcal out

Note: Energy Budget for a "Typical" 4000 Tap Bucket System, Tractor Gathered, With Wood-Fired Evaporator

fabrication was determined. The energy cost of materials for tractors was calculated by assuming that steel comprised 99% of the tractor weight, at 3.15 t/45 hp tractor. Given 554 maple tractors, it was calculated that the total weight of these tractors is approximately 1571 tonnes. At 1.19×10^7 kcal/tonnes for the production of steel, the total kilocalorie input was determined and depreciated to give the yearly figure for tractor materials. Transportation was taken at an estimated 4.87×10^5 kcal/tonnes of agricultural machinery, which, after depreciation, produced the value for tractor transport input.

Inputs for crawler fabrication were calculated by taking the total number of crawlers in use times an average of 38 hp per crawler, giving a total of 12,804.86 crawler horsepower in state maple production. Assuming crawler fabrication to be similar to that of tractors, the 2.65×10^6 kcal/hp was used to compute the total input for the manufacture of all crawlers. By depreciating at a rate equal to that for tractors, the total yearly kilocalorie input for crawler fabrication was determined. The energy cost of materials for crawlers was calculated first by multiplying an average of 4.7 tonnes per crawler by the 337 crawlers in use, to give a total of 1592.7 tonnes of crawler. Assuming again a composition of 99% steel, material tonnage was multiplied by 1.19×10^7 kcal/tonne of steel. This product was then depreciated to approximate the kilocalorie input per year for crawler materials. Transportation input was figured by taking total crawler tonnage times the 4.87×10^5 kcal/t. This value was again depreciated to give yearly input for crawler transport.

Input for snow machine production was calculated by assuming energy input per pound to be similar to the figure given for the production of automobiles. This figure, 17,785 kcal/kg, when multiplied by an average of 159.1 kg per snow machine, gave a figure that, when depreciated at 20% per year, produced a value of 5.66×10^5 kcal/year for the production of a snow machine. When multiplied by the estimated 87 machines in use, this gave the figure recorded for total snow machine production input. Transportation was calculated by assuming 1207 km from the manufacturer multiplied by the total tonnage of snow machines in use. This value was taken times the figure of 396 kcal/tonne-km for trucking, and this total was depreciated over 5 years to give the kcal input for snow machine transport.

Evaporators

Energy input for evaporator materials was calculated next. Average evaporator size was determined to be 1.22×4.27 m, with approximately 1152 evaporators in use in Vermont. From the weight and composition of the average evaporator, it was calculated that it would take about 338.2 tonnes of English tin, 505.5 tonnes of galvanized sheet steel, 138.2 tonnes galvanized steel, and 200 tonnes of other forms of steel to manufacture all wood- and oil-fired evaporators used in Vermont. Using a figure of 4.83×10^7 kcal/tonne of galvanized steel, and assuming this figure to be an approximation of the energy required to produce tin plate, a value of 4.74×10^{10} kcal was calculated. The rolling steel component of evaporators was calculated to have a weight of 200 and, given 1.89×10^7 kcal/t of steel, a figure of 3.78×10^9 kcal was derived. Total kilocalorie input was taken to be 5.50×10^{10} kcal and when this was depreciated over 10 years at 10%/year, the yearly input for evaporator materials was determined. Fabrication input was taken at an arbitrary 10% of the figure for evaporator materials. A value for domestic metals transport was computed by totaling the tonnage of domestic metals used and by figuring 643.7 km from the steel mills to factory and 116.9 kcal/tonne-km by rail. Transportation of English tin was calculated by assuming shipment of 338.2 tonnes over 3218 km by waterway to port at 93 kcal/tonne-km, and by assuming transport 321.9 km from port to factory by rail at 116.9 kcal/tonne-km. An additional 80.5 km was assumed for trucking the fabricated evaporators to the maple producers. Total transportation of materials and evaporators was calculated to be an energy input of 2.15×10^7 kcal/year after depreciation.

Collection and Storage Tanks

The energy cost of collection and storage tanks lack a figure of the fabrication of these vessels. Input for materials processing was figured by assuming two collection tanks per average producer on buckets and taking the total weight as 113.6 kg of galvanized steel to compute a figure of 116.4 tonnes of metal for collection tanks. For storage tanks, it was again assumed two tanks per producer, together weighing a total of 318.2 kg of galvanized steel. This multiplied by the total number of producers gave a figure of 345.5 tonnes of galvanized steel. The total tonnage of galvanized steel for storage and collection vessels was multiplied by 4.83×10^7 kcal/tonne/km and this total was depreciated over 20 years at 5%/year to give the total input for materials processing for tanks. Transporation input was calculated by using the total tonnage for tank metals times an assumed 643.7 km from steel mill to factory. This value for total tonne-kilometers, when taken times the 116.9 kcal/tonne-km, gave transportation input to the factory. An additional 80.5 km was assumed to truck the tanks from factory to producer at 395.8 kcal/tonne-km. Total transportation inputs were tallied and depreciated, yielding the figure given for collection and storage tank input.

Buckets and Covers

The energy cost of buckets and covers was found to be significant and was calculated by first multiplying the total number of buckets in use in the state by 77.4 kcals per bucket for the total energy cost of fabrication. This figure was then depreciated over 30 years at a rate of 3.3% per year, giving the kilocalorie input for bucket fabrication. Fabrication input of covers was assumed to be negligible. The kilocalorie input for materials was derived by multiplying the average number of buckets in use by the weight per bucket. Similarly, the total tonnage of the covers was computed by assuming one cover per bucket and multiplying the weight of 0.355 kg per cover by the total covers used. The total tonnage of galvanized steel in buckets and covers was 1846.4 tonnes. This figure was multiplied by 4.83×10^7 kcal/tonne of galvanized metal, and the product was depreciated at 3.3%/year, giving a yearly figure for energy input attributed to buckets and covers. Transportation input for these was figured by using the total of 1846.6 tonnes of metal multiplied by an assumed 643.7 km from the mill, to give total tonne-km. This figure was taken times 116.9 kcal/tonne-km by rail to produce an estimate of the energy cost from steel mill to factory. An additional 80.5 km was assumed to transport the finished buckets and covers to the sugar makers by truck at a rate of 636.9 kcal/tonne. Distant and local transportation energy inputs were added, and the sum depreciated at the same rate of 3.3%/year to give total energy input for the transportation of buckets and covers.

Tubing

Since no data were available on energy requirements for the manufacturing of tubing, this figure was omitted. Input for the raw materials production and processing was figured by first determining total footage of polyvinyl tubing in use and the average footage per tap for different diameter plastic tubing. These figures were multiplied by the estimate of 286,941 taps on tubing in the state and then by the weights per 30.5 m for the different diameter tubings. A figure of 66.8 t of polyvinyl tubing was calculated for the total tonnage for the state, and this figure was multiplied by 2.84×10^7 kcal/tonne of polyvinyl processed to give a figure of 1.9×10^9 kcal to manufacture plastic materials. After depreciation over 10 years at 10%/year, the energy input per year for tubing materials was derived. Transportation input was calculated for 66.8 tonnes of plastic at an estimated distance of 402.3 km from the manufacturer. At 395.8 kcal/tonne-km for trucking, this worked out to 1.1×10^7 kcal for the transport of plastic tubing. The transportation input for the petroleum and natural gas feedstocks was not determined. However, it is definitely a significant energy input, because a large percentage is imported from outside the U.S.

Packing Containers

The energy cost of syrup packing containers was calculated assuming all metal cans and drums. The inputs for fabrication and materials were calculated together and were derived by first determining the total tonnage of metal cans used per year in the state. Knowing the distribution of syrup sales to be 67% syrup sold retail and wholesale and 33% sold bulk, an estimate of liters sold in cans and gallons sold in drums was calculated. For the 820,194 ℓ sold in cans, it was assumed that 25% was sold in each of the four can sizes. From a total of 54,102 gallon cans at 0.368 kg per can, 108,205 half-gallon cans at 0.268 kg per can, 432,820 quart cans at 0.172 kg per can, and 865,640 pint cans at 0.114 kg per can, it was calculated that 244.2 tonnes of steel were used in can packaging. This figure multiplied by 1.89×10^7 kcal/tonne of metal containers is the energy input per year for the materials and the fabrication process of containers used in packaging. Transporation input was calculated by assuming 222.31 tonnes of steel shipped a distance of 643.7 km by rail to the factory, at a figure of 116.9 kcal/tonne-km. An additional 80.5 km was assumed by truck from factory to sugarmaker at 395.8/tonne-km. The sum of these figures gives the input figure listed for transport of steel-can packaging.

The yearly kilocalorie input for bulk syrup packaging was derived by using the figure of 33% of the total gallons sold in bulk drum or 106,590 gal sold in 30 gal bulk drums. Dividing the total gallons sold in drums by the drum capacity yields the total number of drums used. At an estimated 11.4 kg/30 gal drum, the total weight of galvanized steel in syrup drums came to 20.2 t. This value times 4.83×10^7 kcal/tonne of galvanized metal yielded the input value listed. This figure does not include fabrication cost of drums and assumes that the drums are not reused by the syrup producer. Transportation of metal and drums was calculated by taking the 20.2 tonnes steel times an assumed 643.7 km from the mill to the factory. This value, when multiplied by 116.9 kcal/tonne-km gives energy input for transport to the manufacturer. An additional 80.5 km of transport by truck at 636.9 kcal/tonne were added to give the total kilocalorie value for drum transport.

Petroleum

The energy cost of mining and refining the petroleum used in Vermont's sugarbushes was calculated by taking the figure of 2,594,456 ℓ of petrol used (including heating oil, gasoline, and diesel) times 3.47×10^5 kcal/42 gal barrel for petroleum processing, to give the total input per year for petroleum fuel production. Transportation input for this fuel was determined by calculating the total tonnage of heating oil and gasoline at 0.75 kg/ℓ and the total tonnage of heating oil and diesel fuel at 0.85 kg/ℓ, to give a figure of 2157.4 t petroleum used per season. Assuming transport of a conservative 7242 km from the Middle East at 93 kcal/tonne-km by waterway, a total estimate of 1.45×10^9 kcal input for yearly fuel transport from overseas was figured. An additional 80.5 km at 395.8 kcal/tonne-km by truck was assumed to give the total kilocalorie input listed for petroleum transport.

Wood

Assuming that a chain saw is used to cut wood and given that it takes approximately 1 kcal of gasoline to produce 500 kcal of wood, a total of 7.9×10^7 kcal of fuel was required to cut and buck the wood used in evaporation. This figure plus the labor energy input gives the total input listed for wood production. Solar input was neglected, and transportation of the wood was assumed to be negligible, since it was assumed that most of it is cut near the sugarbush.

Food and Feed

Energy input for the draft animal feeds was calculated in two parts. First, by taking the total weight of oats fed to maple horses, calculated earlier, times 530 kcal/kg for the production of oats, the energy input for oats production was determined. The input for oats processing includes only drying and was derived by multiplying the total weight of oats fed to Vermont maple horses by 124.7 kcal/kg of oats dried. Transportation of oats was assumed to be local — 43.3 km times the tonnage of oats fed. This product, when taken times the figure of 395.8 kcal/tonne/km for trucking, gave the kilocalorie figure for oats transportation.

The yearly energy requirement for hay feed was calculated by taking the total tonnage of hay fed to horses used for sap gathering times the value of 1.53×10^5 kcal/tonne of hay produced. This input for the production of baled hay includes only the energy content of the gasoline used in the total haying process. Transportation of hay was again assumed to be a local — 48.3 km from farm to sugarbush. This figure times the previously calculated tonnage of hay fed per year gives total tonne/km for hay transport. The product was multiplied by the 395.8 kcal/tonne/km value to determine the yearly energy input attributed to hay transport.

Production and transportation costs of human food were not calculated. Average inputs have been estimated at 1 kcal of cultural/production/transportation energy per kilocalorie of food energy. This would double the energy input for human labor, which, with production energy, would still be far less than 1% of the total energy input in maple syrup production.

SUMMARY AND CONCLUSION

The Energy Budget for Maple Production in Vermont presents good empirical evidence that the maple sugaring industry as a whole requires a high absolute energy input for the production of maple syrup. The estimated 19 kcal input for every kilocalorie of food (syrup) yielded indicates that Vermont maple syrup production is a rather energy-inefficient food producing system. The Energy Budget also identifies the specific sugaring subprocess areas that account for the greatest energy inputs to maple production as it exists today in Vermont. Clearly, the energy used for evaporation, wood and oil combined, accounts for more than two thirds of the total energy input for syrup production. Fossil fuel for gathering equipment operation, energy costs for tractor fabrications, and energy input in materials for tanks, evaporators and packaging containers, as well as energy required to transport fossil fuels from the Middle East to Vermont, require further large energy expenditures. Also indicated is the extent to which Vermont maple production is currently dependent upon fossil fuels and, more encouragingly, upon the use of renewable woods.

There is no typical production unit in the Vermont maple industry. We have constructed two economy of scale models, which are as representative as possible of operations of their sizes — 1000 and 4000 taps.

Model I: 1000 Tap Bucket System-Tractor Gathered, with Wood-Fire Evaporator

This model included one 43 hp gas-powered tractor for sap gathering and other chores, a standard 4 ft × 14 ft wood-fired evaporator sufficient to handle sap flow from 1000 taps, and the recommended 16 ft × 20 ft sugarhouse. One sap storage tank and one sap-gathering tank were adequate, and the necessary 1000 buckets and covers were included. Syrup packaging included 63 gallon steel cans, 125 half-gallon cans, 250 quart cans, and 500 pint cans.

Human labor input was 246 hr, 633 ℓ of gasoline were used, and 9.1 cd (@2.27 m^3/cd) of maple, birch, and elm were used in evaporation. A total of 947.5 ℓ of syrup were produced.

Model II: 4000 Tap Bucket System-Tractor Gathered, with Wood-Fired Evaporator

This model included two 43 hp gas-powered tractors for sap gathering and other chores, a standard 6 ft × 16 ft evaporator sufficient to handle sap flow from 400 taps, and the suggested 18 ft × 22 ft sugarhouse. Three sap-gathering tanks and four sap storage tanks were necessary, as well as the 4000 buckets and covers. Syrup packaging included 250 gallon steel cans, 500 half-gallon cans, 1000 quart cans, and 2000 pint cans.

Human labor input was 987 hr, 2520.3 ℓ of gasoline were consumed by tractor, and 36.5 cd (@2.27 m^3/cd) of maple, birch, and elm were used for evaporation. A total of 3790 ℓ of syrup were produced.

One can see that there is little relative energy efficiency advantage between a 1000 tap and 4000 tap operation (approximately 10%). The data indiate that there is probably (on relatively comparable systems) not an economy of scale advantage in terms of energy at any level of commercial maple production.

The construction of models of three current alternative modes of maple producing technologies and methods (Figures 2 and 3) makes it possible to not only compare energy efficiencies across models, but also examine the major energy input variables in a selected subprocess to determine specifically which techniques are more energy intensive in the total sense of the word. The following chart is an example of this type of energy assessment of technologies within a particular system.

A Comparison of Two Major Energy Input Variables in Bucket-Gathering System

(Assuming 500 gal per season)

	Gathering vehicle input	**Evaporator fuel input**	
Horse	2220 kcal/gal syrup, 1.11×10^6 kcal per season attributed to horse	183,800 kcal/gal syrup, 9.19×10^7 kcal per season attributed to wood[3]	**Wood**
Tractor	27,800 kcal/gal syrup, 1.39×10^7 kcal per season attributed to tractor (12.5 times horse)	163,600 kcal/gal syrup, 8.18×10^7 kcal per season attributed to oil[4]	**Oil**

1. Includes energy cost of horse labor, feeds production, processing, and transport.
2. Includes energy cost of tractor fuel, tractor fabrication, transport, fuel mining, and processing.
3. Includes energy cost of wood fuel (content), cutting, splitting, and stacking.
4. Includes energy cost of oil fuel (content), fuel mining, processing, and transport.

This chart gives an energy comparison between the use of the horse as a sap-gathering vehicle and the use of the gas-powered farm tractor. The farm tractor, in addition to operating on fossil fuels that must be transported from distant ecosystems, is also shown to be responsible for "consuming" 12.5 times the energy that the horse "consumes" in the production of a gallon of syrup. The horse "fuel," the plant energy in grain and hay, is or can potentially be produced locally. The comparison between wood and oil as evaporator fuels reveals that wood fuel, with present evaporator rigs, is a slightly less efficient process than oil fuel. But this energy inefficiency can be reduced substantially by local technological modification — the simple addition of an engineered preheater hood to wood evaporators. The "net energy" of wood, to use Odum's term,can be increased. Moreover, wood is locally reproducible. The total efficiency of oil use is beyond the control of local engineering and can only develop into a more capital- and energy-intensive enterprise. The "net energy value" of oil as an evaporator fuel is fixed or decreasing — it is impossible for the sugar maker to tamper with the total eff iciency of the petrol-network.

Presenting the energy input data in this form provides a basis for developing and evaluating new strategies for producing maple syrup that are, on the whole, less energy intensive and rely more completely on indigenous, renewable sources of power. Realistically, we cannot expect maple producers to abandon existing dollar concerns in favor of more energy efficient systems of food production. Only when economies reflect energy efficiency will production become more energy efficient. Practical attempts to maximize energy efficiency in maple production must also be evaluated from the standpoint of dollar economies, ecology, and engineering. It is necessary to make findings available in a form that will have direct meaning and value for the sugar maker, as well as for the state planner.

It is difficult at present to visualize a time in which changes in international relations, energy availability, and national economy will thrust maple sugar and syrup back into prominence (even regionally) as a widely used sweetener. High production costs, particularly for labor, seem to indicate that maple syrup will remain a "specialty" food for at least the near future. Results from this study indicate that maple does offer the potential of being produced using a fairly high percentage of locally-produced renewable resources. Clearly, increased efficiency in utilization of wood in the process of evaporation and integration of draft horses into year-round operations, such as skidding logs as lumber, firewood, or pulping operations, can increase energy efficiency.

Energy Inputs and Outputs for Livestock Production Systems

RESOURCE-CULTURAL ENERGY REQUIREMENTS OF THE DAIRY PRODUCTION SYSTEM

P. A. Oltenacu and M. S. Allen

INTRODUCTION

Over the last 15 years, the U.S. dairy cow population has been declining steadily at an average rate of 0.43 million cows per year. This dramatic decrease in cow numbers, from 17.5 million in 1960 to slightly over 11 million in 1975, was concurrent with a steady increase in milk yield per cow per year. The net result was a modest decline in total milk production, from 60 billion kg in 1960 to the present 52 billion kg per year.

At the farm level, the changes were even more obvious. The average farm size has increased, and farms have become more specialized as a new generation of farmers, better educated and ready to adopt new technology, has changed dairying to an increasingly intensive production system.

The purpose of this chapter is to identify and assess the inputs of cultural energy into the U.S. dairy system and to examine the efficiency of the current dairy production system.

Since the dairy production systems are not uniform across the land, the U.S. was divided into six production regions that represent approximately 85% of the total dairy industry. The regions are defined in Table 1. Where it was impossible or impractical to obtain data for each region as a whole, the representative states, italicized in Table 1, were used to characterize the region.

DAIRY SYSTEM INPUTS

Seven major categories of resource inputs were considered. They are as follows: feed, fuel, insecticide, machinery, equipment, labor, and land. In the feed category, disappearance data for the representative states in each region were used to determine the relative proportion of each feed used, with the actual quantities being calculated to agree with National Research Council (NRC) requirements for a cow unit. All inputs were calculated per cow unit per year, where the cow unit represents a milking cow and her replacement as shown in Table 2. The feed disappearance data included commercially mixed feeds which were further broken down into their components using an average of complete formula feed compositions as estimated by at least one major feed company in each region. All feeds were summarized into six categories as follows: feed grains (corn, sorghum, oats, barley, and other grains [wheat, rye]); soybean meal; alfalfa meal; by-products (wheat mill run, cottonseed meal, other oilseed meals, molasses, wheat middlings, distillers grains, wheat bran, corn gluten, hominy, animal protein, brewer's grains, and beet pulp); roughages (hay, corn silage, sorghum silage, other silage, and other forage); and pasture.

The energy figures for the feeds shown in Table 3 include all cultural energy inputs required for the production of each crop. It was assumed that the primary reason for growing the crop from which the by-products result is not to produce dairy feeds; therefore, no cultural energy required to grow the crop was charged against the milk production system. The energy required for transport to the farm at an average distance of 100 km was the only charge for the by-products.

In the next category, the major fuels used in the dairy system were considered. These are gasoline, liquid propane, diesel fuel, and electricity. They were converted to megacalories with the conversion constants in Table 3.

Table 1
DEFINITION OF GEOGRAPHICAL REGIONS

Region	States
Appalachia	Va., W. Va., *Tenn.*, Ky., Ind.
Lake States	Iowa, Ohio,[a] Ind., *Mich.*, *Wis.*, Ill., *Minn.*, S.D.
Northeast	Maine, N.H., Vt., *N.Y.*, Mass., Conn., R.I., N.J., *Pa.*, Ohio, Md., W. Va., *Va.*
Pacific	*Calif.*, Ore., Wash.
Southeast	Ga., *Fla.*
Southern Plains	*Tex.*, Mo., Ark.

Note: Representative states in each region are italicized

[a] In cases where a state is included in more than one region, the population of that state is divided equally among the regions in which it is included.

Table 2
COMPOSITION OF A DAIRY COW UNIT[1]

Cow unit		Average weight (kg)
Cows =		591
Birth to 12 months =	0.45	159
12 months to freshen =	0.53	386
Total	**1.98**	

The purpose of this paper is to identify and assess the inputs of cultural energy into the U.S. dairy system and to examine the efficiency of the current dairy production system.

The insecticides used were derived from a national figure since no detailed figures for use on dairy cattle were available.

The machinery category includes only the items directly involved with dairy production such as tractors, pickup trucks, manure handling machinery, and manure scrapers. The machinery used for crop production is included in the feed cultural energy figures.

The equipment figures include stanchion barns, freestall barns, loafing barns, dry lots, milk houses, pipelines, milking parlors, bulk tanks, water heaters, feed handling equipment, forage storage equipment, manure handling equipment, etc., with regional variation.

Both machinery and equipment figures are reported as the amount of depreciation per cow unit per year in terms of 1967 dollars and have been converted to megacalories.

Labor figures reported include both family and hired labor.

Cropland is reported as land in crops used to feed one cow unit and does not include land idle or in fallow.

DAIRY SYSTEM OUTPUTS

The U.S. produced 52.5 billion kg of milk from 11 million cows with an average production of 4700 kg in 1975 (Tables 4 and 5). This amounted to a little more than 1.8 billion kg protein from milk. In addition, the dairy system produced 0.2 billion kg of protein as meat from cull cows for a total of 2 billion kg animal protein produced in 1975. The Lake States region has the highest total milk production and 37% of the

Table 3
ENERGY CONVERSION FACTORS

Item	Amount	Mcal	Yield (kg)	Ref.
		Input		
Machinery	1 1967 dollar	18.105	—	2
Equipment	1 1967 dollar	16.900	—	2
Insecticide	1 kg	73.920	—	3
Grain grinding	1 kg	0.038	—	3
Gasoline	1 ℓ	8.256	—	4
LP gas	1 ℓ	6.292	—	4
Diesel	1 ℓ	9.321	—	4
Electricity	1 kWh	3.007[a]	—	
Transportation (truck)	1 kg/km	0.00083	—	5
Feeds				
Corn	1 ha	6,424	5,965	3, 6
Sorghum	1 ha	7,489	3,470	3, 6
Oats	1 ha	2,070	1,826	3, 6
Barley	1 ha	2,724	2,112	3, 6
Other feed grains	1 ha	3,754[b]	1,862	3, 6
Hay (+ alfalfa)	1 ha	2,527	6,052	3, 6
Sorghum silage	1 ha	3,670	19,086	3, 6
Corn silage	1 ha	4,965	22,567	3, 6
Other silage	1 ha	3,670	6,052	3, 6
Other forage	1 ha	2,527	6,052	3, 6
Pasture				
Northeast	1 ha	1,773	5,007	3
Appalachia	1 ha	2,009	5,007	3
Lake States	1 ha	1,836	5,007	3
Southern Plains	1 ha	703	5,007	3
Southeast	1 ha	2,009	5,007	3
Pacific	1 ha	309	5,007	3
Soybean meal	1 kg	0.079		3
Alfalfa meal	1 kg	0.636		3

[a] 1 kWh of electricity contains 859 kcal;[4] this was multiplied by 3.5 in order to account for the fuels used to produce the electricity.

[b] Adapted from figures for wheat and rye.

Table 4
COW POPULATION, MILK YIELD, AND CULTURAL ENERGY EFFICIENCY BY REGION (1975)

Region	Number of cows (10^3)	Average Production (kg/cow/year)	Total Mcal cultural energy input per cow unit	Mcal/ 100 kg milk
Appalachia	709	4,070	8,030	197
Lake States	4,157	4,667	8,460	181
Northeast	2,504	4,813	7,250	151
Pacific	1,072	6,006	6,960	116
Southeast	327	4,381	5,820	133
Southern Plains	731	4,321	7,460	173
Total (85% of U.S.)	**9,500**	**4,776**	**7,820**	**164**
Total U.S.	**11,151**	**4,706**	**7,820**	**166**

Table 5
OUTPUTS AND VARIOUS PARTIAL MEASURES OF EFFICIENCY FOR THE U.S. DAIRY PRODUCTION SYSTEM BY REGION

	U.S.	Appalachia	Lake States	Northeast	Pacific	Southeast	Southern Plains
Total milk production (10^9 kg/year)	52.5	2.9	19.4	12.1	6.4	1.4	3.2
Milk protein production (10^9 kg/year)	1.837	0.101	0.679	0.422	0.225	0.050	0.111
Total protein production (10^9 kg/year)	2.000	0.111	0.740	0.459	0.241	0.055	0.122
Mcal cultural energy per kilogram protein	44	51	48	40	31	35	45
Kilogram protein per hectare land	123	102	112	145	141	172	123
Kilogram protein per hour labor	2.4	2.2	2.2	2.2	5.2	3.5	2.6

U.S. dairy cow population based on 1975 figures (Table 4). The Northeast has the next highest total production with 22% of the U.S. dairy population. The Pacific region has the highest production per cow with 6,000 kg per cow per year, and although it has less than 10% of the U.S. dairy population, over 12% of the total U.S. milk is produced within this region. The last three regions (Appalachia, Southeast, and Southern Plains) account for less than 16% of the U.S. dairy population and slightly over 14% of the total milk produced. All three regions have a lower production figure per cow unit than the national average.

MEASURES OF EFFICIENCY

The efficiency with which cultural energy is used in the dairy system was represented by the amount of cultural energy inputs (in megacalories) per 100 kg milk produced for each region (Table 4). Other partial measures of efficiency included are the cultural energy inputs per kg of protein produced, protein produced per hectare of land, and protein produced per hour of labor (all of which consider protein from cull cows as well as milk protein produced; see Table 5).

The average cultural energy used to produce milk for the U.S. is 166 Mcal/100 kg milk. The regions range from 116 to 197 Mcal/100 kg with the Pacific region being the most efficient and the Appalachian region being the least. The two regions with the highest efficiency (Pacific and Southeast) are also the regions with the greatest number of cows per farm (Tables 4 and 6). The Southeast uses less cultural energy per cow unit, but the efficiency of cultural energy use for milk production is higher in the Pacific region because of its higher production per cow. At the other end of the scale, the region with the greatest cultural energy input per cow unit is the Lake States, although Appalachia has the lowest efficiency because of its lower production per cow.

The regions follow the same pattern for efficiency in terms of cultural energy per kilogram of protein produced as would be expected. However, the regions are ranked differently when efficiency is measured in terms of protein produced per hectare of land or hour of labor.

The U.S. averages 123 kg of protein from dairy sources per hectare of land used to

Table 6
THE REGIONAL VARIATION FOR VARIOUS INPUTS IN DAIRY PRODUCTION

Region	Number of cows per farm	Inputs (as % of total cultural energy) Feed	Fuel	Machinery and equipment	Hours of labor
Appalachia	54	65	28	6	71
Lake States	38	66	27	6	81
Northeast	47	62	29	8	82
Pacific	218	70	24	5	43
Southeast	418	69	25	4	48
Southern Plains	96	69	25	5	65
U.S.	50	65	27	6	74

grow feed for these purposes. The Southeastern states are the most efficient (172 kg/ha) and the Northeast and Pacific states are next with 145 and 141 kg/ha, respectively. The Southern Plains, Lake States, and Appalachia follow with 123, 112, and 102 kg/ha, respectively (Table 5).

When efficiency is represented by the amount of protein produced per hour of labor, there is great variation between some regions. The U.S. averages 2.4 kg/hr, and the regions range from 2.2 to 5.2 kg/hr. The highest output per hour is from the Pacific states, and the Southeast and Southern Plains are next with 3.5 and 2.6 kg/hr, respectively. The other three regions each average 2.2 kg/hr labor (Table 5).

EFFECTS OF VARIOUS INPUTS ON EFFICIENCY

Feeds

The major cultural energy input for all regions is for feed, which accounts for 65% of the total cultural energy used per cow unit nationally and ranges from 62 to 70% (Table 6). These figures are influenced by the amount and type of feed fed and the amount of energy from other inputs. In absolute units, the average amount of cultural energy for feed in the U.S. dairy industry is about 5000 Mcal per cow unit. This figure is also influenced by the amount of feed fed and the proportions of different feeds fed per cow unit.

Roughages are the largest feed category, representing 54% of all the total digestible nutrients (TDN) on a dry matter basis available per cow unit for the U.S. (Table 11). Hay and corn silage are the major feeds in this category with sorghum silage, other silages (haylage), and other forage (green chop) fed at lower levels depending upon region (Table 7). The amount of roughages fed per cow vary greatly between regions. Roughages comprise 59% of the total TDN fed in the Pacific states but constitute only 15% of that fed in the Southeast. The large amounts of alfalfa hay fed in the Pacific states and the very high concentrates (feed grains and by-products) fed in the Southeast account for this large variation. The Lake States and the Northeast feed roughages at comparatively high levels (56 and 58% of total TDN, respectively), and Appalachia and the Southern Plains feed lower levels of roughages (47 and 37% of total TDN, respectively).

Feed grains represent the next largest category of feed at 28% of the total TDN fed. However, they account for 38% of the feed cultural energy, reflecting the energy intensive nature of grain feeding in milk production. The level of feed grains fed vary between regions from 19% of total TDN per cow unit in the Northeast to 38% in the Southeast (Table 8). The Lake States and Southern Plains regions feed relatively high levels (35 and 33% of total TDN, respectively), and the Pacific and Appalachian re-

Table 7
MAJOR ROUGHAGES FED BY REGION AS PERCENTAGE OF TOTAL TDN

Feed	Region					
	Appalachia	*Lake States*	*Northeast*	*Pacific*	*Southeast*	*Southern Plains*
Hay	18	31	30	43	8	31
Corn silage	21	18	23	10	2	2
Sorghum silage	3	—	—	—	2	2
Other silage	4	3	2	2	—	—
Other forage	1	5	3	4	2	1
Total Roughages	**47**	**56**	**58**	**59**	**15**	**37**

Table 8
MAJOR[a] FEED GRAINS FED BY REGION AS PERCENTAGE OF TOTAL TDN

Feed grain	Region					
	Appalachia	*Lake States*	*Northeast*	*Pacific*	*Southeast*	*Southern Plains*
Corn	22	28	16	14	36	13
Sorghum	—	—	—	—	—	18
Oats	—	7	2	—	—	2
Barley	—	—	—	4	—	—
Other feed grains	—	—	—	1	1	1
Total feed grains	**24**	**35**	**19**	**20**	**38**	**33**

[a] Greater than 1% of total TDN.

gions feed comparatively low levels (20 and 24%, respectively). Corn is the major grain fed in all regions except in the Southern Plains where sorghum (milo) comprises 55% of the feed grains. Oats, barley, and other grains (wheat and rye) are fed to a lesser extent. An additional charge for grinding, to make grains suitable as an animal feed, was also made (Table 3).

By-products represent 13% of the total TDN fed nationally. Because only the cultural energy expense of transportation was charged against the dairy production system for them, the level of by-products fed greatly influences the total cultural energy values for feed. This can be seen by comparing the Lake States region with the Southeast. Both regions feed roughly the same amount of feed per cow unit on a TDN basis, and feed grains are fed at nearly the same levels. As shown in Table 9, by-products in the Southeast are fed at five times the levels fed in the Lake States on a per cow unit basis, representing 31% of the total TDN per cow unit in the Southeast and only 6% of the total TDN in the Lake States. The effect of feeding very different levels of by-products is reflected in the total megacalories of cultural energy for feed for the two regions. The Lake States, at 5563 Mcal, require much more energy to produce the same amount of TDN as the Southeast at 4003 Mcal.

Table 9
MAJOR[a] BY-PRODUCTS FED BY REGION AS PERCENTAGE OF TOTAL TDN

By-product	Region					
	Appalachia	Lake States	Northeast	Pacific	Southeast	Southern Plains
Wheat middlings	6.9	3.9	10.3	3.6	13.2	7.3
Hominy	3.2	—	5.1	—	6.1	—
Corn gluten	—	—	—	—	3.2	3.0
Molasses	—	—	—	2.9	2.8	—
Cottonseed meal	—	—	—	—	—	5.5
Wheat mill run	—	—	—	4.8	—	—
Brewers and distillers grains	—	—	—	—	—	2.6
Animal protein	—	—	—	3.2	—	—
Total by-products	14	6	20	13	31	20

[a] By-products that comprise >2.5% of total TDN fed per cow unit.

Table 10
TOTAL BY-PRODUCTS FED BY REGION

Region	By-products (kg × 10^{-9})
Appalachia	0.59
Lake States	1.57
Northeast	3.04
Pacific	1.09
Southeast	0.61
Southern Plains	0.89

Note that although the Lake States feed by-products at only 6% of total TDN fed, the region as a whole feeds a total of 2.6 times the by-products fed by the Southeast (Table 10). The Southeast and Appalachia feed the lowest total levels of by-products, but because of their lower cow populations, they result in feeding more by-products per cow than the national average. Table 16 shows that all regions but the Lake States feed by-products at a level equal to or higher than the average of 15% of the total TDN fed per cow unit nationally.

Table 16 also shows the major by-products fed in the U.S. dairy system by region. Wheat middlings are present in the largest quantity with hominy being the next most abundant. Corn gluten, molasses, cottonseed meal, wheat mill run, brewers and distillers grains, and animal protein are fed at varying levels. The levels of different by-products fed in each region at any one time are very dynamic, being greatly influenced by economic factors, although they stay relatively constant when averaged over time and region. By-products tend to be fed where they are produced since transportation costs limit dispersion.

Besides greatly influencing the total cultural energy values for feed and, thus, efficiency in terms of cultural energy per unit output, by-products are also a major influence on efficiency in terms of kilograms protein produced per hectare of land. Since the land used to grow the primary crops from which the by-products result was not charged against the dairy system, it is obvious that greater amounts of by-products fed will increase output per hectare. The biologial efficiency of the cow will also influence this measure. High-producing cows require less of the total feed (and, therefore, fewer hectares of land to produce feed) for maintenance, and more is then available for production. Evidence of the influence of both of these factors on output per hectare of land can be seen in Table 5. The Southeast has the greatest efficiency, and, although the production per cow is low, this region feeds the highest level of by-products per cow unit (Table 9). The Pacific region also has a higher than average output per hectare. Although the region feeds the national average for proportion of by-products in total feed, the higher production per cow is principally responsible for the high efficiency.

Pasture in the U.S. dairy system provides an average of only 2% of the total TDN fed from 0.14 ha per cow unit. Appalachia has the most pasture at 0.54 ha per cow unit comprising 9% of the total TDN. In the Lake States and Northeast, pasture comprises only 1% of the total TDN at 0.07 ha of pasture per cow unit for each region with all the other regions at intermediate levels.

Alfalfa meal is fed in low quantities in most of the U.S. and averages less than 0.5 % of the total TDN, being fed at higher levels in the Pacific (2% of TDN), Southeastern, and Southern Plains states (1% of TDN).

Soybean meal averages 2% of the total TDN fed per cow unit in the U.S. The highest proportion is fed in the Southeast where it comprises 10% of the total TDN fed. Soybean meal is also fed at comparatively high levels in the Appalachian region (5% of total TDN), Pacific region (4%), and in the Southern Plains (also 4%). The Lake States and the Northeast feed lower levels at 2 and 1% of TDN, respectively.

Fuels

Gasoline,liquid propane, diesel fuel, and electricity are the major fuels used in dairy production and are the second highest category in terms of energy used next to feeds. The figures for electricity were converted from kilowatt hours to megacalories and then adjusted by a factor of 3.5 (for a national average) in order to account for the fossil fuels used to produce the electricity. In terms of megacalories per cow unit, more energy is used in the form of electricity than all other dairy farm fuels combined. Nationally, 1326 Mcal of energy from electricity are consumed per cow unit. Diesel fuel accounts for the next largest amount at 354 Mcal per cow unit. Gasoline and liquid propane gas are consumed at a rate of 248 and 157 Mcal per cow unit, respectively (Table 7). In total, these fuels account for 27% of the cultural energy consumed per cow unit nationally. As shown in Table 6, in regions that have more cows per farm, fuels account for a lower percentage of cultural energy per cow unit.

Machinery and Equipment

Machinery and equipment combined account for 6% of the total cultural energy used per cow unit. These figures also vary by regions, and they are generally lower for the regions with larger farms and higher in regions with fewer cows per farm (Table 6) This is a direct consequence of the fact that, in general, the larger farms make more efficient use of machinery and equipment.

Labor

Nationally, each cow unit requires 74 hr of labor per year. Again, the trend of an indirect relationship between the labor input per cow unit and the size of the farm can be seen when the regions are considered (Table 6). This is mainly due to the predominance of labor-consuming management systems, such as pipeline milking and stan-

Table 11
RESOURCE INPUTS (PER COW AND REPLACEMENT) INTO THE U.S. DAIRY PRODUCTION SYSTEM[a]

Number of Cow Units — 9,500,000; Yield — 4,776 kg Per Cow Per Year, 45,367,000,000 kg Milk Per Year[b]

Inputs	Amount fed (kg)	Amount of protein (kg)	TDN (kg)	% of total TDN	Mcal cultural energy	% of total cultural energy	Ref.
Feed[c]							
Feed grains	1,646	158	1,288	28	1,927	25	8, 9
Soybean meal	145	63	110	2	11	—	
Alfalfa meal	28	5	16	—	18	—	
By-products	819	167	597	13	66	1	
Roughages	8,477	570	2,451	54	2,859	37	
Pasture[d]	722	40	108	2	214	3	10
Total	11,837	1,003	4,570	100	5,095	65	
	Units	**Quantity**	**Mcal**				
Gasoline	ℓ	30	248				10
LP gas	ℓ	25	157				
Diesel	ℓ	38	354				
Electricity	kWh	441	1,326				
Grain grinding			63				3
Insecticide[e]	kg	0.19	14				11
Machinery[f]	$ 1967	4.31	78				10
Equipment[f]	$ 1967	25.34	428				10
Labor	hr	74					10
Cropland per cow unit	ha	1.32					
Cropland and pasture per cow unit	ha	1.46					
Total Mcal/cow unit			7,822				
Total Mcal/100 kg milk			166				

[a] The six regions (Tables 8 to 13) represent approximately 85% of the total U.S. dairy production. Figures for the U.S. represent a weighted average of the figures for the six regions.

[b] Regional production figures are weighted averages of production by states included in each region as defined above.[7]

[c] Total amounts fed of commercially mixed feeds, soybean meal, cottonseed meal, other oilseed meals, urea, molasses, alfalfa meal, animal protein, corn, sorghum (milo), oats, barley, other feed grains, hay, corn silage, sorghum silage, other silage, and other forage; based on representative states. The commercially mixed feeds were further broken down based on estimates of a major feed company in each region.[8] Feed grains, by-products, and roughages were combined in these summarized tables. Since this is disappearance data and no accurate figures of the amount actually fed per animal is available, the quantities were adjusted to agree with NRC requirements for a cow unit. The feeds fed and their proportions were kept constant. Protein and TDN levels of the feeds are expressed on a dry matter basis.[9]

[d] Pasture figures were not adjusted.[10]

[e] No detailed figures for insecticide use on dairy cattle were available.[11]

[f] The Wholesale Price Index was applied to adjust to $ 1967.[10]

chion barns, among the smaller farms vs. freestall barns and milking parlors among the larger farms.

Land

Total land required to grow feed and pasture per cow unit averages 1.46 ha nationally. Of the 1.46 ha, 0.14 ha are in pasture and 1.32 ha are in feed grains, soybean meal, alfalfa meal, hays, silages, and other forages (Tables 11 to 17). The total land

Table 12
RESOURCE INPUTS (PER COW AND REPLACEMENT) INTO THE DAIRY PRODUCTION SYSTEM: APPALACHIA[a]

Number of Cow Units — 709,000; Yield — 4,070 kg Per Cow Per Year, 2,886,000,000 kg Milk Per ayear[b]

Inputs	Amount fed (kg)	Amount of protein (kg)	TDN (kg)	% of total TDN	Mcal cultural energy	% of total cultural energy	Ref.
Feed[c]							
Feed grains	1,281	114	1,030	24	1,446	18	8, 9
Soybean meal	287	126	219	5	23	—	
Alfalfa meal	26	5	15	—	17	—	
By-products	829	168	621	14	66	1	
Roughages	8,159	413	2,013	47	2,558	32	
Pasture[d]	2,700	151	405	9	1,083	14	10
Total	13,282	977	4,303	100	5,193	65	
	Units	**Quantity**	**Mcal**				
Gasoline	ℓ	33	272				10
LP gas	ℓ	30	189				
Diesel	ℓ	45	419				
Electricity	kWh	458	1,377				
Grain grinding			49				3
Insecticide[e]	kg	0.19	14				11
Machinery[f]	$ 1967	4.71	85				10
Equipment[f]	$ 1967	25.44	430				10
Labor	hr	71					10
Cropland per cow unit	ha	0.99					
Cropland and pasture per cow unit	ha	1.53					
Total Mcal/cow unit			8,031				
Total Mcal/100 kg milk			197				

[a] The six regions (Tables 8 to 13) represent approximately 85% of the total U.S. dairy production. Figures for the U.S. represent a weighted average of the figures for the six regions.
[b] Regional production figures are weighted averages of production by states included in each region as defined above.[7]
[c] Total amounts fed of commerically mixed feeds, soybean meal, cottonseed meal, other oilseed meals, urea, molasses, alfalfa meal, animal protein, corn, sorghum (milo), oats, barley, other feed grains, hay, corn silage, sorghum silage, other silage, and other forage; based on representative states. The commercially mixed feeds were further broken down based on estimates of a major feed company in each region.[8] Feed grains, by-products, and roughages were combined in these summarized tables. Since this is disappearance data and no accurate figures of the amount actually fed per animal is available, the quantities were adjusted to agree with NRC requirements for a cow unit. The feeds fed and their proportions were kept constant. Protein and TDN levels of the feeds are expressed on a dry matter basis.[9]
[d] Pasture figures were not adjusted.[10]
[e] No detailed figures for insecticide use on dairy cattle were available.[11]
[f] The Wholesale Price Index was applied to adjust to $ 1967.[10]

required varies from 1.60 ha per cow unit for the Pacific region to 0.98 ha for the Southeastern states. This figure is influenced by the type of feed, the amount of feed fed per cow unit, and yield per hectare. Appalachia and the Lake States also require more land per cow unit than the average, at 1.53 and 1.59 ha, respectively. The Northeast and Southern Plains fall below the national average with 1.26 and 1.36 ha per cow unit, respectively.

Table 13
RESOURCE INPUTS (PER COW AND REPLACEMENT) INTO THE DAIRY PRODUCTION SYSTEM: LAKE STATES[a]

Number of Cow Units — 4,156,500; Yield — 4,667 kg Per Cow Per Year, 19,398,000,000 kg Milk Per Year[b]

Inputs	Amount fed (kg)	Amount of protein (kg)	TDN (kg)	% of Total TDN	Mcal cultural energy	% of total cultural energy	Ref.
Feed[c]							
Feed grains	2,053	197	1,600	35	2,270	27	8, 9
Soybean meal	92	40	70	2	7	—	
Alfalfa meal	2	0	1	—	1	—	
By-products	378	71	270	6	30	—	
Roughages	9,119	601	2,570	56	3,128	37	
Pasture[d]	345	19	52	1	127	2	10
Total	11,989	928	4,563	100	5,563	66	
	Units	**Quantity**	**Mcal**				
Gasoline	ℓ	34	281				10
LP gas	ℓ	21	132				
Diesel	ℓ	36	336				
Electricity	kWh	502	1,510				
Grain grinding			78				3
Insecticide[e]	kg	0.19	14				11
Machinery[f]	$ 1967	4.70	85				10
Equipment[f]	$ 1967	26.94	455				10
Labor	hr	81					10
Cropland per cow unit	ha	1.52					
Cropland and pasture per cow unit	ha	1.59					
Total Mcal/cow unit			8,458				
Total Mcal/100 kg milk			181				

[a] The six regions (Tables 8 to 13) represent approximately 85% of the total U.S. dairy production. Figures for the U.S. represent a weighted average of the figures for the six regions.

[b] Regional production figures are weighted averages of production by states included in each region as defined above.[7]

[c] Total amounts fed of commercially mixed feeds, soybean meal, cottonseed meal, other oilseed meals, urea, molasses, alfalfa meal, animal protein, corn, sorghum (milo), oats, barley, other feed grains, hay, corn silage, sorghum silage, other silage, and other forage; based on representative states. The commerciall- mixed feeds were further broken down based on estimates of a major feed company in each region.[8] Feed grains, by-products, and roughages were combined in these summarized tables. Since this is disappearance data and no accurate figures of the amount actually fed per animal is available, the quantities were adjusted to agree with NRC requirements for a cow unit. The feeds fed and their proportions were kept constant. Protein and TDN levels of the feeds are expressed on a dry matter basis.[9]

[d] Pasture figures were not adjusted.[10]

[e] No detailed figures for insecticide use on dairy cattle were available.[11]

[f] The Wholesale Price Index was applied to adjust to $ 1967.[10]

SUMMARY

The amount of cultural energy used in the dairy system includes all of the inputs in terms of megacalories needed for dairy production per year. Feed, fuels, and machin-

Table 14
RESOURCE INPUTS (PER COW AND REPLACEMENT) INTO THE DAIRY PRODUCTION SYSTEM: NORTHEAST[a]

Number of Cow Units — 2,504,400; Yield — 4,813 kg Per Cow Per Year, 12,054,000,000 kg Milk Per Year[b]

Inputs	Amount fed (kg)	Amount of protein (kg)	TDN (kg)	% of total TDN	Mcal cultural energy	% of total cultural energy	Ref.
Feed[c]							
Feed grains	1,112	105	875	19	1,242	17	8, 9
Soybean meal	73	32	55	1	6	—	
Alfalfa meal	3	1	2	—	2	—	
By-products	1,215	227	921	20	97	1	
Roughages	9,550	584	2,659	58	3,035	42	
Pasture[d]	336	19	50	1	119	2	10
Total	12,289	968	4,562	100	4,501	62	
	Units	Quantity	Mcal				
Gasoline	ℓ	32	264				10
LP gas	ℓ	27	170				
Diesel	ℓ	38	354				
Electricity	kWh	437	1,314				
Grain grinding			42				3
Insecticide[e]	kg	0.19	14				11
Machinery[f]	$ 1967	4.55	82				10
Equipment[f]	$ 1967	29.72	502				10
Labor	hr	82					10
Cropland per cow unit	ha	1.19					
Cropland and pasture per cow unit	ha	1.26					
Total Mcal/cow unit			7,247				
Total Mcal/100 kg milk			151				

[a] The six regions (Tables 8 to 13) represent approximately 85% of the total U.S. dairy production. Figures for the U.S. represent a weighted average of the figures for the six regions.

[b] Regional production figures are weighted averages of production by states included in each region as defined above.[7]

[c] Total amounts fed of commercially mixed feeds, soybean meal, cottonseed meal, other oilseed meals, urea, molasses, alfalfa meal, animal protein, corn sorghum (milo), oats, barley, other feed grains, hay, corn silage, sorghum silage, other silage, and other forage; based on representative states. The commercially mixed feeds were further broken down based on estimates of a major feed company in each region.[8] Feed grains, by-products, and roughages were combined in these summarized tables. Since this is disappearance data and no accurate figures of the amount actually fed per animal is available, the quantities were adjusted to agree with NRC requirements for a cow unit. The feeds fed and their proportions were kept constant. Protein and TDN levels of the feeds are expressed on a dry matter basis.[9]

[d] Pasture figures were not adjusted.[10]

[e] No detailed figures for insecticide use on dairy cattle were available.[11]

[f] The Wholesale Price Index was applied to adjust to $ 1967.[10]

ery and equipment are the largest inputs which comprise 65, 27, and 6% of all cultural energy consumed by the dairy system, respectively.

A total of 87.2 × 10⁹ Mcal is used per year in the U.S. dairy production system. Other inputs include labor, which amounts of 825 × 10⁶ hr, and land in feed (including pasture), of which 16.3 × 10⁶ ha are used.

Table 15

RESOURCE INPUTS (PER COW AND REPLACEMENT) INTO THE DAIRY PRODUCTION SYSTEM: PACIFIC[a]

Number of Cow Units — 1,072,000; Yield — 6,006 kg Per Cow Per Year, 6,438,000,000 kg Milk Per Year[b]

Inputs	Amount fed (kg)	Amount of protein (kg)	TDN (kg)	% of total TDN	Mcal cultural energy	% of total cultural energy	Ref.
Feed[c]							
Feed grains	1,240	119	977	20	1,498	22	8, 9
Soybean meal	237	103	181	4	19	—	
Alfalfa meal	155	28	90	2	99	1	
By-products	1,013	218	659	13	81	1	
Roughages	8,488	743	2,915	59	3,135	46	
Pastures	964	54	145	3	59	1	10
Total	12,097	1,265	4,967	100	4,891	70	
	Units	**Quantity**	**Mcal**				
Gasoline	ℓ	19	157				10
LP gas	ℓ	27	170				
Diesel	ℓ	42	391				
Electricity	kWh	309	929				
Grain grinding			47				3
Insecticide[e]	kg	0.19	14				11
Machinery[f]	$ 1967	2.93	53				10
Equipment[f]	$ 1967	1 8.19	307				10
Labor	hr	43					10
Cropland per cow unit	ha	1.41					
Cropland and pasture per cow unit	ha	1.60					
Total Mcal/cow unit			6,960				
Total Mcal/100 kg milk			116				k

[a] The six regions (Tables 8 to 13) represent approximately 85% of the total U.S. dairy production. Figures for the U.S. represent a weighted average of the figures for the six regions.

[b] Regional production figures are weighted averages of production by states included in each region as defined above.[7]

[c] Total amounts fed of commercially mixed feeds, soybean meal, cottonseed meal, other oilseed meals, urea, molasses, alfalfa meal, animal protein, corn, sorghum (milo), oats, barley, other feed grains, hay, corn silage, sorghum silage, other silage, and other forage; based on representative states. The commercially mixed feeds were further broken down based on estimates of a major feed company in each region.[8] Feed grains, by-products, and roughages were combined in these summarized tables. Since this is disappearance data and no accurate figures of the amount actually fed per animal is available, the quantities were adjusted to agree with NRC requirements for a cow unit. The feeds fed and their proportions were kept constant. Protein and TDN levels of the feeds are expressed on a dry matter basis .[9]

[d] Pasture figures were not adjusted.[10]

[e] No detailed figures for insecticide use on dairy cattle were available.[11]

[f] The Wholesale Price Index was applied to adjust to $ 1967.[10]

Feed represents 5000 Mcal per cow unit per year as a national average. This figure varies with region and is influenced greatly by the amount of by-products fed. Most of the by-products available are in the Northeast, Lake States, and Pacific regions.

Table 16
RESOURCE INPUTS (PER COW AND REPLACEMENT) INTO THE DAIRY PRODUCTION SYSTEM: SOUTHEAST[a]

Number of Cow Units — 327,000; Yield — 4,381 kg Per Cow Per Year, 1,433,000,000 kg Milk Per Year[b]

Inputs	Amount fed (kg)	Amount of protein (kg)	TDN (kg)	% of total TDN	Mcal cultural energy	% of total cultural energy	Ref.
Feed[c]							
Feed grains	2063	186	1663	38	2299	39	8, 9
Soybean meal	578	253	442	10	46	1	
Alfalfa meal	63	10	37	1	40	1	
By-products	1860	393	1382	31	149	3	
Roughages	2321	156	646	15	780	13	
Pasture	1718	96	258	6	689	12	10
Total	8603	1094	4428	100	4003	69	
	Units	**Quantity**	**Mcal**				
Gasoline	ℓ	17	140				10
LP gas	ℓ	20	126				
Diesel	ℓ	28	261				
Electricity	kWh	313	941				
Grain grinding			78				3
Insecticide[e]	kg	0.19	14				11
Machinery[f]	$ 1967	2.32	42				10
Equipment[f]	$1967	11.80	199				10
Labor	hr	48					10
Cropland per cow unit	ha	0.64					
Cropland and pasture per cow unit	ha	0.98					
Total Mcal/cow unit			5824				
Total Mcal/100 kg milk			133				

[a] The six regions (Tables 8 to 13) represent approximately 85% of the total U.S. dairy production. Figures for the U.S. represent a weighted average of the figures for the six regions.

[b] Regional production figres are weighted averages of production by states included in each region as defined above.[7]

[c] Total amounts fed of commercially mixed feeds, soybean meal, cottonseed meal, other oilseed meals, urea, molasses, alfalfa meal, animal protein, corn, sorghum (milo), oats, barley, other feed grains, hay, corn silage, sorghum silage, other silage, and other forage; based on representative states. The commercially mixed feeds were further broken down based on estimates of a major feed company in each region.[8] Feed grains, by-products, and roughages were combined in these summarized tables. Since this is disappearance data and no accurate figures of the amount actually fed per animal is available, the quantities were adjusted to agree with NRC requirements for a cow unit. The feeds fed and their proportions were kept constant. Protein and TDN levels of the feeds are expressed on a dry matter basis.[9]

[d] Pasture figures were not adjusted.[10]

[e] No detailed figures for insecticide use on dairy cattle were available.[11]

[f] The Wholesale Price Index was applied to adjust to $ 1967.[10]

Due to their lower cow populations, however, the Southeast, Southern Plains, and Appalachian regions feed more by-products per cow unit.

Fuels used amount to 2100 Mcal per cow unit. This figure varies by region and seems to be principally influenced by the average size of the farms in each region. Machinery and equipment figures translate to 500 Mcal per cow unit and also tend to be lower in regions with a large number of cows per farm.

Table 17
RESOURCE INPUTS (PER COW AND REPLACEMENT) INTO THE DAIRY PRODUCTION SYSTEM: SOUTHERN PLAINS[a]

Number of Cow Units — 731,000; Yield — 4,321 kg Per Cow Per Year, 3,159,000,000 kg Milk Per Year[b]

Inputs	Amount fed (kg)	Amount of protein (kg)	TDN (kg)	% of total TDN	Mcal cultural energy	% of total cultural energy	Ref.
Feed[c]							
Feed grains	1919	199	1462	33	3256	44	8, 9
Soybean meal	224	99	172	4	18	—	
Alfalfa meal	66	12	38	1	42	1	
By-products	1212	333	885	20	97	1	
Roughages	4197	424	1613	37	1546	21	
Pasture	1464	82	220	5	206	3	10
Total	9082	1149	4390	100	5165	69	
	Units	**Quantity**	**Mcal**				
Gasoline	ℓ	25	206				10
LP gas	ℓ	32	201				
Diesel	ℓ	44	410				
Electricity	kWh	338	1016				
Grain grinding			73				3
Insecticide	kg	0.19	14				11
Machinery	$ 1967	3.80	69				10
Equipment	$ 1967	17.68	299				10
Labor	$ 1967	65					10
Cropland per cow unit	ha	1.07					
Cropland and pasture per cow unit	ha	1.36					
Total Mcal/cow unit			7459				
Total Mcal/100 kg milk			173				

[a] The six regions (Tables 8 to 13) represent approximately 85% of the total U.S. dairy production. Figures for the U.S. represent a weighted average of the figures for the six regions.

[b] Regional production figures are weighted averages of production by states included in each region as defined above.[7]

[c] Total amounts fed of commercially mixed feeds, soybean meal, cottonseed meal, other oilseed meals, urea, molasses, alfalfa meal, animal protein, corn, sorghum (milo), oats, barley, other feed grains, hay, corn silage, sorghum silage, other silage, and other forage; based on representative states. The commercially mixed feeds were further broken down based on estimates of a major feed company in each region.[8] Feed grains, by-products, and roughages were combined in these summarized tables. Since this is disappearance data and no accurate figures of the amount actually fed per animal is available, the quantities were adjusted to agree with NRC requirements for a cow unit. The feeds fed and their proportions were kept constant. Protein and TDN levels of the feeds are expressed on a dry matter basis.[9]

[d] Pasture figures were not adjusted.[10]

[e] No detailed figures for insecticide use on dairy cattle were available.[11]

[f] The Wholesale Price Index was applied to adjust to $ 1967.[10]

Labor averages 74 hr per cow unit and is influenced by herd size because of the labor-saving management systems possible with larger farms.

In terms of cultural energy efficiency, the U.S. averages 166 Mcal/100 kg milk. There are large regional variations (116 to 197 Mcal/100 kg) which are due mainly to the differences in production per cow.

When cull cows are included as outputs and total production is in terms of protein produced, the U.S. averages 44 Mcal/kg. In terms of land used, 123 kg protein are produced per hectare in feed and pasture. Proein is produced at a national rate of 2.4 kg/hr of labor.

Efficiency in terms of cultural energy used can be increased in the future by increasing milk production per cow and by increasing herd size to an optimum level. Increasing the quantity of silages fed (up to the biological limits of the cow's intake) may be an effective way of increasing efficiency as silage gives large yields per hectare and is less energy intensive than are grains.

REFERENCES

1. **Nott, S. B., Ricks, D. S., Speicher, J. A., and Brown, L. H.,** A Dairy Budget Guide, Extension Bulletin E-1018, Cooperative Extension Service, Michigan State University, East Lansing, August 1976.
2. **Herendeen, R. A. and Bullard, C. W.,** Energy Costs of Goods and Services, 1963 and 1967, CAC Doc. No. 140, University of Illinois, Urbana-Champaign, 1974.
3. **Pimental, D.,** personal communication, 1978.
4. **Cervinka, V., Chancellor, W. J., Coffelt, R. J., Curley, R. G., and Dobie, J. B.,** Energy Requirements for Agriculture in California, California Department of Food and Agriculture and University of California, Davis, 1974.
5. **Pimental, D. and Pimental, M.,** *Food, Energy and Society,* Arnold Press, England, in press, 1978.
6. Agricultural Statistics, U.S. Department of Agricuture, Washington, D.C., 1976.
7. Agricultural Statistics, U.S. Department of Agriculture, Washington, D.C., 1975.
8. U.S. Census of Agriculture, Vol. 5, U.S. Department of Agriculture, Washington, D.C., 1971.
9. National Research Council, Nutrient Requirements of Dairy Cattle, 1971.
10. Firm Enterprise Data System, Dairy Report, Economic Research Service, U.S. Department of Agriculture, and Oklahoma State University, Stillwater, 1976.
11. Agricultural Economics Rep. 252, Economic Research Service, U.S. Department of Agriculture, Washington, D.C., 1971.

ENERGY USE IN AGRICULTURE POULTRY

Charles E. Ostrander

INTRODUCTION

The poultry industry has been one of the fastest moving and drastically changing industries in agriculture during the past two decades. It has evolved from small flocks of nondescript chickens on almost every farm to a sophisticated, highly technical, and specialized industry. No longer does each farm maintain a flock of chickens for eggs and meat. Poultry production has become more specialized and sophisticated, and there are fewer but larger poultry operations. It is not uncommon to have flocks of many thousands of laying hens, and there are several with over 1 million layers on one operation.

Through technology, environmental housing, and mechanization, poultry operations have become very efficient in the use of labor and land. Through improved technology in breeding, feeding, housing, and management, the industry is supplying the egg demand with fewer layers today than 20 years ago, even with the increase in population.

Most of the broiler industry is in the southern states, but as can be seen in Table 1,[1] there are some broilers grown in other areas. The broiler industry started, on a commercial basis, on the Del-Mar-Va peninsula (peninsula east of Chesapeake Bay, made up of parts of Maryland, Delaware, and Virginia) and later developed in large integrated operations in the southern states. In 1977, Arkansas was the largest producer of broilers. There continues to be a substantial broiler industry in Maine, Del-Mar-Va, and California.

This industry continues to expand, producing over 3 billion head of broilers in 1977.[2] It is one of the most efficient converters of grains to animal protein.

Turkeys are produced in all regions of the country except the East South Central region. The West North Central region continues to be the leading turkey-producing area. The turkey industry has also made great strides in improving efficiency through better breeding and management.

The duck industry, while small compared to broilers and turkeys, is a relatively important industry in some areas. Long Island, N.Y. continues to be an important producer of ducks, but larger numbers are now being produced in Indiana, Wisconsin, and some other areas. Ducks grow very rapidly and are reasonably efficient converters of grain to animal protein.

The poultry industry produced over $7 billion of gross income in 1976.[2] This is nearly double the value in 1965. Eggs accounted for 45, broilers 42, and turkeys 12% of the total gross income.

Energy use per head of poultry for production activities has declined slightly since the mid-1960s, and considerably since the 1940s, as more efficient use of heating fuels, particularly for broilers, was more than sufficient to offset larger requirements for electricity, particularly for layers and turkeys.[1] The efficiency of energy use in marketing improved between the 1960s and 1974 due to economics of scale, larger units and equipment, more direct marketing channels, and increased density of supply areas.

Substantial quantities of various types of energy are required by the poultry and egg industry. The energy expenditures in 1974 amounted to almost $550 million for 36.9

trillion kcal of energy used in the total industry, including production and marketing:[1]

Propane	738,000,000ℓ
Natural gas	130,800,000 m^3
Fuel oil	33,300,000 ℓ
Coal	29,600 t
Gasoline and diesel fuel	120,100,000ℓ
Electricity	1,415,900,000 kWh

These costs of energy sources for production were approximately $126 million, or 2% of gross farm income from poultry and eggs.[1]

Rates of energy use per 1000 birds varies greatly with types of poultry, among regions and among producers. Climatic conditions and seasonal variations in temperatures caused some differences, but these were often modified by housing design, types of equipment, and alternative management practices. The open type, uninsulated poultry houses often used in the more southern areas of the country use less total energy in construction and materials, but this is often balanced by the fact that this type of housing uses labor energy and feed energy less efficiently than the more elaborate, insulated houses in the more northern areas. The broiler industry uses considerably more energy than other phases of the poultry industry.

ENERGY USE BY THE POULTRY INDUSTRY

The energy used by the various types of poultry varies as does the amount used in various regions of the U.S. The average energy used per 1000 birds in the U.S. in 1974 is shown in Table 3.[1] Turkeys are the largest users of energy per 1000 birds. Layers are the largest users of electricity.

The amount of energy used by various regions producing poultry is also affected by the type of housing used. The monthly use varies considerably by the type of house used (Table 7).[1]

Considerable energy is used in marketing poultry and poultry products. During 1974 this was about $421 million, or 9% of the difference between farm and retail values. This is considerably greater than the $126 million for production during the same period.

Efficiencies in the use of energy for marketing have actually been improving, but the costs of energy have increased so greatly since the mid-1960s that total energy costs have increased greatly.

There have been many changes in the type of product, particularly in the meat production phase of the poultry industry. There is more emphasis in cutup, prepacked, and precooked products than in years past. This affects the energy used.

The energy used for processing chickens and eggs is rather constant throughout the year, but for turkeys this increases dramatically during the late summer and fall since most turkeys continue to be marketed during the Thanksgiving and Christmas holiday period (Table 7).[1]

Energy is used for heating and cooling in the processing industry, and the ratio of use changes drastically in the north and south (Table 7).

The energy productivity in marketing has changed dramatically over the years. Using the period 1959 to 1969 as 100, Table 8 points out that some phases of marketing have changed considerably in energy productivity while others have remained rather constant. Some of this is due to further processing of some products; some is due to improved transportation, better distribution practices, larger and more efficient processing facilities, and hauling larger loads to points of distribution.

More direct marketing and the decline in traditional wholesaling has improved the wholesaling efficiency in the use of energy. Energy costs have continued to rise, and the amount of further processing has increased; therefore, energy costs have about doubled in the period 1965 to 1974 (Table 9).

Chick and poult hatcheries have improved energy efficiency by using better and larger equipment. While costs have continued to rise, the number of BTUs required to hatch chicks has tended to improve.

Feed mills have tended to improve energy efficiency. A 1964 study[1] showed 83,166 to 98,413 kcal required to mill a metric ton of feed. Recent studies show 58,216 to 59,602 kcal/t of pelleted poultry feed and under 27,722 kcal/t of mash.

Larger loads have tended to reduce the energy required to deliver feed to the farm. Increased density, particularly in the broiler industry, has led to improved energy efficiency in delivery. Even though improvements have been made, it is suggested that it might be possible to effect a 10 to 20% savings in embodied energy in poultry rations, but with current prices of feed ingredients, costs per ton to producers might increase by a smaller to a larger ratio.[1]

There is considerable room for improvement in the use of energy in the poultry industry. It is not unusual to find some producers using twice as much energy per 1000 birds as other producers. These variations may be due to types of housing, mechanization, elevation, windbreaks, management, and perhaps other factors. Better insulation by use of new types of insulating materials, better control of ventilation, reuse of heat dissipated from ventilation and processing, a movement back to more basic foods with less further processing, a reexamination of irradiation preservation, newer shipping and transportation procedures to improve shelf-life, and changes in retail marketing procedures could possibly save large amounts of energy.

Tables 10 to 14 attempt to reveal the approximate energy used in inputs to production in egg production, broiler production, and turkey production in various areas of the U.S. These will vary with individual producers, but this provides a picture of the energy efficiency of the poultry industry.

The poultry industry has improved efficiency over the years and consistently strives for improved efficiency. Poultry is one of the most efficient converters of dietary energy inputs to protein produced of the animal agriculture species (Table 15).[4]

The turkey industry is the largest user of energy per 1000 birds, followed by egg production. However, the egg production industry is consistently the most efficient in relation of inputs to outputs. Broilers are the least efficient when comparing inputs to outputs.

Table 1
POULTRY PRODUCED, BY REGION AND U.S. — 1974

Item	New England	Middle Atlantic	East North Central	West North Central	South Atlantic	East South Central	West South Central	Mountain	Pacific	U.S.
Broilers produced	82.1	68.0	36.7	38.6	1,280.5	707.0	766.1	0	125.6	3,104.6
Turkeys produced	0.3	3.4	17.0	43.2	28.5	0	19.2	7.5	20.7	139.8
Layers, average on hand	14.5	26.6	38.0	33.5	60.9	27.9	30.8	6.5	45.9	284.6
Other chickens produced	14.5	25.8	37.7	37.5	62.5	27.8	39.6	6.3	32.8	284.5
Other poultry raised	0.6	8.4	7.9	1.5	4.1	0.6	3.0	0.2	2.9	29.2

Note: Figures given represent million head.

Table 2
ENERGY USE IN POULTRY PRODUCTION, BY TYPE, U.S. — 1974

Item[a]	Propane (million ℓ)	Natural gas (million m^3)	Fuel oil (million ℓ)	Coal (1000 t)	Gasoline (million ℓ)	Diesel (million ℓ)	Electricity (million kWh)
Broilers	462.9	63.4	24.2	17.1	23.5	—	504.1
Turkeys	160.5	32.9	1.9	5.6	20.1	6.1	66.9
Hens	19.3	7.1	1.9	—	36.3	6.8	828.8
Chickens	89.3	15.4	4.5	5.5	21.6	3.4	14.3
Other poultry	6.1	12.0	0.8	1.4	2.3	—	1.8
Total	738.1	130.8	33.3	29.6	103.8	16.3	1,415.9
kcal (billion)	5,687	1,545	380	213	1,049	186	4,054
							13,144 Total[b]

[a] Turkeys and hens include breeders; chickens mainly include laying flock replacements, but also some nonbroiler meat chickens; and other poultry includes ducks, geese, guineas, game birds, etc.

[b] For production only — does not include marketing.

Table 3
U.S. AVERAGE ENERGY USE PER 1000 BIRDS — 1974[a]

Type of poultry	Propane (ℓ)	Natural gas (m^3)	Fuel oil (ℓ)	Coal (t)	Gasoline (ℓ)	Diesel fuel (ℓ)	Electricity (kWh)
Broilers	154.8	21.2	8.1	0.006	7.85	—	168.5
Turkeys[b]	1222.3	250.6	14.5	0.04	153.1	46.5	509.5
Hens	67.4	24.8	6.6	—	91.8	23.7	2893.1
Chickens[c]	312.7	53.9	15.8	0.02	75.6	11.9	50.1
Other poultry[d]	208.9	410.9	27.4	0.05	78.8	—	61.6

[a] These are true averages used per 1,000 birds. Some producers may not use a particular fuel; therefore, these figures do not depict the actual amount of any one material used by a particular producer.

[b] Includes breeders.

[c] Mainly laying flock replacements, but includes some nonbroiler meat chickens.

[d] Ducks, geese, guineas, game birds, etc.

Table 4
ENERGY USE IN POULTRY PRODUCTION BY REGIONAL PRECENTAGE OF U.S. TOTAL — 1974

Region	Propane	Natural gas	Fuel oil	Coal	Gasoline and diesel	Electricity
New England	0.4	0.0	47.2	0.0	3.4	5.4
Middle Atlantic	3.5	6.2	6.0	15.6	6.6	8.1
East North Central	6.7	1.4	5.2	6.1	11.3	8.9
West North Central	16.2	0.1	2.8	2.3	14.0	8.5
South Atlantic[a]	34.1	8.4	18.7	42.5	24.5	26.9
East South Central	15.9	8.1	0.3	26.8	10.0	11.9
West South Central	18.5	39.5	0.0	0.0	14.9	13.6
Mountain	1.4	2.7	0.0	6.7	2.5	1.7
Pacific	3.3	33.6	19.8	0.0	12.9	15.1

Note: The type of energy used by various regions varies considerably.

[a] The South Atlantic is the largest user of energy because of the broiler industry.

Table 5
MONTHLY ENERGY NEEDS IN POULTRY PRODUCTION FOR THE U.S., BY TYPE — 1974

Month	Propane (million ℓ)	Natural gas (million m^3)	Fuel oil (million ℓ)	Coal (t)	Gasoline and diesel (million ℓ)	Electricity (million kWh)
January	127.2	19.2	5.3	5.1	7.95	110.3
February	108.4	15.4	4.2	4.3	11	103.0
March	110.7	15.2	4.2	4.4	11	106.6
April	71.6	10.0	3.0	2.8	12.9	103.2
May	48.9	7.0	2.3	1.8	14	119.2
June	24.6	4.1	1.1	0.9	12.1	135.2
July	12.9	2.6	0.8	0.5	9.84	164.2
August	11.8	2.1	1.1	0.5	7.95	159.6
September	18.6	2.8	1.1	0.8	8.71	112.4
October	36.8	5.1	2.3	1.6	9.84	99.1
November	65.2	8.1	3.0	2.7	9.5	98.7
December	102.0	14.1	4.6	4.3	7.95	103.4
Total	738.8	105.7	33.0	29.7	122.7	1,415.9
Average	61.6	8.8	2.75	2.5	10.2	118.0

Note: The amount of energy used by the poultry industry varies considerably during the year, particularly the use of propane. Propane use during the winter months is considerably greater than during the summer months, as shown. This is primarily due to the fact that propane is used extensively for brooding broiler chicks.

Table 6

MONTHLY USE OF ENERGY IN DIFFERENT REGIONS, BY TYPE OF POULTRY AND POULTRY HOUSING CONDITIONS — 1974

Example	Energy type[a]	Jan.	Feb.	Mar.	Apr.	May	June	July	Aug.	Sept.	Oct.	Nov.	Dec.	Annual use per 1000 birds
		Monthly rate of use as percent of average monthly rate (%)												
Broilers														
Northern State, semi- and environmentally controlled housing	Propane	204.0	178.0	153.7	95.5	55.4	30.6	19.5	24.3	42.2	77.0	130.3	189.6	56.4
	Electricity	99.0	94.0	90.0	88.0	91.0	101.0	118.0	122.0	104.0	99.0	96.0	98.0	226
Southern State, 10% environmentally controlled housing	Propane	219.3	180.5	151.7	81.5	52.3	19.9	8.5	12.5	35.3	80.1	146.9	211.3	44.0
	Electricity	92.0	83.0	78.0	78.0	90.0	110.0	140.0	135.0	115.0	98.0	90.0	91.0	45
Turkeys														
Midwestern State, 70% confinement rearing, fair housing	Propane	192.1	156.9	124.9	99.4	76.9	56.4	37.3	37.7	52.0	82.4	121.1	162.9	700
	Electricity	96.4	96.4	84.3	78.3	78.3	102.4	144.6	144.6	96.4	90.4	93.4	96.4	599
West Coast State, 5% confinement rearing	Propane	174.9	140.7	114.1	83.6	60.8	55.1	60.8	72.2	83.6	98.9	117.9	138.8	264
	Electricity	93.0	93.0	81.0	74.0	74.0	110.0	165.0	160.0	100.0	80.0	85.0	85.0	442
Hens														
Northern State, closed housing, high mechanization	Electricity	88.0	87.0	85.0	84.0	94.0	105.0	135.0	140.0	104.0	94.0	92.0	90.0	3829
Southern State, largely open housing, medium mechanization	Electricity	83.0	82.0	77.0	75.0	95.0	118.0	160.0	155.0	100.0	86.0	85.0	84.0	2400
Northern State	Gasoline/	78.0	75.8	85.0	106.8	107.0	107.8	93.0	93.3	112.0	124.5	124.5	92.5	40.0
Southern State	diesel	86.8	84.5	93.7	103.0	103.3	98.8	96.8	97.0	112.5	115.7	115.7	92.5	40.0
Turkeys														
Northern State, 80% range reared,	Gasoline/ diesel	47.5	49.2	54.1	68.9	104.9	113.1	136.1	134.4	159.0	170.5	103.3	59.0	61.0
20% range reared	Gasoline/ diesel	69.0	71.4	78.6	100.0	114.3	107.1	92.9	90.5	126.2	142.9	121.4	85.7	42.0

[a] Propane, gasoline, and diesel units are in gallons and electricity is in kilowatthours.

Table 7
FACTORS AFFECTING MONTHLY USE OF ENERGY IN POULTRY AND EGG MARKETING

	Chickens (% of annual volume)		Turkeys (% of annual volume)		Eggs[a] (% of annual volume)		South[b]		North[c]	
Month	**Slaughter**	**Further processed**	**Slaughter**	**Further processed**	**Farm production**	**Commer-cially broken**	**Number of heating degree days[d]**	**Number of cooling degree days[e]**	**Number of heating degree days[d]**	**Number of cooling degree days[e]**
January	8.7	9.3	4.6	5.8	8.6	6.5	599	5	1281	0
February	7.5	8.3	2.8	4.9	8.0	7.1	451	8	1094	0
March	8.1	9.2	2.8	5.5	8.7	8.2	339	25	945	0
April		8.9	8328	5.1	8.5	8.3	105	73	545	0
May	9.1	8.8	5.5	6.1	8.7	9.9	18	212	262	22
June	8.8	7.7	8.3	7.5	8.2	10.1	2	391	67	112
July	8.8	7.2	10.9	9.3	8.3	9.6	0	518	14	233
August	9.0	8.1	13.0	11.8	8.2	9.2	0	495	25	191
Sept-ember	7.8	8.0	12.0	10.9	7.9	8.2	1	318	144	63
October	9.0	9.2	14.7	13.9	8.2	9.1	85	114	412	7
November	7.5	8.2	13.4	11.3	8.3	7.5	317	13	773	0
December	7.5	7.1	8.2	7.9	8.4	6.3	552	4	1157	0
Total	**100.0**	**100.0**	**100.0**	**100.0**	**100.0**	**100.0**	**2469**	**2176**	**6719**	**628**

[a] Averages for 1973 to 1974.
[b] Combined averages for California, Texas, Arkansas, Georgia, and Florida.
[c] Combined averages for Maine, Minnesota, Delaware-Maryland, Pennsylvania, and Colorado.
[d] Weather Bureau data, 1941-1970, average degrees below 65°.
[e] Weather Bureau data, 1941-1970, average degrees above 65°.

Table 8
MEASUREMENTS OF ENERGY PRODUCTIVITY IN MARKETING IN THE U.S. BY SELECTED PERIODS (1965 to 1969 = 100)

Commodity and function	1955 to 1959	1960 to 1964	1965 to 1969	1970 to 1973	1974[a]
Grade A large eggs					
Assembly	52	70	100	111	101
Packing	118	105	100	102	116
Long-distance transportation	75	91	100	115	130
Wholesaling	75	89	100	166	185
Retailing	113	110	100	116	125
Total[b]	**81**	**92**	**100**	**122**	**132**
Broilers (ready-to-cook)					
Assembly	62	96	100	120	120
Processing	111	104	100	107	108
Long-distance transporation	80	91	100	110	119
Wholesaling	77	90	100	161	182
Retailing	113	110	100	117	125
Total[b]	**85**	**98**	**100**	**122**	**128**
Turkeys (ready-to-cook)					
Assembly	81	97	100	131	132
Processing[c]	121	110	100	105	94
Long-distance transportation	73	92	100	118	132
Wholesaling	95	96	100	140	187
Retailing[c]	135	122	100	97	92
Total[b,c]	**106**	**106**	**100**	**111**	**109**

[a] Preliminary.
[b] All functions.
[c] The period 1970 to 1974 reflects the effects of changes in product form toward more whole bird processed and cutup packs. For whole ready-to-cook birds, productivity has probably risen.

Table 9
VOLUMES OF OUTPUT AND ESTIMATED COSTS OF ENERGY IN POULTRY AND EGG PRODUCTION AND MARKETING, U.S., 1965 and 1974

Item	Unit	1965[a]	1974[a]
Farm production volumes			
Average of layers on farms	Number	301	286pT
Commercial broilers produced	Dozen	2334	2993
Turkeys produced	Dozen	105	131
Nonbroiler chickens and other poultry raised	Dozen	266	295
Costs for energy in production	Dollars	52.6	126.5
Marketing volumes			
Farm egg production	Cases	182.1	183.0
Less: eggs used for hatching	Dozen	11.1	12.3
Less: eggs used for farm consumption	Dozen	0.6	0.2
Eggs processed	Dozen	15.9	21.0
Shell eggs marketed commercially 1/[b]	Dozen	154.5	149.5
Poultry meat sold ready-to-cook 2/[c]	Pounds	7271	9263
Poultry meat further processed	Dozen	745	1518
Energy costs in marketing			
Assembly	Dollars	31.9	65.2
Primary processing	Dozen	47.7	110.8
Further processing	Dozen	5.4	18.7
Long-distance transportation 3/[d]	Dozen	35.1	73.3
Wholesaling, retailing	Dozen	90.2	153.1
Total	Dozen	210.3	421.1
Energy, costs, production, and marketing	Dozen	262.9	547.6
Index of energy prices	(1967 = 100)	99	206

[a] Figures given represent millions.
[b] Excludes hatching and farm consumption.
[c] Excludes farm consumption.
[d] Includes delivery of further processed.

Table 10
ENERGY INPUTS PER 1000 LAYING HENS: EGG PRODUCTION — NORTHERN AREAS

Inputs	Amount	kcal	Ref.
Buildings	30.93 m^{2}[a]	832,500	13, 14
Labor	400 hr		5
Equipment	11.35 kg	234,605	6
Gasoline and diesel	150 ℓ	1,516,350	1
Electricity	3,829 kWh	10,962,427	1
Laying ration	38,636 kg	105,500,000[b]	7
Transporation	38,759.85	9,961,282	
Total		**129,007,164**	

[a] In personal communication, Hannon stated 38,000 kcal/ft² for all outbuildings. He suggested we use 50,000 kcal/ft² for insulated and environment control laying and brooding houses. Depreciated over 20 years.

[b] Amount of kilocalories required to produce and process amount of ration shown.

Table 11
ENERGY INPUTS PER 1000 LAYING HENS: EGG PRODUCTION — SOUTHERN AREAS

Inputs	Amount	kcal	Ref.
Buildings	46.45 m^2	950,000	
Labor	450 hr		5
Equipment	8.5 kg	155,025	6
Gasoline and diesel	150 ℓ	1,516,350	1
Electricity	2,400 kWh	6,871,200	1
Laying ration	38,590 kg	105,200,000	1, 7
Transportation[a]	38,710 kg	9,948,470	
Total		**124,641,045**	

[a] Transportation = [equipment (kg) + fuel (kg) + feed (kg) × 257 kcal/kg]

Table 12
ENERGY INPUTS PER 1000 REPLACEMENT PULLETS: REARING

Inputs	Amount	kcal	Ref.
	Northern Areas		
Buildings	92.9 m^2	2,500,000	
Labor	61 hr		5
Equipment	5.68 kg	117,406	6
Brooding (propane)	211.5 ℓ	1,629,565	1
Electricity	226 kWh	647,038	1
Rearing ration	6,810 kg	22,020,000	1, 7
Transportation	6,940.48 kg	1,783,703	
Total		**28,697,712**	
	Southern Areas		
Buildings	92.9 m^2	2,500,000	
Labor	61 hr		5
Equipment	5.68 kg	117,406	6
Brooding (propane)	165.0 ℓ	1,271,325	1
Electricity	45 kWh	128,835	1
Rearing station	6,810 kg	22,020,000	1, 7
Transportation	6,913.0 kg	1,766,649	
Total		**27,894,215**	

Table 13
ENERGY INPUTS PER 1000 BROILERS

Inputs	Amount	kcal	Ref.
	Northern Areas		
Buildings	69.7 m^2	1,875,000	
Labor	9 hr		12
Equipment	3.78 kg	78,133	6
Brooding (propane)	211.5 ℓ	1,629,565	1
Electricity	226 kWh	873,038	1
Broiler ration	3,182 kg	9,240,000	1, 9
Transportation	3,311 kg	850,927	
Total		**14,546,663**	
	Southern Areas		
Buildings	69.7 m^2	1,425,000	
Labor	12 hr		12
Equipment	3.78 kg	78,133	6
Brooding (propane)	165 ℓ	1,271,325	1
Electricity	45kWh	128,835	1
Broiler ration	3,182 kg	9,240,000	1, 9
Transportation	3,283 kg	843,764	
Total		**12,987,057**	

Table 14
ENERGY INPUTS PER 1000 TURKEYS

Inputs	Amount	kcal	Ref.
	Midwest		
Buildings	418.1 m^2	8,550,000	
Labor	125 hr		8
Equipment	6.0 kg	124,020	6
Brooding (propane)	2,625 ℓ	20,225,625	1
Electricity	599 kWh	1,714,937	1
Gasoline and diesel	158 ℓ	1,597,222	1
Feed	34,700 kg	110,780,000	1, 10
Transportation	36,374 kg	9,348,118	
Total		**152,339,922**	
	West Coast States		
Buildings	418.1 m^2	8,550,000	
Labor	124 hr		8
Equipment	6.0	124,020	6
Brooding (propane)	990 ℓ	7,627,950	1
Electricity	442 kWh	1,265,446	1
Gasoline and diesel	229 ℓ	2,314,961	1
Feed	34,700 kg	110,780,000	1, 10
Transportation	35,462 kg	9,113,734	
Total		**139,776,111**	

Table 15
OVERALL EFFICIENCY WITH WHICH FARM ANIMALS CONVERT DIETARY ENERGY TO PROTEIN[a]

Food product	Level and/or rate of output	Grams of protein produced per million calories of digestible energy input
Eggs	200 eggs per year	10.1
	250 eggs per year	13.7
Broiler	3.5 lb/12 weeks; 3 lb feed per 1 lb gain	11.9
	3.5 lb/10 weeks; 2.5 lb feed per 1 lb gain	13.7
	3.5 lb/8 weeks; 2.1 lb feed per 1 lb gain	15.9
Pork	200 lb; 6 lb feed per 1 lb gain	5.0
	200 lb/6 months; 4 lb feed per 1 lb gain	6.1
	200 lb; 2.5 lb feed per 1 lg gain	8.7
	Biological limit (?); 2 lb feed per 1 lb gain; no losses	12.1
Milk	7,936 lb/year; No concentrates[b]	10.5
	11,905 lb/year; 25% of energy as concentrates	12.8
	20,000 lb/year; 50% of energy as concentrates	16.3
	30,000 lb/year; 65% of energy as concentrates	20.5
Beef	1,100 lb/15 months; 8 lb feed per 1 lb gain	2.3
	1,100 lb/12 months; 5 lb feed per 1 lb gain	3.2
	Highly intensive system; no losses	4.1

[a] These data include the energy cost of reproduction, rearing of breeding stock, and mortality as well as that of production.

[b] Concentrates are high-energy feeds, e.g., grains and oilseed meals.

From Reid, T., Proceedings of Cornell Nutrition Conference, Cornell University, Ithaca, N.Y., 1975, 16. With permission.

REFERENCES

1. **Rogers, G. B., Benson, V. W., and Van Dyne, D. L.,** Energy Use and Conservation in the Poultry and Egg Industry, Agricultural Economics Rep. No. 354, Economic Research Service, U. S. Department of Agriculture, Washington, D.C., 1976.
2. Agriculture situation and outlook, in New York Economic Handbook, 1978, Department of Agricultural Economics, New York State College of Agriculture and Life Sciences Extension 77-29, Cornell University, Ithaca, N.Y., 1977.
3. **Benson, V. W.,** A Guide to Energy Savings — for the Poultry Producer, U.S. Department of Agriculture, Federal Energy Administration, Washington, D.C., 1977.
4. Energy Use in Agriculture — Now and for the Future, Council for Agricultural Science and Technology Rep. No. 68, Department of Agronomy, Iowa State University, Ames, August, 1977.
5. **Bratton, C. A. and Thacker, G. H.,** Poultry Farm Business Summary, New York, 1976, Department of Agricultural Economics, College of Agriculture and Life Sciences, AE Extension 77-20, Cornell University, Ithaca, N.Y., 1977.
6. **Davis, H. R.,** personal communication.
7. **Scott, M. L. and Austic, R. E.,** Recommendations for Formulating Poultry Rations, Department of Poultry Science, New York State College of Agriculture and Life Sciences, Cornell University, Ithaca, N.Y., 1975.
8. **Gallimore, F. and Irvin, R. J.,** The Turkey Industry: Structure, Practices and Costs, Marketing Rsearch Rep. No. 1000, Economic Research Service, U.S. Department of Agriculture, Washington, 1973.
9. **Scott, M. L., Young, R. J., and Nesheim, M. C.,** Broiler Rations, Department of Poultry Science, New York State College of Agriculture and Life Sciences, Cornell University, Ithaca, N.Y., 1967.
10. **Scott, M. L., Young, R. J., and Nesheim, M. C.,** Turkey Rations, Department of Poultry Science, New York State College of Agriculture and Life Sciences, Cornell University, Ithaca, N.Y., 1967.
11. **Marsden, S. J.,** Turkey Production, Agricultural Handbook No. 393, Agricultural Research Service, U.S. Department of Agriculture, Washington, D. C., 1971.
12. **Parkhurst, R. T.,** Commercial Broiler Production, Agricultural Handbook No. 320, Agricultural Research Service, U. S. Department of Agriculture, Washington, D.C., 1967.
13. **Hannon, B., Stein, R., Segal, B., and Serber, D.,** Energy Use in Building Construction, CAC Doc. No. 228, Center for Advanced Computations, University of Illinois, Urbana, 1977, 84.
14. **Hannon, B.,** personal communication.

CULTURAL ENERGY, LAND, AND LABOR REQUIREMENTS OF SWINE PRODUCTION SYSTEMS IN THE U.S.

J. T. Reid, P. A. Oltenacu, M. S. Allen, and Ottilie D. White

CHRONOLOGY OF CHANGES IN SWINE PRODUCTION IN THE U.S.

Pork Consumption

Since 1950, the average amount of pork consumed per capita per year in the U.S. has varied less from year to year than that of any other meat or of milk or eggs. From 1950 through 1976, the mean consumption varied from 26.4 to 31.4 kg of carcass per caput per year.[1] Prior to that period, the level of consumption was generally greater, reaching 33. 3 kg in 1940.

Swine Population

During the period 1955 through 1976, the number of hogs slaughtered varied irregularly from 75.3×10^6 to 87.6×10^6, except for 1975 during which 74.0×10^6 were slaughtered and 1971 during which 95.5×10^6 were killed for meat purposes.[2,3] Between 1960 and 1977, the number of sows which farrowed varied from 9.9×10^6 (1975) to 14.0×10^6 (1970) per year.[4] Accordingly, the annual pig crop during that period varied from 71.3×10^6 (1975) to 101.9×10^6 (1970) per year.[4] The reduced number of sows farrowed and pigs produced in 1975 appears to reflect the backlash of the exceedingly high prices of corn and soybeans during 1973 and 1974 that resulted from large exports of U.S. grain. From 1960 to 1971, grains were produced generally in excess of the demand. The average number of pigs per litter has varied between 6.99 (1960) and 7.36 (1968) during the period 1960 to 1976, with no trend of perceptible improvement in this characteristic.[4] Unquestionably, one of the major means by which the monetary, energetic, and land efficiency of swine production could be improved is that of increasing the number of pigs produced per sow.

Although there appears to be no adequate long-term census of the number of pigs weaned per sow on a national basis, Curtis and Craig[5] reported 5.5 in 1900 in Iowa, and Wiley[6] cited 6.1 as the average number raised by 343 sows on 40 Indiana farms registered in a production-record program in 1921. If it is assumed that 85% of the pigs farrowed were weaned, the U.S. sow would have produced an average of only 6.14 weaned pigs per litter in 1975 and 1976. This indicates that no improvement has been made in the number of weaned pigs produced per sow during the past 55 years.

Feeding and Management

A century ago, legume pastures or hay, corn grain, and water were the main constituents of the diets of pigs.[7] However, it was recognized that the addition of skim milk improved performance. By 1900, a variety of feedstuffs comprised the diet because performance was improved over that of the simple diets,[7] and Henry[8] in 1898 reported the increased growth rate which resulted from the supplementation of high-corn diets with bone meal. In 1912 Dietrich[9] stated that although corn and soybeans met the needs of pigs for satisfactory growth, the performance was not as good as that on diets containing a variety of feeds. Morrison[10] and colleagues in 1922 formulated their "trio mixture" consisting of 50% tankage or meat scraps, 25% ground legume hay, and 25% linseed meal as a supplement for pigs not on pasture. This was a major advance preceding the 1930 to 1950 era during which the B-vitamin, amino-acid, and trace-element needs of pigs became recognized and feed sources or supplements to meet them were identified. These developments made it possible to raise pigs in drylots.

Nevertheless, soybean meal was not a major feed ingredient in swine rations until after 1940, even though Robison[11] had shown in 1924 that heat treatment of soybeans greatly improved their nutritive value for pigs. With the knowledge gained concerning amino-acid, B-vitamin, and mineral nutrition, as well as that of antibiotic feeding, during the period 1950 to 1960, soybean meal became the dominant high-protein feed for pigs in all regions of the country. However, in the southwestern region, pigs currently are fed for economic reasons 44% as much cottonseed meal as they consume of soybean meal. Other regions utilize very little cottonseed meal as a protein supplement.

For most of the country, corn is the main low-protein, high-carbohydrate feed in swine rations; however, in the southwestern region, 77% of the feed consumed by pigs is grain sorghum. Although corn comprises 47% of the total feed consumed by pigs in the Mountain-Pacific region, 34% of the feed provided to pigs in that region consists of barley and oats.

Since 1950, swine production pronouncedly changed from the use of pasture and other feeds, such as corn, to more complete diets comprised of corn and soybean meal (or of sorghum grain and soybean meal plus cottonseed meal), with various supplements fed under conditions of total confinement on solid or slotted concrete floors.[12] Although the degree of improvement resulting therefrom is not known, such dietary adjustments and total confinement made it possible to manage many more pigs per unit of labor and land in 1976 than in 1950.[12] Such changes also have resulted in average daily gains of body weight of 0.95 kg in 1976 as compared with only 0.84 kg in 1950,[12] over the range of 18 to 100 kg body weight. This means that the time required to produce a 100-kg pig has been greatly reduced since 1950.

Year-Round Farrowing

Also, seasonal production (e.g., spring or fall farrowing) has been largely eliminated, with sows being bred to farrow at any time of the year.[12] For example, in 1976 the percentages of the 10.0×10^6 sows that farrowed during the following quarter-year periods, were: 20.5, December, 1975 to February, 1976; 29.1 March to May; 25.2, June to August; and 25.2, September to November.[13]

Carcass Changes

The nature of the pig produced today is decidedly different from that produced in 1950. In 1950, 14.5% of the liveweight was trimmable fat; by 1975, this was reduced to 6.5%. Thus, the weight of trimmed fat was reduced by 8 kg in a 100-kg pig.[12] However, this does not necessarily mean that today's pig is that much leaner than the 1950 version because there might have occurred a redistribution of fat inter- and intramuscularly that could well account for a considerable part of the reduction in trimmable fat.[14]

OBJECTIVES

It is the purpose of this chapter to identify and quantify the various inputs of cultural energy and labor into the present swine production system in the U.S. and to examine the land requirements of feed production for the swine enterprise in 1975. Also, an attempt will be made to assess the efficiency of pork production in various regions of the U.S. per unit of input of cultural energy, land, and labor.

PORK PRODUCTION SYSTEM INPUTS

The resource inputs included in this study are: feed, fuel, machinery, equipment, land, and labor. Data concerned with the disappearance of feed in the respective states

Table 1
STATES INCLUDED IN GEOGRAPHICAL REGIONS

Region	States
Corn Belt	Ill., Ind., Iowa, Kan., Mich., Minn., Mo., Wis., Ohio
Northeast	Md., N.Y., N.J., Pa., Vt., Conn., Mass., R.I , N.H., Maine, Del.
Southeast	Va., W.Va., Ky., N.C., S.C., Tenn., Ga., Fla., Ala., Miss., La.
Mountain-Pacific	Calif. Colo., Idaho, Mont., Nebr., Nev., N.D., Ore., S.D., Utah, Wash., Wyo.
Southwest	Ark., Ariz., N.M., Okla., Tex.

Table 2
INPUTS PER 100 KG OF LIVEWEIGHT OF PIGS SLAUGHTERED IN THE CORN BELT REGION

Inputs		Quantity	Energy (Mcal)	Land (acres)
Feed				
Corn	(kg)	307	330.7	0.13
Soybean meal	(kg)	46	52.9	0.07
Alfalfa meal	(kg)	3	8.8	
Skim milk, dried	(kg)	1		
Tankage	(kg)	13		
Grain grinding			13.2	
Total		**370**	**405.6**	**0.20**
Fuel				
Gasoline	(gal)	0.84	26.2	
Liquid propane	(gal)	0.88	21.0	
Diesel	(gal)	0.79	28.0	
Electricity	(kWh)	27.97	84.1	
Total			**159.3**	
Depreciation				
Machinery	(1967 $)	0.82	14.8	
Equipment	(1967 $)	2.75	45.6	
Total		**3.57**	**61.4**	
Labor	(hrs)	1.74		
Total			**626.3**	**0.20**

of each of five regions (the Corn-belt, Northeast, Southeast, Mountain-Pacific, and the Southwest), including all of the 48 contiguous states, were employed to determine the relative proportion of each feed used in each of the five regions.[15] (The states included in each region are in Table 1.) These values in combination with the requirements of swine as outlined by the National Research Council[16] provided the quantities of each feed consumed. These were reduced to the amounts of feeds required to produce 100 kg of pig liveweight per year as in Tables 2 to 6. The feeds are summarized in the following categories: corn, grain sorghum, barley, and oats; soybean meal; alfalfa meal; and by-products (skim milk, tankage, fish meal, and cottonseed meal).

The cultural energy requirements to produce the various feeds and their yields per hectare are in Table 7. It should be noted here that no cultural energy was charged against pork production for the by-product inputs on the assumption that the energy cost should be appropriately assigned to the major product from which the by-product was derived. However, the energy cost of transporting by-products an average distance of 100 km was charged as in Table 8. Also, the energy cost of feed grinding was charged to the pork production system.

Among the forms of energy used for power or other purposes, electricity was the dominant one, diesel fuel and gasoline were intermediate, and liquid propane was em-

Table 3
INPUTS PER 100 KG OF LIVEWEIGHT OF PIGS SLAUGHTERED IN THE NORTHEASTERN REGION

Inputs		Quantity	Energy (Mcal)	Land (acres)
Feed				
Corn	(kg)	332	348.3	0.13
Soybean meal	(kg)	48	50.7	0.07
Alfalfa meal	(kg)	3	8.8	
Skim milk, dried	(kg)	1		
Tankage	(kg)	13		
Grain grinding			13.2	
Total		**387**	**421.0**	0.20
Fuel				
Gasoline	(gal)	0.84	26.2	
Liquid propane	(gal)	0.88	21.0	
Diesel	(gal)	0.79	28.0	
Electricity	(kWh)	27.97	84.1	
Total			**159.3**	
Depreciation				
Machinery	(1967 $)	0.82	14.8	
Equipment	(1967 $)	2.76	46.6	
Total		**3.58**	**61.4**	
Labor	(hr)	1.74		
Total			**641.7**	0.20

Table 4
INPUTS PER 100 KG OF LIVEWEIGHT OF PIGS SLAUGHTERED IN THE SOUTHEASTERN REGION

Inputs		Quantity	Energy (Mcal)	Land (acres)
Feed				
Corn	(kg)	301	324.1	0.13
Oats	(kg)	20	22.0	0.02
Soybean meal	(kg)	44	50.7	0.07
Cottenseed meal	(kg)	5		
Alfalfa meal	(kg)	3	8.8	
Fish meal	(kg)	5		
Skim milk, dried	(kg)	1		
Tankage	(kg)	9		
Grain grinding			13.2	
Total		**388**	**418.8**	**0.22**
Fuel				
Gasoline	(gal)	1.10	34.4	
Liquid propane	(gal)	0.46	11.0	
Diesel	(gal)	0.71	25.0	
Electricity	(kWh)	35.45	106.6	
Total			**177.0**	
Depreciation				
Machinery	(1967 $)	1.04	18.8	
Equipment	(1967 $)	3.44	58.1	
Total		**4.48**	77.0	
Labor	(hr)	3.11		
Total			**672.8**	0.22

Table 5
INPUTS PER 100 KG OF LIVEWEIGHT OF PIGS SLAUGHTERED IN THE MOUNTAIN-PACIFIC REGION

Inputs		Quantity	Energy (Mcal)	Land (acres)
Feed				
Corn	(kg)	172	185.2	0.07
Oats	(kg)	21	24.2	0.02
Barley	(kg)	104	143.3	0.13
Soybean meal	(kg)	47	52.9	0.07
Cottonseed meal	(kg)	2		
Alfalfa meal	(kg)	3	8.8	
Fish meal	(kg)	6		
Skim milk, dried	(kg)	1		
Tankage	(kg)	8		
Grain grinding			13.2	
Total		**364**	**427.6**	**0.29**
Fuel				
Gasoline	(gal)	0.95	29.6	
Liquid propane	(gal)	0.64	15.2	
Diesel	(gal)	0.68	24.1	
Electricity	(kWh)	30.51	91.8	
Total			**160.7**	
Depreciation				
Machinery	(1967 $)	0.90	16.37	
Equipment	(1967 $)	3.24	54.76	
Total		**4.14**	**71.13**	
Labor	(hr)	3.15		
Total			**659.43**	**0.29**

Table 6
INPUTS PER 100 KG OF LIVEWEIGHT OF PIGS SLAUGHTERED IN THE SOUTHWESTERN REGION

Inputs		Quantity	Energy (Mcal)	Land (acres)
Feed				
Grain Sorghum	(kg)	295	637.1	0.22
Oats	(kg)	18	19.8	0.02
Soybean meal	(kg)	36	61.7	0.05
Cottonseed meal	(kg)	16		
Alfalfa meal	(kg)	3	8.8	
Skim milk, dried	(kg)	1		
Tankage	(kg)	14		
Grain grinding			13.2	
Total		**383**	**740.6**	**0.29**
Fuel				
Gasoline	(gal)	0.93	28.9	
Liquid propane	(gal)	0.59	14.2	
Diesel	(gal)	0.55	19.4	
Electricity	(kWh)	28.09	84.4	
Total			**146.9**	
Depreciation				
Machinery	(1967 $)	0.84	15.2	
Equipment	(1967 $)	3.66	60.4	
Total		**4.50**	**75.6**	
Labor	(hr)	4.59		
Total			**963.1**	**0.29**

Table 7
YIELDS AND ENERGY COSTS OF CERTAIN FEEDS USED IN SWINE DIETS[17]

Feed	Yield(kg/ha)	Energy requirement (Mcal/kg)
Corn	5965	1.08
Grain sorghum	3470	2.16
Barley	2112	1.29
Oats	1826	1.13
By-products	—	0.00

Table 8
ENERGY CONVERSION FACTORS AND YIELDS OF CERTAIN CROPS

	Amount	Energy (Mcal)	Yield (kg)	Ref.
Feeds				17
Corn	1 ha	6424	5965	
Grain sorghum	1 ha	7489	3470	
Oats	1 ha	2070	1826	
Barley	1 ha	2724	2112	
Soybean meal	1 kg	0.079	—	
Alfalfa meal	1 kg	0.636	—	
Fuels				
Gasoline	1 gal	31.251	—	18
Liquid propane	1 gal	23.817	—	18
Diesel	1 gal	35.283	—	18
Electricity[a]	1 kWh	3.007	—	
Machinery	$1 (1967)	18.105		19
Equipment	$1 (1967)	16.900		19
Transportation (truck)	1 kg/km	0.00083		20
Insecticide	1 kg	73.920		17
Grain grinding	1 kg	0.038		17

[a] 1 kWh of electricity represents 859 kcal (Cervinka et al.[18]); this value was multiplied by 3.5 to account for the fuels used to produce the electricity.

ployed in smallest quantity (Tables 2 to 6). The energy equivalents of these forms of power or fuel are in Table 8.

The machinery employed directly in pork production includes tractors, pickup trucks, grinders, manure-handling devices, and feed distribution units. The machinery employed to produce crops is accounted for in the cultural energy costs of feed production. Energy costs of equipment include those represented by barns, dry lots, farrowing pens, water systems, feed-handling equipment, manure-handling equipment, and heating, cooling, or ventilating systems. The values for both machinery and equipment cultural energy costs are in Tables 2 to 6 as the amount of depreciation per 100 kg of pig liveweight produced per year in terms of 1967 dollars as in Table 8.

Labor includes that which is hired and that which is provided by the family, and it is expressed in terms of hours per 100 kg of pig liveweight produced (Tables 2 to 6). The labor charge is that expended directly in the swine production function, but it excludes that required in feed production and similar activities.

Table 9
SUMMARY OF SWINE ENTERPRISE OUTPUTS BY REGION IN 1975

Region	Pigs slaughtered (Number/10^6)	Average liveweight of pigs (kg)	Total production of carcass: Weight (Kg/10^6)	Total production of carcass: Protein (Kg/10^6)
Corn Belt	45.178	110.2	3386.2	440.2
Northeast	3.669	105.2	262.6	34.1
Southeast	13.461	106.1	971.6	126.3
Mountain-Pacific	9.163	105.2	655.7	85.2
Southwest	2.267	107.9	166.4	21.6
Total U.S.	**73.738**	**108.5**	**5442.5**	**707.4**

Table 10
SUMMARY OF CULTURAL ENERGY, LABOR, AND LAND REQUIREMENTS OF PRODUCING 100 KG OF PIG LIVEWEIGHT IN VARIOUS REGIONS OF THE U.S.

Inputs	Corn Belt	Northeast	Southeast	Mountain-Pacific	Southwest	Total U.S.
Energy (Mcal)						
Feed	405.6	421.0	418.8	427.6	740.6	421.6
Fuel	159.3	159.3	177.0	160.7	146.9	163.1
Machinery and Equipment	61.4	61.4	77.0	71.1	75.6	65.8
Total	**626.3**	**641.7**	**672.8**	**659.4**	**963.1**	**650.5**
Labor (hr)	1.74	1.74	3.11	3.15	4.59	2.24
Land (acres)	0.20	0.20	0.22	0.29	0.29	0.22

SWINE SYSTEM OUTPUTS

The U.S. produced 73.74 × 10^6 slaughtered pigs in 1975 having an average liveweight of 108.5 kg (Table 9). On the assumption that the average dressing percentage was 68, this represented 5.44 × 10^9 kg of pork carcasses. Thus, there was available to the average U.S. consumer 25.5 kg of pork carcass or 3.3 kg of carcass protein. Pork carcass weight can be converted to cooked pork weight by multiplying by the factor, 0.35.

The numbers of slaughter pigs and carcass weights produced in the five regions are also summarized in Table 9. Of the total pork produced in the U.S. in 1975, the proportions produced in the various regions were: Corn Belt, 62.2%; Southeast, 17.8%; Mountain-Pacific, 12.1%; Northeast, 4.8%; and the Southwest, 3.1%.

EFFICIENCY OF PORK PRODUCTION

Cultural Energy, Labor, and Land Costs of Swine Production

The efficiency of pork production in the various regions and the entire country was examined principally in terms of the amounts of cultural energy, labor, and land required to produce 100 kg of pig liveweight. These data are in Table 10. It will be noted that the energy cost of production is not much different among the regions of the country, with the exception of the Southwest in which it was 48% greater than the average for the total country in 1975. Also, in the Southwest, the requirement for labor is more than twice as much as, and that for land is 33% greater than, that of the average for the entire U.S.

The higher total cultural energy cost of swine production in the Southwest than in any other region is chiefly attributable to the higher energy requirement of feed pro-

Table 11
CONTRIBUTION OF VARIOUS FEEDS TO SWINE DIETS IN VARIOUS REGIONS

Feed component	Corn Belt	Northeast	Southeast	Mountain-Pacific	Southwest
			Percent of total diet		
Corn	83.0	83.2	77.6	47.2	—
Grain sorghum	—	—	—	—	77.0
Barley	—	—	—	28.5	—
Oats	—	—	5.2	5.7	4.7
Soybean meal	12.4	12.4	11.3	12.9	9.4
Alfalfa meal	0.8	0.8	0.8	1.0	0.8
By-products	3.2	3.6	4.9	4.7	8.1
Other feeds	0.6	0.0	0.2	0.0	0.0

duction. This is the result of the large proportion (77%) of the total diet consisting of grain sorghum rather than of corn or of corn and other grains (Table 11). In Table 7, the average yield of grain sorghum is 58% of that of corn per unit of land, and grain sorghum requires two times the input of energy per unit weight of grain as does corn. In addition, the nutritive value of grain sorghum for pigs is 90 to 95% of that of corn.[10]

The amount of land required to produce feed for swine in the Corn Belt and the Northeast is only 69% of that needed in the Southwest per unit of pig liveweight (Table 10). If the southwestern region did not use 2.3 and 2.5 times as much by-products in swine rations as do the northeastern and Corn Belt regions, the difference in land requirements would be still greater.

Also in the Mountain-Pacific region, the land requirement of feed production per unit of pig liveweight is 44% greater than that in the Corn Belt or the Northeast. This reflects the use of barley and oats in the Mountain-Pacific region (Table 11) and the low yield per land unit of these grains (Table 7). Since the energy cost per unit weight of producing barley and oats is not much greater than that of corn and considerably more corn than barley and oats is employed in swine rations in the Mountain-Pacific region, the energy cost of feed production per unit of weight of pig is about the same as that of all regions except the Southwest.

The total requirement of fuel for power and other purposes in the swine production unit (i.e., excluding that required in feed production) in 1975 was not much different from region to region, and the average amount for the U.S. was 163 Mcal per 100 kg of pig liveweight produced. Of the total energy used in the five regions, electricity contributed 52.1 to 60%; gasoline, 16.4 to 20%; diesel fuel, 13.6 to 17.8%, and liquid propane gas, 6.2 to 13.7%.

Inexplicable differences in the labor requirement of swine production exist. In the Corn Belt and Northeast, the labor requirement was the least; in the southeastern and Mountain-Pacific regions, it was intermediate, but almost twice that of the former regions; but in the Southwest, it is inordinately high (4.59 hr/100 kg of liveweight produced). The weighted average labor requirement for the U.S. is 2.24 hr/100 kg pig liveweight. These values represent the amount of labor exerted in operating the swine enterprise itself, and they do not include that of feed production.

Cultural Energy, Land, and Labor Costs of Producing Carcass Protein

A summary of the amounts of cultural energy, land, and labor required to produce 1 kg of pig carcass protein is in Table 12. The data for the southwestern region are

Table 12
ENERGY, LAND, AND LABOR REQUIREMENTS OF THE PRODUCTION OF CARCASS PROTEIN BY PIGS

	Cost of producing 1 kg carcass protein		
Region of U.S.	**Energy (Mcal)**	**Land (acres 10^{-3})**	**Labor (min)**
Corn Belt	71	22.4	11.8
Northeast	73	22.4	11.8
Southeast	76	24.9	21.1
Mountain-Pacific	75	32.4	21.4
Southwest	109	32.4	31.1
Total U.S.	**74**	**24.4**	**15.2**

Table 13
PROPORTIONS OF TOTAL NONSOLAR ENERGY AS CULTURAL AND DIETARY ENERGY REQUIRED BY SWINE PRODUCTION

	Partition of energy inputs	
Regions	**Cultural**	**Dietary**
	Percent of nonsolar energy	
Corn Belt	27.8	72.2
Northeast	27.4	72.6
Southeast	28.3	71.7
Mountain-Pacific	29.2	70.8
Southwest	36.4	63.6
Total U.S.	**28.3**	**71.7**

somewhat higher than those for the other regions except for the land requirement which is the same as that of the Mountain-Pacific region. These values reflect the same relative patterns as the costs of the same inputs of producing pig liveweight as in Table 10 because the protein content of the carcass is highly correlated with the liveweight.[14]

The average cultural energy cost of producing 1 kg of pork-carcass protein in the U.S. is 74 Mcal. In another study based on the national average production conditions, Reid[21] estimated a value of approximately 67 Mcal, based upon the feeding of a corn-soybean meal diet. That value did not include the energy cost of waste disposal, but it assumed that the energy value of the extracted soybean oil would be equivalent to the energy cost of feed processing and distribution to and from storage on the farm.

Proportions of Nonsolar Energy as Cultural and Dietary Energy Required for Swine Production

Nonsolar energy was partitioned between the cultural and dietary energy requirements. The total cultural energy requirements for the various regions and the entire country are those in Table 10. The dietary energy costs were computed from the amounts of feed in each region (Tables 2 to 6) on the basis that 1 kg of feed provides 4.4 Mcal of dietary energy.

In Table 13, in the U.S., of the total nonsolar energy required, 28% is cultural energy and 72% is dietary energy. The major deviation from these proportions is the Southwest where the partition is 36 and 64%, respectively.

SUMMARY

1. The average requirements of producing 100 kg of pig liveweight in the U.S. in 1975 were 651 Mcal of cultural energy, 2.24 hr of labor, and 0.22 acres of land. Of the total cultural energy required, the percentages comprised by the following inputs were feed production, 64.8; fuel, 25.1; and machinery and equipment used directly in pork production, 10.1.
2. The cultural energy requirements of swine production are considerably greater in the southwestern region than in any other region. This is the result largely of feed production costs. In the Southwest, the major feed provided to swine is grain sorghum (77% of the total diet). Grain sorghum yields considerably less per acre, and it requires two times as much cultural energy per unit weight to produce as does corn which is the major ration ingredient in all other regions.
3. In 1975, the average inputs of cultural energy, land, and labor required to produce 1 kg of pork carcass protein in the U.S. were 74 Mcal, 24.4×10^{-3} acres, and 15.2 min of labor, respectively. The main regional deviations from these values were those of the Southwest in which the requirements were 109 Mcal, 32.4×10^{-3} acres, and 31.1 min of labor, respectively. In the Mountain-Pacific region, 32.4×10^{-3} acres and 21.4 min were required. The most efficient production of pork protein (1 kg) occurred in the Corn Belt and northeastern regions: 71 to 73 Mcal, 22.4×10^{-3} acres, and 11.8 min of labor.
4. The major energy cost of swine production in the U.S. is that represented by the heat of combustion value of the diet. Of the total nonsolar energy required, 72% is dietary energy, and 28% is cultural energy.
5. The major gains to be made in the efficiency of swine production are those associated with increased reproductive rates and decreased mortality rates. These changes resulting in more pigs being generated and reared per sow might not be effected without additional cultural and dietary energy inputs, e.g., in providing milk substitute diets, early weaning, etc. Nevertheless, such changes would reduce the overhead costs of rearing and maintaining the brood sow. Whether a net gain would result or at what level of change the net gain would be improved needs to be determined.

REFERENCES

1. **Anon.,** National Food Situation, Economic Research Service, U.S. Department of Agriculture, November issues in respective years.
2. **Randall, C. K.,** 1973 Handbook of Agricultural Charts, Agricultural Handbook No. 455, U.S. Department of Agriculture, Washington, D.C., 1973.
3. **Hoffman, G. and Ball, E.,** Livestock and Meat Situation, Economic Research Service, Bull. 210, U.S. Department of Agriculture, Washington, D. C., 1976.
4. **Koehn, M. L. and Wood, E. L.,** Agricultural Statistics, 1976, Statistical Report Service, U.S. Department of Agriculture, 1976, 312.
5. **Curtis, C. F. and Craig, J. A.,** A study of pork production from the standpoint of the farm and the market, *Iowa Exp. Stn. Bull.*, 48, 1900.
6. **Wiley, J. R.,** Hog production record demonstrations in Indiana, in *Proc. Am. Soc. Anim. Prod.*, 1922, 55.
7. **Hays, V. W.,** A century of progress in swine production, in *Agricultural and Food Chemistry: Past, Present, Future,* Teranishi, R., Ed., AVI Publishing, Westport, Conn., 1978, chap. 15.
8. **Henry, W. A.,** *Feeds and Feeding,* 1st ed., W. A. Henry Publishing, Madison, Wisc., 1898.

9. **Dietrich, W.,** *Swine Breeding, Feeding, and Management,* Saunders, Chicago, 1912.
10. **Morrison, F. B.,** *Feeds and Feeding,* 22nd ed., Morrison, Ithaca, N. Y., 1959, 867.
11. **Robison, W. L.,** The influence of the method of oil extraction on the feeding value of soybean oil meal, in *Proc. Am. Soc. Anim. Prod.,* 1924, 60.
12. Committee. Food from Animals, Council Agr. Sci. Techn., Rpt. No. 1979. (To be published)
13. **Cochran, P. and Sikora, B.,** Livestock and Meat Statistics, Stat. Bull. 522, Economic Research Service, U.S. Department of Agriculture, Washington, D.C., 1977, 23.
14. **Reid, J. T., Bensadoun, A., Bull, L. S., Burton, J. H., Gleeson, P. A., Han, I. K., Joo, Y. D., Johnson, D. E., McManus, W. R., Paladines, O. L., Stroud, J. W., Tyrrell, H. F., Van Niekerk, B. D. H., Wellington, G. H., and Wood, J. D.,** Changes in body composition and meat characteristics accompanying growth of animals, in *Proc. Cornell Nutr. Conf.,* 1968, 18.
15. **Anon.,**Unpublished swine data, Firm Enterprise Data System, Economic Research Service, U. S. Department of Agriculture, and Oklahoma State University, Stillwater, 1976.
16. Subcommittee on Swine Nutrition, *Nutrient Requirements of Swine,* 7th ed., National Academy of Science, Washington, D.C., 1973.
17. **Pimentel, D.,** personal communication, 1977.
18. **Cervinka, V., Chancellor, W. J., Coffelt, R. J., Curley, R. G., and Dobie, J. B.,** Energy requirements for agriculture in California, California Department of Food and Agriculture and University of California Report, 1974.
19. **Herendeen, R. A. and Bullard, C. W.,** Energy Cost of Goods and Services 1963 and 1967, Center for Advanced Computation, Doc. No. 140, University of Illinois, Urbana, 1974.
20. **Pimentel, D. and Pimentel, M.,** *Food, Energy and Society,* ed. Arnold Press, London, 1978.
21. **Reid, J. T.,** Comparative efficiency of animals in the conversion of feedstuffs to human foods, in *Proc. Cornell Nutr. Conf.,* 1975, 16.

CULTURAL ENERGY IN U.S. BEEF PRODUCTION

C. Wayne Cook, John J. Combs, and Gerald M. Ward

INTRODUCTION

Beef cattle production is the largest single agricultural enterprise in the U.S. Beef cattle account for approximately 30% of farm cash receipts. Nix et al.[13] estimated retail sales of beef from fed steers and heifers to be about $14.9 billion. Another $1 billion was derived from beef by-products. Over a period of 20 years, American consumers spent an average of 2.5% of their disposable income for beef, the largest single item in the food budget. The great significance of beef, the complexities of the industry and the fact that grain is consumed in large quantities has raised many questions about fossil fuel use in beef production.

RANGELAND

Approximately 50% of the surface land area of the earth is classified as rangeland and has no alternate use for food production other than from grazing animals. The use of these rangelands can be integrated with the use of cropland aftermath, and grazeable forage crops to produce acceptable table meat with a minimum of fossil fuel expenditure.

There are 483,387,000 ha (1.2 billion acres) of rangeland in the U.S. that have an estimated forage potential equivalent to a yearlong supply for 35 million cows or a quantity capable of feeding 85% of the total current beef cattle population in the U.S. This large resource represents a stable foundation from which to formulate many grazing systems for red meat production for the future.

About two thirds of the land area in the 11 western states is rangeland, about 46% of the six plains states is rangeland and about 44% of the total land area in the U.S. is rangeland. Pasture and hay lands occupy only 2.7% of the land area of the 11 western states, only 10.6% of the six plains states and only 8.6% of the total land area of the U.S. Rangeland forage is a natural renewable resource that regrows each year to be processed into red meat. This renewable resource represents a tremendous source of animal feed if managed properly. When grazed for the production of meat, it does indeed represent wealth, but when preserved merely for viewing, this resource is wasted and either deprives people of food because of a scarcity or because the cost from other intensive feeding systems prices it beyond the means of most people.

Value of the Range Livestock Industry

There were approximately 43 million beef stock cows in the U.S. in 1976, with about 20% of these in the 11 western states and 36% in the six plains states (Figure 1). The total value of production for beef is approximately $14 billion annually.

The annual value of production of the sheep and cattle industry in the 11 western range states is about $3.1 billion and for the six plains states about $5.3 billion.[18]

The value of grazed range and forage crops is not generally recognized because they are taken for granted as a renewable crop that will be there infinitely for animal production. These forage crops contain large quantities of cellulose and hemicellulose, along with nonstructural carbohydrates and proteins. The ruminant is one of the few animals that can convert structural carbohydrates such as cellulose and hemicellulose into meat protein for human consumption.

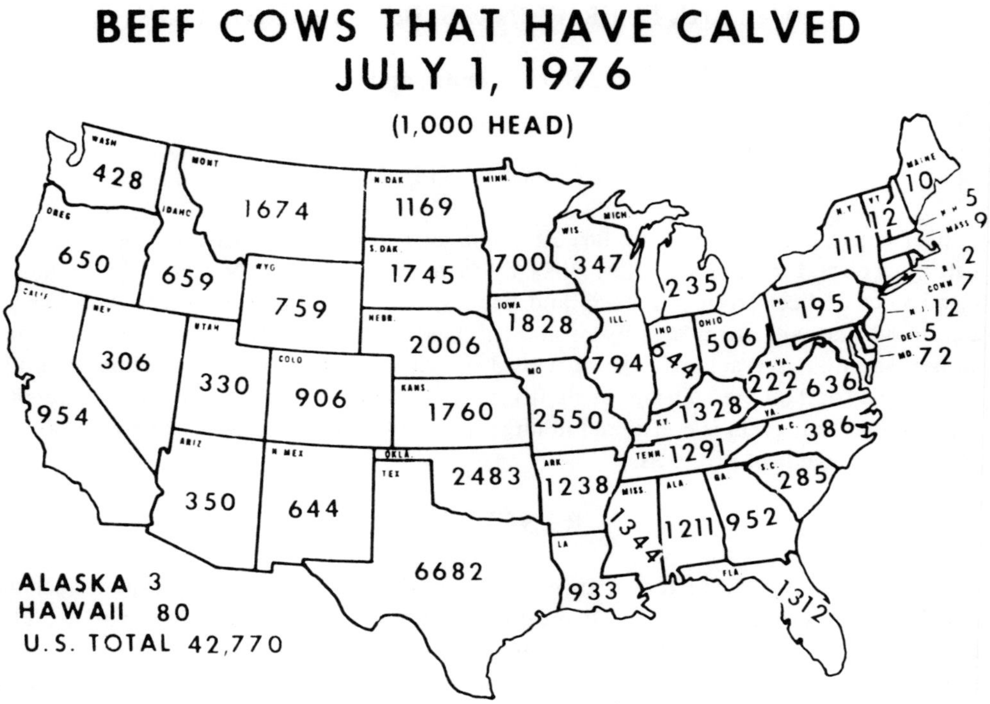

FIGURE 1. Distribution of beef cows and calves.

The ruminant is an animal capable of harvesting its own food from the forage crops grown upon the land. The ruminant has evolved as a consumer of native vegetation that monogastric animals cannot utilize. Products from the utilization of grazing animals provide meat, milk, and fiber for human sustenance. Of great importance is the form and kind of protein in people's present day diet. The current importance of ruminants in the agricultural economy is suggested by their cash receipts that represent about 43% of total receipts for all agricultural commodities.[5]

Even in 1976, when grain was inexpensive because of surplus production, 80% of the feed units for beef production and about 90% of the feed units for sheep production were provided by forages.[10]

Beef and dairy products raised largely from forages supply about one half of the protein and one third of the energy in the diets of people in the U.S.[17]

Yields of forage per hectare either directly in terms of dry matter or indirectly in terms of animal gain are frequently higher than comparable values for cereals. Thus, an integration of plant and animal agriculture appears to have great promise in furnishing meat protein to maintain an acceptable standard of living. There is indeed a tremendous potential for processing cellulosic feedstuffs through ruminant animals in the future.

It has been found that on a range and herd sustaining basis that every 11.6 kg of useable forage eaten by a cow-calf unit will yield about ½ kg of weaner calf. However, every 5 kg of range forage eaten by a growing steer will produce 0.5 kg of gain.[2]

Rangelands provide a diversity of plant species and topographic features. Therefore, a combination of grazing animal species gives greater uniformity of forage resource utilization because different animals use different plant species and different portions of the plant. In addition, animal species prefer certain topographic aspects to others and therefore a combination of animal species use the range more uniformly.[1]

FEEDLOT

The feedlot developed as a system to market feed grains. In times when the major grain market was livestock feed, it became profitable to use large amounts of grain to speed up the growth and fattening process which resulted in beef that was uniformly tender, juicy, and generally very acceptable to consumers. In marked contrast to the cow-calf operations where only minimal inputs are economically justified, feedlots are highly mechanized, capital intensive operations. Feeding periods run from 60 to 200 days so decisions on cattle feeding can be made very rapidly as compared to the cow-calf operation where decisions about size of operation are made mostly only once a year.

Cattle feeding and feedlots first developed in the Corn Belt, but more recently have expanded in western states in response to increased feed grain production under irrigation. Cattle feeders can be divided into farmer feeders and the larger commercial feedlots. Farmer feeders produce their feed and the objective is to maximize return from his crops. Commercial feeders must maximize returns on capital. Leading states in cattle feeding are in order: Texas, Iowa, Nebraska, Kansas, California, and Colorado. These six states accounted for 62% of the cattle on feed in 1976.

The cultural energy expended in the production of range forage, grazing dry land crops and red meat are presented in Table 1.

Most items of input of cultural energy were taken from the Advisory Board's suggestions referenced herein as Pimentel.[16] Energy input for all items include raw materials, manufacturing, distribution, and maintenance and depreciation when applicable. Cultural energy expenditures for operation of machinery and trucks include the power, fuel, depreciation, and maintenance for operation per hectare or kilometer. The total values for cultural energy expenditures for liquid fuels in Table 1 were calculated on the basis of total production and distribution costs which were derived by adding 699 kcal to the production costs presented by the Advisory Board.[16]

These items include the management of the forage produced in place on the land and the care and management of the grazing animals.

The cultural energy expenditures for common feeds used in range supplements and rations in feedlot finishing are presented in Table 2. In like fashion, cultural energy expenditures for special seeded forages (crested wheatgrass and sorghums) along with the common broad range types of mixed and shortgrass and mountain range, along with the production of both plant and animal energy and protein, are presented in Table 3. The total expenditure of cultural energy for typical ranching operations under several management systems is presented in Table 4.

Range sheep and cattle production in the Intermountain Great Basin area is typical of ranching operations of the states of Colorado, Wyoming, Idaho, Utah, and Nevada where animals graze desert ranges during winter, foothills during the spring and fall, and mountain ranges during the summer (Table 4). Crested wheatgrass is frequently used during the early spring for calving and lambing pastures. Supplements are fed on the winter range from December until April.[4]

Range cattle production on the Central Great Plains area consists of grazing yearlong with minimum concentrate supplements during the winter. Sorghum forages are planted for late summer and autumn grazing, especially for yearling steers (Table 4).

To determine weights of feedlot animals, composition of ration and average feeding period, questionnaires were sent to various feedlot specialists around the U.S.[14] Average weight of feeder cattle coming into western feedlots was 227 kg and are slaughtered at an average weight of 468 kg. Average daily gain was 1.27 kg/day, requiring 190 days on feed to reach slaughter weight.

Table 5 shows the machinery usage for various size feedlots during the 190-day feeding period. Cultural energy used for the various facets of feedlot operations are presented in Table 6. For comparison, Table 7 lists various estimates for energy use in feedlot operations. Feed processing is a variable energy input, depending upon the particular process utilized. Equipment complement and energy consumption for feed grinding, flaking, and pelleting are shown in Tables 8, 9, and 10, respectively. Table 11 depicts a composite ration fed to feedlot animals in the western states. The total energy input for a 10,000-head capacity feedlot using various feed processing methods is presented in Table 12.

Table 1
AVERAGE CULTURAL ENERGY EXPENDED PER UNIT FOR THE VARIOUS ITEMS USED IN CROP AND MEAT PRODUCTION IN WESTERN LIVESTOCK

Item	Unit	Cultural energy expenditure (kcal)	Ref.
Machinery			8
Offset disc	hr	403,866	
One-way	hr	424,866	
Rod weeder	hr	324,654	
Prilled fertilizer spreader	hr	293,829	
Anhydrous applicator	hr	337,654	
Chisel or sweep	hr	388,866	
Grain drill	hr	283,829	
Cultivator	hr	333,645	
Plow or lister	hr	448,878	
Swather	hr	246,923	
Baler	hr	284,036	
Stacker	hr	393,829	
Combine corn or wheat	hr	333,533	
Chopper	hr	372,290	
Tractor 108DB	hr	312,792	
D-5 Cat	hr	598,096	
Pickup ¾ ton	km	4,183	8
Truck 2 ton	km	3,090	8
Labor	hr	2,500	4
Fuel[a]			16
Gasoline	1	10,808	
Diesel	1	12,113	
Electricity			
Steam[1]	kwh	2,863	16
Hydro[a]	kwh	900	
Fertilizer			16
Ammonium nitrate prilled	kg (N)	14,700	
Anhydrous ammonia	kg (N)	12,000	
Triple super phosphate	kg (P_2O_5)	2,700	
Irrigation			
Sprinkler CP	ha, cm	162,770	8
Ditch gravity[a]	ha	40,525	16
Insecticides			16
Toxaphene	kg (AI)	38,100	
Parathion	kg (AI)	13,810	
Herbicides			16
Atrazine	kg (AI)	45,240	
Paraquat	kg (AI)	109,520	
2,4-D	kg (AI)	24,200	
2,4,5-T	kg (AI)	56,700	
Dicamba	kg (AI)	70,240	
Seed			16
Alfalfa	kg	62,000	
Sorghum	kg	10,000	
Crested wheat	kg	4,000	
Wheat	kg	3,000	
Hybrid corn	kg	25,000	
Barley	kg	3,000	

Table 1 (continued)
AVERAGE CULTURAL ENERGY EXPENDED PER UNIT FOR THE VARIOUS ITEMS USED IN CROP AND MEAT PRODUCTION IN WESTERN LIVESTOCK

Item	Unit	Cultural energy expenditure (kcal)	Ref.
Feed			
Wheat[a]	kg	732	
Corn[a]	kg	1,654	
Soybean	kg	1,462	4
Barley[a]	kg	732	
Milo[a]	kg	737	
Silage corn[a]	kg	261	
Hay			4
Domestic	kg	269	
Native	kg	232	
Alfalfa			
Sun cured			
Irrigated[a]	kg	194	
Dry	kg	180	4
Dehydrated and pelleted	kg	4,000	9
Irrigated pasture[a]	kg	1,112	
Dryland pasture	kg	194	8
Drying Feed	kg	108	16
Transportation			4
To and from market	kg × km	2.9	
Buildings			
Residence	m^2	1,544,444	
Service	m^2	442,222	
Fence[a]			6
Barbed	km	435,695	
Mesh	km	440,893	
Water development[a]			
Wind mill	aum	7,950	
Pond	aum	2,930	

[a] Calculated by authors.

Table 2

QUANTITY AND KILOCALORIES PER HECTARE FOR VARIOUS FEEDS USED AS SUPPLEMENTS FOR RANGE LIVESTOCK OR AS RATION MIXTURES IN FINISHING ANIMALS

	Feed											
	Corn grain (irrigated)		Corn silage (irrigated)		Alfalfa (irrigated)		Wheat or barley (dry)		Milo (dry)		Domestic pasture (irrigated)	
Items of input	Quantity	kcal	Quantity	kcal	Quantity	kcal	Quantity	kcal	Quantity	kcal	Quantity	kcal
Labor (hr)	11.4	28,500	112.6	281,500	11.4	28,500	5.15	12,875	5.35	13,375	1.97	4,925
Machinery (hr)												
Offset disc	.635	256,454	.635	256,455	.24	96,927			.63	254,436	.07	28,270
Plow/lister	.538	241,496	.538	241,496					.538	241,496		
One-way							.63	267,666	.32	135,957		
Rod weeder							.22	71,422				
Sweeps							.95	369,412				
Anhydrus fertilizer applicator	.382	128,984	.382	128,983								
Prilled fertilizer applicator	.190	55,827	.190	55,827	.063	18,511					.21	61,704
Planter	.340	96,501	.340	96,501								
Drill					.085	24,125	.26	73,795			.04	11,353
Cultivator	.645	215,201	.645	215,201								
Swather					1.402	346,186						
Stacker					.385	151,624						
Bailer					1.60	454,457						
Combine							.407	135,747	.407	135,747		
Corn picker	.947	315,855										
Silage chopper			.824	306,767								
Pickup truck (km)	.55	2,301	.55	2,301	.55	2,301	.82	3,430	.82	3,430	.82	3,430
Irrigation												
Center sprinkler (cm)	38.2	6,217,814	35.1	5,713,227							41.6	6,771,232
Ditch gravity (ha)					1.0	40,525						
Transportation (kg × km)	23,361	67,747	317,387	920,422	237,947	690,046	1,529	4,434	1,805	5,234		
Fertilizer (kg)												
Nitrogen	206	2,472,000	206	2,472,000							195	2,866,500
Phosphorus	56	151,200	56	151,200	123	332,100					84	226,800
Herbicide (kg)	2.24	54,208	2.24	54,208								
Insecticide (kg)	7.84	108,270			.56	7,734						
Seed (kg)	25	625,000	25	625,000	1.10	68,200	67	201,000	45	135,000	1.50	22,500
Fence (km)	.016	6,971	.016	6,971	.016	6,971	.05	21,785	.05	21,785	.03	13,071
Total		**11,044,329**		**11,528,059**		**2,268,207**		**1,161,576**		**946,460**		**10,009,785**
Yield (kg/ha)		6,678		42,500		11,700		1,586		1,285		9,000
Produce (kcal/kg)		1,654		271		194		732		737		1,112

Table 3
INPUT OF CULTURAL ENERGY AND OUTPUT IN PLANT AND ANIMAL BIOMASS AND KILOCALORIES FOR PLANTS AND ANIMALS PER HECTARE IN DRYLAND FOR SOME SEEDED PASTURAGE AND COMMON RANGE TYPES IN THE WESTERN U.S.

	Crested Wheatgrass (April to June 15)		Sorghum Pasture (July 15 to October 1)		Mixed Range (May 15 to October 1)		Shortgrass Range (May 15 to October 1)		Mountain Range (June 15 to September 1)	
Production	Quantity	kcal/ha	Quantity	kcal/ha	Quantity	kcal/ha	Quantity	kcal/ha	Quantity	kcal/ha
Primary										
Labor (hr)	1.93	4,825	8.00	20,000	0.51	1,275	0.49	1,225	0.61	1,525
Machinery (hr)	0.22	77,727	1.10	387,345	0.03	10,599	0.02	7,066	0.03	10,599
Fence (km)	0.05	21,785	0.05	21,785	0.02	8,714	0.01	4,357	0.03	13,071
Water, mill (AUM)	0.75	5,966	0.35	2,784	1.00	7,955	1.50	11,932	1.50	11,932
Seeds (kg)	0.80	3,200	22.32	42,408	—	—	—	—	—	—
Herbicides	0.50	12,100	0.89	65,476	—	—	—	—	0.01	242
Pickup (km)	0.90	3,765	0.82	3,430	0.50	2,301	0.50	2,091	0.95	3,974
Transportation (kg × km)	87	252	92	267	21	61	20	58	98	284
Total		**129,620**		**543,495**		**30,905**		**26,729**		**41,627**
Yield (kg)	350[a]	1,516,900	735[a]	3,655,155	201[a]	84,440	102[a]	44,880	161[a]	70,035
kcal output/kcal input	11.70		6.72		2.73		1.67		1.68	
Protein yield (kg)	35		81		22		12		19	
Secondary										
Labor (hr)	0.75	1,875	0.75	1,875	0.63	1,575	0.70	1,750	0.80	2,000
Machinery (hr)	0.12	42,608	0.12	42,608	0.05	17,753	0.04	14,203	0.06	21,304
Feeds (kg)	—	—	—	—	0.13	—	0.10	—	—	—
Insecticides (kg)	0.05	773	0.05	773	0.05	773	0.05	773	0.05	773
Transportation (kg × km)	187	542	200	580	187	542	180	522	200	580
Pickup (km)	0.65	2,719	0.93	3,890	0.65	2,719	0.50	2,092	0.95	3,974
Total		**48,517**		**49,726**		**23,362**		**19,340**		**28,631**
Meat yield (kg)[b]	30.00		49.20		13.39		6.03		10.71	
kcal/ha	35,250		59,220		15,733		7,085		12,584	
kcal output/kcal input	0.198		0.098		0.290		0.154		0.179	
Protein yield (kg)	2.79		4.57		1.49		0.65		1.15	

[a] Useable forage.
[b] Live weight

Table 4
AVERAGE CULTURAL ENERGY EXPENDED FOR VARIOUS RANGE SHEEP AND CATTLE OPERATIONS

Item	Intermountain Great Basin				Central Great Plains					
	Range sheep[a] 2000 female herd		Range cows[a] 250 female herd		Range-Sorghums[a] 250 cow operation		Range cows[a] 250 female herd		Range-Sorghums 400 steer operation	
	Quantity	Mcal	Quantity	Mcal	Quantity	Mcal	Quantity	Mcal	Quantity	Mcal
Labor (hr)	6,080	15,200	3,170	7,925	2,737	6,842	2,737	6,842	3,780	9,450
Machinery (hr)	1,500	67,500	1,200	54,000	1,744	78,480	1,744	78,480	1,900	85,500
Pickup	25,000	104,575	20,000	83,660	15,000	62,745	15,000	62,745	18,000	75,294
Transportation (kg × km)	555	1,609	573	1,662	7	20	7	20	47	136
Feed (kg)	20,089	19,350	20,089	22,500	30,134	23,490	30,134	23,490	80,357	
Sorghum pasture (ha)	—	—	—	—	—	54,035	—	—	—	84,769
Fence (km)										
Mesh	40	17,636	—	—	—	—	—	—	—	—
Barbed			80	34,856	35	15,249	30	13,071	35	15,249
Water (aum)										
Mill	1,200	9,546	1,000	7,955	2,000	15,910	1,500	11,932	2,000	15,910
Pond	3,600	10,548	2,000	5,860	1,000	2,930	1,500	4,395	1,000	2,930
Total		**245,974**		**218,418**		**259,701**		**200,975**		**351,878**
Market weight or gain (kg)	38.2[b]		108.8[c]		198.7[d]		184.8		176.8	
Calf or lamb crop (%)	103.5		93		93		92		98[e]	
Total energy in carcass (mcal)	121,382		49,498		54,423		50,578		138,138	
Number marketed	2,070		233				23230		392	
Live weight marketed (kg)	79,011		42,127		46,288		42,509		69,300	
Cultural energy/food energy	2.03		4.41		4.77		3.97		2.54	
Mcal cultural energy/kg of live weight	3.11		5.18		5.61		4.73		5.08	

[a] These calculations include the expenditures of cultural energy required to operate the entire ranching enterprise.

[b] Sold as choice fat lambs at 90#, 70%, and sold as feeder lambs at 75#, 30%.

[c] All sold as good grass fat calves.

[d] Sold as choice grass fat calves, 70%.

[e] Percent death loss was 2%, therefore, 98% were marketed; all sold as good grade grass fat steers weighing 810#.

Table 5
MACHINERY ENERGY USE 1000, 5000, AND 10,000 HEAD CAPACITY[5]

	1000 Head			5000 Head			10,000 Head		
	Hr Use/ head	Mcal depreciation/head	Mcal fuel/head	Hr Use/ head	Mcal depreciation/head	Mcal fuel/head	Hr use/ head	Mcal depreciation/head	Mcal fuel/head
Receive cattle									
Platform scale	.03	4.91	0	.03	.98	0	.03	.47	0
Feed distribution									
Auger — 6''	.30	.30	1.84	.16	15.04	13.61	.09	11.49	11.06
Augur — 4''	.13	.25	.53	.02	.81	1.12	.02	.41	1.12
Silage conveyor	.35	.35	1.45	.05	2.42	5.89	.05	1.19	5.89
Platform scale	.82	4.84	0	.22	5.32	27.48	.22	4.43	27.48
Silage loader	.02	4.07	1.12	.05	.85	4.56	.05	.76	4.56
Silage wagon	.41	.82	0						
Feed truck	1.06	37.13	131.52	.51	19.81	45.24	.26	4.81	23.46
Tractor	.05	3.32	4.56	.08	4.64	7.42	.11	2.55	13.25
Waste removal									
Tractor	.24	18.65	21.23	.24	4.87	17.41	.24	4.80	24.47
Dump truck	.09	8.07	11.78	.09	2.37	11.78	.09	2.56	12.24
Total		**82.73**	**174.03**		**57.11**	**134.51**		**33.47**	**123.53**

Table 6
CULTURAL ENERGY REQUIREMENTS FOR FEEDLOT OPERATION[5a]

Item	Re-ceive cattle	Prepare feed order	Feeding	Inspection	Vet care	Waste removal	Loading out	Overhead	Total
1000 Head									
Labor									
Hr	.17	.17	1.11	.16	.08	1.9	.08	.47	
Mcal	.43	.43	2.78	.40	.20	.47	.20	1.19	6.10
Fuel-Mcal	0	0	141.02	0	0	33.00	0	90.93	264.95
Depr-Mcal	4.91	0	51.09	0	0	26.73	0	0	82.73
Total (Mcal)	5.34	.43	194.89	.40	.20	60.20	.20	92.12	353.78
5000 Head									
Labor									
Hr	.21	.13	.59	.16	.08	.24	.11	.47	
Mcal	.51	.32	1.46	.40	.20	.59	.28	1.19	4.95
Fuel-Mcal	0	0	52.66	0	0	29.19	0	90.93	172.78
Depr-Mcal	.98	0	24.45	0	0	7.24	0	0	32.67
Total (Mcal)	1.49	.32	78.57	.40	.20	37.02	.28	92.12	210.40
10,000 Head									
Labor									
Hr	.25	.11	.46	.16	.08	.25	.16	.47	
Mcal	.63	.28	1.15	.40	.20	.63	.40	1.19	4.88
Fuel-Mcal	0	0	50.11	0	0	36.71	0	90.93	177.75
Depr-Mcal	.47	0	18.30	0	0	7.36	0	0	26.13
Total (Mcal)	1.10	.28	69.56	.40	.20	44.70	.40	92.12	208.76

[a] Mcal/head.

Table 7
CULTURAL ENERGY USED FOR FEEDLOT OPERATIONS

	Mcal/head/day	Ref.
Average for Colorado	1.00	8
Average for California	0.36	7
New York estimate	0.52	15
New Mexico survey	9.90	15
U.S. survey	0.46	12
Survey 14 Kansas feedlots	1.20	11

Table 8
EQUIPMENT COMPLEMENT AND ENERGY CONSUMPTION[5] OF A 91,000 kg/DAY FEED GRINDING FACILITY

Unit	Installed hp or capacity	To produce 91,000 kg[a] (Mcal)	Mcal/kg
Front end loader	30 hp-gasoline	96	.0011
Receiving conveyor	5 hp-hr/36,000 kg	25	.00027
Receiving elevator	15 hp-hr/36,000 kg	75	.00082
Grain cleaner	13 hp-hr/36,000 kg	65	.00071
Receiving drag conveyor	2 hp-hr/36,000 kg	5	.00005
2 drag conveyors	@ 2 hp	40	.00044
Hammermill	180 kg/hp-hr	1000	.011
Pneumatic conveyor	40 hp	400	.0044
Rotary airlock	1.5 hp	15	.00016
24 discharge conveyors	@ 2 hp	470	.0052
2 mixers (@ 2 ton)	@ 20 hp	400	.0044
Feed elevator	7.5 hp	75	.00082
Scalper	1 hp	10	.00011
3 finished feed transfer conveyors	@ 3 hp	90	.00099
Total		**2800**	**.030**
Total + 25% for labor, machinery, and miscellaneous		3500	.038

[a] Electrical generating factor of 32% is assumed.

Table 9
ENERGY CONSUMPTION IN GRAIN FLAKING[5]

Unit	Installed capacity	To produce 91,000 kg	Mcal/kg
Grain steamer	.056 kg steam/kg corn	4,600	0.05
Roller mill	40 hp-hr/3600 kg	2,000	0.022
Support facilities (Table 8 total less hammermill)		1,800	0.020
Total		**7,400**	**0.093**
Total + 25% for machinery, labor and miscellaneous		9,300	0.12

Table 10
ENERGY CONSUMPTION IN PELLETING[5]

Unit	Installed capacity	Mcal to produce 91,000 kg	Mcal/kg
Pellet mill	68 kg/hp-hr	850	.0093
Pellet steaming	2500 lbs steam/40 tons English	2,600	.028
Pellet cooler	20 hp-hr/9 metric tons	130	.0014
Elevator	1.5 hp-hr/9 tons	9.6	.0001
Grader	2 hp-hr/10 tons	11	.0001
Feeder and fat applicator	3 hp-hr/10 tons	17	.0002
Total, specialized equipment		**3,300**	**.0039**
+ supporting equipment (Table 8)		2,800	.030
Total		6,100	.069
Total including 25% for labor, machinery, and miscellaneous		**7,600**	**.084**

Table 11
COMPOSITE RATION FOR WESTERN CATTLE FEEDING (FEED FOR ENTIRE FEEDING PERIOD FOR ONE ANIMAL)[a]

Ingredient	% of ration	kg required	kcal/kg[b]	kcal
Corn	18.0	338.4	1,889	639,238
Corn silage	16.5	310.2	761	236,062
Barley	16.0	300.8	754	226,803
Sorghum milo	13.0	244.4	755	184,522
Alfalfa hay	11.0	206.8	643	132,972
Wheat	10.0	188.0	754	141,752
Molasses	6.1	114.6	—	—
Native hay	3.7	69.6	932	64,867
Tallow	2.5	47.0	—	—
Beet pulp	2.2	41.4	3,825	158,355
Cottonseed meal	1.0	18.8	—	—
	100.0	1,880.0		1,784,571

[a] 9,392 kcal/hd/day.
[b] From Table 2.

Table 12
TOTAL ENERGY INPUT FOR 10,000 CAPACITY FEEDLOT

	Type of feed processing					
	Grinding		Flaking		Pelleting	
	kcal/head	kcal/head/day	kcal/head	kcal/head/day	kcal/head	kcal/head/day
Feedlot operation	208,760	1,099	208,760	1,099	208,760	1,099
Ration	1,784,571	9,392	1,784,571	9,392	1,784,571	9,392
Processing	71,400	376	22,560	119	157,920	831
Total	**2,064,731**	**10,867**	**2,015,891**	**10,610**	**2,151,251**	**11,322**
Cultural energy input from birth to slaughter						
Mcal cultural energy/Mcal carcass energy				= 3.8	3.7	3.9
Mcal cultural energy/kg protein				= 65.6	64.6	67.4

[a] Prorated from 120 to 190 days.[9]

REFERENCES

1. **Cook, C. W.,** *J. Range Manage.*, 7, 10, 1954.
2. **Cook, C. W.,** *Colo. State Univ. Exp. Stu. Bull.*, TB109, 1970.
3. **Cook, C. W.,** *J. Range Manage.*, 29, 226, 1976.
4. **Cook, C. W., Denham, A. H., Bartlett, E. T., and Child, R. D.,** *J. Range Manage.*, 29, 186, 1976.
5. CAST Special Publication No. 4, Council for Agricultural Sciences and Technology, 1975.
6. **Hayes, E. T.,** *Science*, 191, 661, 1976.
7. **Hertlein, K. H.,** preliminary paper, mimeograph, University of California, Davis, 1974.
8. **Hobson, B.,** Report on Grant No. SIA-14125, NSF-RANN, Department of Animal Science, Colorado State University, 1976.
9. **Hobson, B. and Yorks, T. P.,** Report on Grant No. SIA-14125, NSF-RANN, Department of Animal Science, Colorado State University, 1977.
10. **Hodgson, H. J.,** *BioScience*, 26, 625, 1976.
11. **Lipper, R. E., Anschutz, J. A., and Weller, J. C.,** Paper presented at 1976 Mid-Central Meeting of the American Society of Agricultural Engineers, Kansas State University, March, 1976.
12. **Lockeretz, W.,** Center for Biology of Natural Systems, Washington University, St. Louis, Missouri, NSF RANN Grant No. GI-043890, 1975.
13. **Nix, J., Ballard, E., and Arata, J.,** LMS-218, U.S. Department of Agriculture, December, 1977.
14. **Old, C. A.,** unpublished data.
15. **Patrick, N. A.,** *Agriculture and Energy*, W. Lockeretz, Ed., Academic Press, New York, 1977.
16. **Pimentel, D.,** personal communication.
17. Agricultural Statistics, U.S. Department of Agriculture, Washington, D.C. , 1974.
18. Agricultural Statistics, U.S. Department of Agriculture, Washington, D.C., 1977.

ENERGY INPUTS FOR BEEF CATTLE PRODUCTION ON PASTURE

Carl S. Hoveland

Forages are the basis of cattle production in the U.S. Beef cattle obtain over three fourths of their feed units from forages. Dairy cattle, because of heavy grain feeding, obtain about two thirds of their feed units from forages. Grasslands, producing forage for cattle, sheep, and goats, occupy about one half the total land area of the 48 contiguous states.[2] Beef animals outnumber dairy animals by over four to one, and the ratio is widening.

Changes are occurring in the U.S. beef industry. The beef cattle population is moving eastward into the humid area of the country. Western range carrying capacity may be increased slightly with improved management, but low precipitation will restrict growth in beef cattle numbers. Future increases in beef cattle numbers and productivity will be concentrated in humid areas, particularly on land not well adapted to arable crops. The greatest potential for growth in the beef cattle industry is the southeastern U.S., where an abundance of relatively cheap nonarable crop land, combined with favorable temperatures and rainfall, offer pasture over most of the year. The development of improved quality grasses and legumes for this region should improve animal performance and reduce the need for fertilizer nitrogen. In recent years, there has been a trend toward less grain feeding of beef cattle. As energy costs escalate and world demand for U.S. grain increases in the future, it is likely that pasture will provide an even higher percentage of the nutrients consumed by beef cattle. Stocker cattle are being grown to heavier weights on pasture alone before finishing in feedlots. Numerous opportunities exist, especially in the southern U.S., for expanded use of finishing beef on pasture with little or no grain feeding.[3]

A built-in inefficiency of beef production is maintenance of a brood cow needed to produce one calf annually. In countries such as Ireland, where two thirds of calves for beef production come from a large dairy industry, energy inputs are substantially lower than in the U.S.[1] Long-lived perennial pastures are essential to reduce energy inputs. In Alabama, cow-calf production, even with the use of legumes to supply nitrogen and with a long grazing season (March 11 to October 5), the energy inputs are substantial (Table 1). In addition, over 1 metric ton of hay and some protein supplement would be needed per cow for winter maintenance. Cow-calf production on western rangeland would have lower energy inputs on pasture, but the shorter grazing season would result in a much longer hay feeding period.

Growing weaned steers on pasture heavily fertilized with nitrogen can produce high yields of beef per hectare because of a high animal stocking rate (Table 2). Energy inputs into such a pasture system are high, mainly because of the nitrogen fertilizer. However, daily gain per steer on many perennial grass pastures, such as Coastal bermudagrass (*Cynodon dactylon*), bahiagrass (*Paspalum notatum*), and tall fescue (*Festuca arundinacea*), are generally low unless legumes are present in the sward.

A low energy input system is illustrated by Kentucky bluegrass (*Poa pratensis*) — birdsfoot trefoil (*Lotus corniculatus*) pasture in Indiana (Table 3). Energy inputs were extremely low, since no fertilizer was required. It is doubtful if a pasture could remain this productive over an extended period of time without adding fertilizers to replace that lost by removal of animal product each year. Again, daily gains of steers were relatively low. This results in additional high energy cost feeding of stored roughage and concentrate at the end of the grazing season.

The productivity of western rangeland can be improved by fertilizing and seeding

Table 1
ENERGY INPUTS PER HECTARE FOR BEEF COWS AND CALVES ON COASTAL BERMUDAGRASS — WINTER ANNUAL CLOVER PASTURE IN SOUTHEASTERN ALABAMA

Input

Item	Quantity/ha	kcal/ha
Pasture establishment		
Labor	12 hr	6,528
Machinery	5.5 kg[a]	142,026
Diesel fuel	64 ℓ[b]	730,496
Phosphorus	15 kg	32,940
Potassium	50 kg	110,100
Lime	250 kg[c]	3,862
Bermudagrass planting stock	400 kg	400,000
Total		**1,425,952**
Energy input per year prorated over 30-year life of stand		*47,532*
Annual maintenance		
Labor	1 hr	544
Machinery	3.5 kg[d]	90,380
Diesel fuel	4 ℓ[b]	45,656
Phosphorus	15 kg	32,940
Potassium	50 kg	110,100
Lime	250 kg[c]	3,862
Clover seed (arrowleaf and crimson)	15 kg	465,000
Total		**748,482**
Prorated energy/ha/year for pasture establishment		*47,532*
Total energy input/ha		**796,014**

Output

Daily gain per cow	410 g	—
Daily gain per calf	870 g	—
Cow gain	325 kg	—
Calf gain	460 kg[e]	—
kcal input/kg calf gain output		1,730
kcal input/kg cow and calf gain		1,014

[a] Assumes that 5.5 metric tons of machinery were used for establishment of pasture (tractor, disk, plow, bermuda planter, fertilizer spreader) and that this machinery serves an average of 100 ha and that the machinery is depreciated over 10 years.
[b] Diesel fuel for operation of tractor.
[c] Lime applied at rate of 1000 kg/ha every 4 years.
[d] Assumes that 3.5 tons of machinery were used annually to overseed the clovers and apply fertilizer.
[e] Calves born in November are weaned in September.

an improved grass, as illustrated for North Dakota (Table 4). Energy inputs per unit of beef are lower than the grass-nitrogen system in Alabama, but higher than the grass-legume system in Indiana. Although daily gain per steer was relatively good, production of beef per hectare was low because of the low stocking rate and short grazing season.

Table 2
ENERGY INPUTS PER HECTARE FOR BEEF STEERS ON COASTAL BERMUDAGRASS PASTURE IN SOUTH EASTERN ALABAMA[5]

Item	Quantity/ha	kcal/ha
Input		
Pasture establishment		
Labor	12 hr	6,528
Machinery	5.5 kg[a]	142,026
Diesel fuel	64 ℓ[b]	730,496
Nitrogen	100 kg	1,470,000
Phosphorus	15 kg	32,940
Potassium	50 kg	110,100
Lime	250 kg[c]	3,862
Bermudagrass planting stock	400 kg	400,000
Total		**2,895,952**
Energy input per year prorated over 30-year life of stand		*96,532*
Annual maintenance		
Labor	1 hr	544
Machinery	3.5 kg[d]	90,380
Diesel fuel	4 ℓ[b]	45,656
Nitrogen	360 kg	5,292,000
Phosphorus	15 kg	32,940
Potassium	80 kg	176,160
Lime	250 kg[c]	3,862
Total		**5,641,542**
Prorated energy/ha/year for pasture establishment		*96,532*
Total energy input/ha		**5,738,074**
Output		
Daily gain per steer	568 g/day	—
Beef yield	695 kg/ha	—
kcal input/kg beef output	—	8,256
Protein yield	111 kg/ha	—
kcal input/ kg protein out	—	51,694

[a] Assumes that 5.5 metric tons of machinery were used for establishment of pasture (tractor, plow, disk, bermuda planter, fertilizer spreader) and that this machinery serves an average of 100 ha and that the machinery is depreciated over 10 years.
[b] Diesel fuel for operation of tractor.
[c] Lime applied at rate of 1000 kg/ha every 4 years.
[d] Assumes that 3.5 tons of machinery were used annually to apply fertilizer.

Finishing of steers for slaughter on pasture without the use of grain requires high quality forage over a long grazing season. Winter annual pasture, planted in September and grazed from November until June in the lower South, can furnish high quality forage for high daily gains and sufficient finish to grade high Good at slaughter (Table 5). As shown, this is a high energy input system. However, where this annual pasture is grown on clay soils year after year, ryegrass and clover seed inputs can be eliminated because of natural reseeding and the nitrogen inputs can be reduced by at least one half. This would result is an energy input of 6473 kcal/kg beef. Although this appears

Table 3
ENERGY INPUTS PER HECTARE FOR BEEF STEERS ON KENTUCKY BLUEGRASS-BIRDSFOOT TREFOIL PASTURE IN INDIANA

Item	Quantity/ha	kcal/ha
Input		
Pasture establishment		
Labor	8 hr	4,352
Machinery	5.8 kg[a]	149,773
Diesel fuel	65 ℓ[b]	741,910
Seed		
Kentucky bluegrass	3 kg	153,000
Birdsfoot trefoil	4 kg	316,000
Total		**1,365,035**
Total energy input per year prorated over 15-year life of stand[c]		*91,002*
Output		
Daily gain per steer	495 g/day	—
Beef yield	380 kg/ha	—
kcal input/kg beef output	—	239
Protein yield	68 kg/ha	—
kcal input/kg protein output	—	1,338

[a] Assumes that 5.8 metric tons of machinery were used for establishment of the pasture (tractor, disk, plow, drill, cultipacker) and that this machinery serves an average of 100 ha and that the machinery is depreciated over 10 years.

[b] Diesel fuel for operation of tractor.

[c] Length of stand assumed to be at least 15 years, even when fertilizer is not applied.[7]

to be a high'energy input system, it does permit the use of relatively poor land unsuited for other crop production to produce satisfactory finished beef without feeding grain.

Energy inputs for beef production are relatively high in the U.S. Some factors responsible for this are (1) maintenance requirements for beef brood cows, (2) feeding large amounts of stored roughages, (3) heavy grain feeding, (4) reliance on synthetic nitrogen fertilizer rather than legumes in pastures, and (5) low nutritive quality of many otherwise adapted perennial forage grasses. Possible ways to improve energy efficiency of U.S. beef production are (1) multiple births of calves, (2) perennial grass cultivars with improved nutritive quality, (3) greater use of legumes as a source of nitrogen in pastures, and (4) greater reliance on well-managed pastures over as much of the year as possible.

Table 4
ENERGY INPUTS PER HECTARE FOR BEEF STEERS ON RUSSIAN WILDRYE PASTURE IN NORTH DAKOTA [8]

Item	Quantity/ha	kcal/ha
Input		
Pasture establishment		
Labor	8 hr	4,352
Machinery	5.8 kg[a]	149,773
Diesel fuel	65 ℓ[b]	741,910
Nitrogen	45 kg	661,500
Phosphorus	30 kg	65,880
Seed	9 kg	108,000
Total		**1,731,415**
Energy input per year prorated over 20-year life of stand[c]		**86,571**
Annual maintenance		
Labor	1 hr	544
Machinery	3.5 kg[d]	90,380
Diesel fuel	4 ℓ[b]	45,656
Nitrogen	45 kg	661,500
Phosphorus	30 kg	13,500
Total		**929,840**
Prorated energy/ha/year for pasture establishment		*86,571*
Total energy input per ha/year		**1,016,411**
Output		
Daily gain per steer	850 g/day	—
Beef yield	183 kg/ha	—
kcal input/kg beef output	—	5,554
Protein yield	33 kg/ha	—
kcal input/kg protein output	—	30,800

[a] Assumes that 5.8 metric tons of machinery were used for establishment of pasture (tractor, disk, plow, drill, fertilizer spreader) and that this machinery serves an average of 100 ha and that the machinery is depreciated over 10 years.

[b] Diesel fuel for operation of tractor.

[c] Personal communication, R. J. Lorenz.

[d] Assumes that 3.5 metric tons of machinery were used for annual application of fertilizer to pasture and that the machinery serves an average of 100 hectares and that the machinery is depreciated over 10 years.

Table 5
ENERGY INPUTS PER HECTARE FOR BEEF STEERS ON WINTER ANNUAL PASTURE IN CENTRAL ALABAMA[9]

Item	Quantity/ha	kcal/ha
Input		
Labor	8.8 hr	4,787
Machinery	5.8 kg[a]	741,910
Diesel fuel	69 ℓ[b]	787,566
Nitrogen	112 kg	1,646,400
Phosphorus	67 kg	147,132
Potassium	67 kg	147,534
Lime	250 kg[c]	3,862
Seeds		
Rye	94 kg	282,600
Ryegrass	11 kg	134,400
Arrowleaf clover	9.0	279,000
Total		**4,175,191**
Output		
Daily gain per steer	931 g/day	—
Beef yield	454 kg/ha	—
kcal input/kg beef output	—	9,196
Protein yield	82 kg/ha	—
kcal input/kg protein output	—	50,917

[a] Assumes that 5.8 metric tons of machinery were used for establishment and fertilization of pasture (tractor, plow, disk, drill, fertilizer spreader) and that this machinery serves an average of 100 ha and that the machinery is depreciated over 10 years.

[b] Diesel fuel for operation of tractor.

[c] Lime applied at rate of 1000 kg/ha every 4 years.

REFERENCES

1. **Collins, D. P., Drennan, M. J., and Flynn, A. V.,** Potential of Irish grassland for beef production, in *Proc. International Meeting on Animal Production from Temperate Grassland,* Irish grassland and Animal Production Association, Dublin, Ireland, 1977, 12.
2. **Sprague, H. B.,** Significance of our grasslands, *Grasslands of the United States, Their Economic and Ecologic Importance,* Sprague, H. B., Ed., Iowa State University Press, Ames, 1974, 4.
3. **Stuedemann, J. A., Huffman, D. L., and Purcell, J. C.,** Forage-feed beef: production and marketing alternatives in the South, Southern Cooperative Series Bulletin 220, Stuedemann, J. A., Huffman, D. L., and Purcell, J. C., Eds., North Carolina Agricultural Experiment Station, Raleigh, 1977, xvii.
4. **Hoveland, C. S., Anthony, W. B., McQuire, J. A., and Starling, J. G.,** Beef cow-calf performance on Coastal Bermuda grass overseeded with winter annual clovers and grasses, *Agron. J.,* 70, 418, 1978.
5. **Hoveland, C. S., King, C. C., Evans, E. M., Harris, R. R., and Anthony, W. B.,** Bermuda grass for forage in Alabama, Agriculture Experiment Station Bulletin 328, Auburn University, Auburn, Ala., 1971.
6. **Lechtenberg, V. L., Rhykerd, C. L., Mott, G. O., and Huber, D. A.,** Beef production on brome grass (*Bromus inermis* Leys.) pastures fertilize anhydrous ammonia, *Agron. J.,* 66, 47, 1974.
7. **Lechtenberg, V. L.,** personal communication.
8. **Rogler, G. A. and Lorenz, R. J.,** Beef production from Russian wildrye (*Elymus gunceus* Fisch.) in the United States, in *Proc. 11th International Grassland Congr.,* Surfers Paradise, Queensland, Australia, 835.
9. **Anthony, W. B., Hoveland, C. S., Mayton, E. L., and Burgess, H. E.,** Rye-rye grass-yucha arrowleaf clover for production of slaughter cattle; Agriculture Experimental Station Circular 182, Auburn University, Auburn, Ala., 1971.

CULTURAL ENERGY IN SHEEP PRODUCTION

C. Kerry Gee*

WESTERN RANGE SHEEP PRODUCTION

The western sheep-producing region has long supplied the major proportion of mutton, lamb, and wool of the nation. Consisting of the 17 most western of the 48 contiguous states, the region in 1976 had 9.2 million stock sheep or about 80% of the U.S. total of 11.5 million. Much of this region is characterized by vast areas of arid or semiarid rangeland vegetated by grasses, forbs, shrubs, and brush. Part of the land is more suitable to sheep grazing than cattle grazing, for sheep utilize forbs and certain species of shrubs more readily than do cattle. They also require less frequent access to water and have greater ability to traverse rough or steep terrain. Cash receipts from sales of sheep, lambs and wool, although providing all or most of the income for many producers, are much less important in the total agricultural picture in the western states than receipts from sales of crops and other livestock products. Sales of sheep, lambs, and wool in the 17 western states are less than 1% of the cash receipts from all farm and ranch commodities. In comparison, sales of cattle and calves account for 30% of all cash agricultural receipts; hogs, 3%; and all dairy products, 6%. The relative contribution of sheep, lamb, and wool sales to total agricultural commodity sales is more significant in some states than others. Their contribution is nearly 7.8% in Wyoming, 6.6% in Utah, and 4.8% in Nevada, but only 0.1% in Oklahoma.

Only about 41% of sheep producers of the West have commercial-scale operations of 50 head or more sheep, but they own nearly 93% of the sheep of the region. About one third of these commercial producers have specialized in sheep, while two thirds have diversified livestock operations. More than two thirds operate as sole proprietors, while the rest have formed partnerships and family corporations. Many have substantial equity positions which indicate past profitability. About half of the feed requirement for commercial sheep comes from private range, while public range supplies about one fifth. The remainder is usually hay, concentrate, or crop residue. Over half of the commercial sheep are grazed under the care of herders, usually on open (unfenced) range. Most lambing occurs in late winter and early spring. More commercial producers practice shed lambing than range lambing, but the number of sheep involved is less — 36 compared to 62%.

Three energy budgets represent sheep production in this region. One describes enterprises in the Texas-New Mexico area which have about 25% of all stock sheep in the region. Range is all privately owned and fenced. The climate is mild in winter and hot and dry in summer. Little shed lambing occurs. Sheep receive a minimum of care, and production per ewe is lower than in other parts of the region. During dry seasons, supplementary feeding is done. Two budgets depict production in the Intermountain and Great Basin areas, where about 50% of western states sheep are found. Except for differences in lambing practices (shed vs. range), operations are similar. Much use is made of open range, and sheep are moved long distances between feed sources. A minimum amount of supplementary feed is used. Operations that shed lamb use more feed from this source than those that range lamb. Shed lambing enterprises usually have more output of lamb per ewe than those that range lamb.

About 65% of lambs marketed in this area are for slaughter. The remainder go as feeder lambs.

* Economics, Statistics and Cooperatives Service, U.S. Department of Agriculture, Ft. Collins, Colorado.

NORTHCENTRAL STATES FARM FLOCK PRODUCTION

Sheep production in the northcentral states* is usually a supplementary farm enterprise. Flocks are small, averaging less than 100 head. About 8% of farms in the region have sheep. In 1976, stock sheep in these states numbered 1.9 million head or 14.5% of total U.S. inventories. Farms reporting sheep in 1975 totaled 66.5 thousand which is about one half of all farms with sheep in the U.S. Despite the large number of farms with this enterprise, sheep contribute only 0.1% to receipts from all agricultural commodities in the northcentral states.

States in this region with largest stock sheep populations are Ohio, Iowa, and Minnesota. Arkansas has the smallest population. Of particular importance is Ohio which ranks tenth among all states in the nation in stock sheep numbers and has nearly one fourth of the region's total stock sheep.

One budget represents farm flock production. Sheep are kept in pastures in summer and corrals and sheltered areas in winter. Shed lambing is common. Supplementary feeds are fed several months of the year. Lambs are all marketed at slaughter weights and grades which require creep or supplementary feeds during the growing period for at least part of the lambs.

SOURCE OF DATA

Input requirements and production for sheep were taken from budgets prepared by the Economics, Statistics and Cooperatives (ESCS) — U.S. Department of Agriculture (USDA) firm enterprise data system which in turn are based on 1975 to 1976 sample surveys of sheep producers.[1,2]

Energy estimates per unit of input are the same numbers used by Cook et al. in their report Cultural Energy in U.S. Beef Production.[3] Adjustments were made in irrigated pasture energy requirements to reflect ditch gravity irrigation instead of center sprinkler irrigation which reduces energy required per kilogram of forage from 1112 to 364 kcal. Pasture costs for the northcentral states sheep budget assume no irrigation which reduces energy inputs to 359 kcal/kg of forage produced. Energy associated with fences was also subtracted from the pasture budgets and entered as a separate input item in the sheep budgets.

* Northcentral states include Arkansas, Illinois, Indiana, Iowa, Michigan, Minnesota, Missouri, Ohio, and Wisconsin.

Table 1
ENERGY INPUTS PER BREEDING EWE FOR SOUTHWESTERN RANGE SHEEP PRODUCTION, ALL SIZES

Item	Quantity[a]	kcal/unit	Total (kcal)
	Input		
Labor (hr)	0.83		
Machinery (hr)	0.01	312,792	3,128
Pickup (km)	19.4	4,183	81,150
Truck (kg × km)	1.2	3,090	3,708
Electricity (kWh)	0.43	2,863	1,231
Range forage (kg)[b]	769.0	—	—
Alfalfa hay (kg)	54.0	473	25,542
Corn (kg)	42.0	1,889	79,338
Soybean meal (kg)	14.0	2,833	39,662
Fence (km)	0.03	440,893	13,227
Water[c]			
Pond (Aum)	3.32	2,930	9,728
Total energy input			**256,714**
	Output		
Lambs (kg)			
Liveweight[d]	20.9		
Carcass weight	10.65	2,630	28,009
Cull ewes (kg)[e]			
Liveweight	7.1		
Carcass weight	3.41	2,470	8,423
Total energy in carcass			**36,432**
kcal input/kcal output			7.0

[a] Based on sheep budget estimates by FEDS.[1]

[b] It is assumed that there is no range improvement other than fencing and some water development and maintenance. Energy associated with these items are included in the appropriate inputs in the budget.

[c] Livestock water structures are measured in animal unit months required.

[d] Assumes 0.59 feeder lamb sold per ewe in January 1 inventory at 78 lb.

[e] Assumes 0.12 culls sold per ewe in January 1 inventory at 130 lb.

Table 2
ENERGY INPUTS PER BREEDING EWE FOR INTERMOUNTAIN AND GREAT BASIN RANGE SHEEP PRODUCTION, RANGE LAMBING, 2500 HEAD AND OVER

Item	Quantity[a]	kcal/unit	Total (kcal)
	Input		
Labor (hr)	2.91		
Machinery (hr)	0.03	312,792	9,384
Pickup (km)	25.1	4,183	104,993
Truck (kg × km)	5.6	3,090	17,304
Electricity (kWh)	0.31	2,863	888
Pasture forage (kg)[b]	20	364	7,280
Range forage (kg)[b]	386	—	—
Alfalfa hay (kg)	48	473	22,704
Barley (kg)	9	754	6,786
Soybean meal (kg)	1.4	2,833	3,966
Fence (km)	0.01	440,893	4,409
Water[c]			
Pond (Aum)	1.0	2,930	2,930
Windmill (Aum)	0.4	7,950	3,180
Total energy input			**183,824**
	Output		
Lambs (kg)[d]			
Liveweight	34.9		
Carcass weight	17.8	2,630	46,815
Cull ewes (kg)[e]			
Liveweight	10.1		
Carcass weight	4.84	2,470	11,955
Total energy in carcass			**58,769**
kcal input/kcal output			3.1

[a] Based on sheep budget estimates by FEDS.[1]

[b] Energy for fences and water development and maintenance are included in other input items. It is assumed that there is no range improvement.

[c] Livestock water structures are measured in animal unit months required.

[d] Assumes 0.79 lambs sold per ewe in January 1 inventory with 75% at a slaughter weight of 102 lb and 25% at a feeder weight of 84 lb.

[e] Assumes 0.16 culls sold per ewe in January 1 inventory at 138 lb.

Table 3
ENERGY INPUTS PER BREEDING EWE FOR INTERMOUNTAIN AND GREAT BASIN RANGE SHEEP PRODUCTION, SHED LAMBING, 2500 HEAD AND OVER

Item	Quantity[a]	kcal/unit	Total (kcal)
	Input		
Labor (hr)	3.91		
Machinery (hr)	0.13	312,792	40,663
Pickup (km)	17.6	4,183	73,621
Truck (kg × km)	11.7	3,090	36,153
Buildings (ft_2[b])	4.4	38,000	167,200
Electricity (kWh)	3.7	2,863	10,593
Pasture forage (kg)[b]	35.0	364	12,740
Range forage (kg)[b]	678.0	—	—
Alfalfa hay (kg)	82.0	473	38,786
Barley (kg)	21.0	754	15,834
Soybean meal (kg)	7.0	2,833	19,831
Fence (km)	0.01	440,893	4,409
Water[c]			
Pond (Aum)	0.9	2,930	2,637
Windmill (Aum)	0.5	7,950	3,975
Total energy input			**426,442**
	Output		
Lambs (kg)[d]			
Liveweight	46.0		
Carcass weight	23.5	2,630	61,805
Cull ewes (kg)[e]			
Liveweight	10.1		
Carcass weight	4.84	2,470	11,955
Total energy in carcass			**73,760**
kcal input/kcal output			5.8

[a] Based on sheep budget estimates by FEDS.[1]

[b] Energy for fences and water development and maintenance are included in other input items. It is assumed that there is no range improvement.

[c] Livestock water structures are measured in animal unit months required.

[d] Assumes 0.89 lambs sold per ewe in January 1 inventory with 65% at a slaughter weight of 107 lb and 35% at feeder weight of 90 lb.

[e] Assumes 0.16 culls sold per ewe in the January 1 inventory at 138 lb.

Table 4
ENERGY INPUTS PER BREEDING EWE FOR NORTHCENTRAL STATES FARM FLOCK PRODUCTION, ALL SIZES

Item	Quantity[a]	kcal/unit	Total (kcal)
	Input		
Labor (hr)	4.0		
Machinery (hr)	0.53	312,792	165,780
Pickup (km)	21.9	4,183	91,608
Buildings (ft_2[b])	6.5	38,000	247,000
Electricity (kWh)	10.8	2,863	30,290
Pasture forage (kg)[b]	471	359	169,089
Alfalfa hay (kg)	262	473	123,926
Corn (kg)	117	1,889	221,013
Soybean meal (kg)	17	2,833	48,161
Fence (km)	0.02	440,893	8,818
Water Pond[c] (Aum)	1.8	2,930	5,274
Total energy input			**1,111,589**
	Output		
Lambs (kg)[d]			
Liveweight	42.7		
Carcass weight	21.8	2,630	57,334
Cull ewes (kg)[e]			
Liveweight	9.7		
Carcass weight	4.8	2,470	11,856
Total energy in carcass			**69,190**
kcal input/kcal output			16.1

[a] Based on sheep budget estimates by FEDS.[2]
[b] Energy for fences and water development and maintenance are included in other input items.
[c] Livestock water facilities are measured in animal unit months required.
[d] Assumes 0.95 slaughter lambs sold per ewe in January 1 inventory at a weight of 99 lb.
[e] Assumes 0.15 culls sold per ewe in January 1 inventory at 143 lb.

REFERENCES

1. Firm Enterprise Data System, 1974, Economic Research Service, U.S. Department of Agriculture, and Oklahoma State University, Stillwater, 1976.
2. Firm Enterprise Data System, 1976, Economic Research Service, U.S. Department of Agriculture, and Oklahoma State University, Stillwater, 1978.
3. **Cook, C. W., Combs, J. J., and Ward, G. M.,** Cultural Energy in U.S. Beef Production, mimeo report, Colorado State University, Fort Collins, undated.

AQUACULTURE

J. Bardach

Aquaculture is the farming and husbanding of fish, shellfish, and other aquatic organisms in fresh or salt water, mainly for food, but also for commercial use (pearls, algal colloids) and for recreation. High-grade animal protein derived from aquatic farming plays an important role in the nutrition of more than a quarter of humanity, in the Chinese Peoples' Republic, in Southeast Asia, in Israel, Japan, and the Soviet Union.

In contrast to world fisheries which exhibited a near-static harvest over the last decade, aquaculture has increased rapidly. Production now stands at over 6 million metric tons world wide and is expected to increase further in importance especially if overfishing, pollution, and high energy costs of certain fishing operations were to curb global fishery yields.[1,2]

The peculiar properties of water and of the organisms which inhabit it make for generic rather than specific resemblance in production methods of aquaculture and land animal husbandry. Aquaculture differs from fisheries, however, by production site and stock being controlled by the aquatic farmer. Notably, aquacultural practices range from being very intensive to extensive with husbandry measures being restricted, in certain species to portions of an animals' life cycle. Such semiculture[3] prevails in many sites for oyster — and other mollusk — farming with fixed and variable inputs consisting of providing attachment devices for naturally occuring seed and the labor and equipment (including boats of various sizes) for limited tending as well as harvest; no fertilizer or feed is required. Similarly low production inputs prevail in salmon ranching where aquaculture only provides fry or juveniles to be sent out to graze on free ocean pastures.

An important difference between aquaculture and the raising of avian or some mammalian stock is that several important, intensively grown, aquacultural species are carnivores (e.g., trout and catfish), requiring extraneous feeding with high-protein rations, the fashioning of which demands substantial inputs of energy. Counteracting this apparent ecological, hence bioenergetic, disadvantage is the superior food converting ability of carnivorous fishes[4] explaining their relatively low-energy use for production of a gram of protein as compared to range and feed-lot beef, lobster, lambs, and shrimp.[2] However, aquaculture also relies on large amounts of water, hence on pumping, and often on the use of cement for containing devices. Thus aquaculture can exhibit widely disparate levels of energy inputs.

Examples of husbandry practices for fish or shellfish with favorable input/output ratios of energy are found among those for filter feeders of plant plankton like the milkfish of the Indo-Pacific or in raft culture for mussels where there is, albeit highly site-specific, reliance on natural seeding and tidal advection of food such as occurs in bays near Vigo, in Spain.[3] Full use of the water column for growing species with different food habits can also enhance overall efficiency of flesh production from ponds and use of the same water for growing ducks or pigs as well as fish, even including the fertilization of adjacent market garden crops, leads to important energy savings in the combined operation. Such practices have been developed in China, India, Malaysia, and Hungary; in them the energy inputs to produce one gram of protein may even approach those for wheat or rice, but they are not as amenable to energy accounting, as are monocultures.

In developed nations where high labor costs prevail, the products of aquaculture tend to be luxury foods. Even though the total energy demand for growing them has not received the attention it will deserve, aquaculture competes favorably with other modes of animal protein production even now: the energy used to produce 1g of protein in catfish farming is equal to this value for eggs, but far lower than for beef or lamb.[2] The analyses which follow indicate that prawns, the price of which exceeds that of catfish by a factor of at least three, requires under 2.5 times the amount of energy input per gram of unprocessed protein as does catfish. Aquaculture management measures are being developed that rely on sound ecologic as well as economic principles such as polyculture, including the growing in one unit aquatic and terrestrial crops and the recycling of nutrients. The analysis for growing two species of carp, tilapia, and a small percentage of mullet in one pond, using also some chicken manure, but employing a fair amount of machinery and based on the average of 80 farms, underlines these advantages: the kcal requirements per gram of unprocessed protein in such fish are lower than those for several table fish as landed in the U.S.[2] and of a comparable order of magnitude to protein in rice, i.e., 40 kcal/g for rice[2] grown in the U.S. and 65 kcal/g for the fish from Israeli fishponds grown under intensive culture. Furthermore, domestication of aquatic species has hardly begun; as more of them are subjected to selective breeding with attendant improvements in growth characteristics, aquaculture will be able to compete even more advantageously with the growing of avian and mammalian stock and with fishing. All three production modes of animal protein will, no doubt, have to be undertaken in future under more energy- and materials-saving conditions than prevail today.

Table 1
FARM PRODUCTION OF MALAYSIAN PRAWN *(MACROBRACHIUM ROSENBERGI)* ON OAHU, HAWAII[a]

Item	Amount	Land (ha)	kcal	Ref.
	Inputs			
Fixed				
Pond construction[b]		3.5		
Scraper				
D 8				
Grader	27.5 days/year		1,922,291	
Pipes	350 m 6 in. PVC 125 psi		36,120	6, 7
Cement flumes	120 ft² total (8 flumes)		238,286	6, 7
Wooden building	2000 ft²		217,143	
Labor	122 days/year			
Annual operating				
Water[c]	130 l/ha/min			
Labor; manual/miscellaneous	72 days/year			
Machinery use				
Running maintenance[d]				7
Harvesting (50 hp vehicle/tractor)	91 days/year	3.5	10,102,698	
Materials				
Net	Shared, with other farms, used 1 day/week; 4 cm mesh, 135 m/2 m nylon		7,000	6

Table 1 (continued)
FARM PRODUCTION OF MALAYSIAN PRAWN *(MACROBRACHIUM ROSENBERGI)* ON OAHU, HAWAII[a]

Item	Amount	Land (ha)	kcal	Ref.
	Inputs			
Fertilizer				8
Sodium nitrate	14 kg/ha		17,250	
Triplesuperphosphate	5 kg/ha		9,000	
Feed[e]				
Broilerchicken mash	4500 kg/ha		9,000,00	
Larva for planting (seed)[f]	50,000 larva/ton of production		19,333	
Total			**21,569,092**	
	Output[g]			
Live malaysian prawn[h]	3000 kcal/ha		324,000	9
Edible portion about 45%				
Caloric content 720 kcal/kg				
Protein content in prawn Flesh 14.6%; 65.7 kg protein from 450 kg of prawn that is edible in 1 metric ton				
kcal/g protein			328.3	
kcal input/kcal output			66.6	
kcal output/labor hour			129.6	

[a] The farm is Aquatic Farms Ltd. in Waikane, Oahu where both prawns and oysters are grown with partial joint use of facilities. The farm has 3.5 ha under water for prawns and 0.45 ha for oysters; it covers an area of about 10 ha. Investments are planned to be ammortized after 10 years.[5] All kcal values adjusted to 1 ton of product/year.

[b] The land would best be used for taro if an alternative crop were to be raised. Taro yields per ha are about 30,000 kg under labor intensive cultivation.

[c] Malaysian prawn farming relies on steady water use supplied by a permanent surface flow; alternative water use would be for taro irrigation which requires 40,000 m^3 per ha as opposed to 68,000 m^3 for prawn.

[d] Energy costs of machinery operation includes ammortization.

[e] The energy content of this food does not include costs of compounding it; the caloric value is derived from experimental calorimetry in feeding trials at 6 kcal/g oven dried. The value is therefore low.

[f] Prawn larva are supplied by a State Hatchery; production figures obtained from that source did not include ammortization. Value entered is therefore low.

[g] It was assumed that the calorie and protein values of prawn are equal to those given for crayfish to which the prawns are closely related.

[h] It is not unreasonable to take the whole animal as output in these calculations as Malaysian prawn are often marketed live or iced and heads-on.

Table 2
INTENSIVE OYSTER PRODUCTION ON LAND OAHU, HAWAII[a]

(Annual basis)

Item	Amount	Land(ha)	kcal	Ref.
Water area under production[b]		0.45		
Inputs				
Fixed				
Building farm, prorated			2,884,436	
Machinery (tractor, grader, dredger)	5 days			
Labor	26 days		72,200	
Pipes and cement flumes[c]	3 level, 30 m³, 1000 m 6 in. and 4 in. PVC		34,700,000	
Plastic trays for oysters[c]	3400 kg		129,956	
Operating				
Seed[d]				
Labor	1095 days			
Electricity, water pumping[e]	10,000 kWh/ month		343,560,000	
Fertilizer				8
Triplesuperphosphate	5 kg/day/ha		7,391,250	
Sodium nitrate	20 kg/day/ha		109,500,000	
Water[f]				
Total			**498,237,842**	
Output				
Oysters[g] *(Grassostrea gigas)*	13,636 kg/ha		5,583,760	
kcal input/kcal output	6,136 kg produced on farm with 910 kcal/kg oyster flesh		89.2	9
kcal in /g protein[b]	Protein content in oyster flesh 10.6%; 650 kg from farm		766.5	9
kcal output/labor hour			619.9	

[a] The farm is Aquatic Farms Ltd. at Waikane on Oahu; on it are grown both oysters and Malaysian Prawn with partial joint use of facilities. Investments are planned to be ammortized in 10 years. Such culture facilities are of interest in view of vanishing natural oyster grounds.[5]

[b] Total land for oyster production is 1 ha with 0.45 ha under water; production figures are based on the water area. The land may be used best for taro production if an alternative crop were to be grown on it. Average taro production per ha in Hawaii is 30,000 kg under labor intensive cultivation.

[c] The trays are made of polyethylene plastic and have a useful life of 10 years.

[d] The oysters reproduce on the farm and seed can be collected.

[e] From Pimentel (mimeographed), after Cervinka.

[f] The water used for the oysters is sea water with some admixture of prawn farm water; it has no direct costs.

[g] Values for shucked oyster flesh are used; weight of shells can be disregarded.

Table 3
EXPERIMENTAL PRODUCTION OF LAKE PERCH *(PERCA FLAVESCENS)* IN WISCONSIN[a]

Item	Amount	Land(ha)	kcal	Ref.
	Input			
Fixed				
Land[b]		2 08 (0.2/ ton)		
Containment structures				
Machinery 50 hp	15.3 days		596,277	
Pipes, conduits	1200m 4 in. PVC pipe 125 psi		765,217	
Buildings	Estimated		1,200,000	
Water[c]	3,400 m^3/ton			
Labor	16 days			
Operating				
Labor				
Maintenance	250 days			
Operation	1095 days/year			
Harvest (for farm)	95 days			
Nets, pails, etc.[d]	30/lm seine, wt about 20 kg Dipnets (for farm), 10/20 l plastic pails		11,000	
Stocking material				
Fingerlings[e]	50 kg/ha		4,634,200	
Fertilizer				
Dried alfalfa	200 kg/ha		3,440	
Medication[f]	5 kg/ha		350,000	
Feed				
40% protein dry pellets[g]	1750 kg/ton of fish		5,250,000	
Direct energy inputs[h]	2862 l fuel oil		32,666,868	
Pumping	20,190 kWh		57,803,970	
Total			**103,280,922**	
	Output			
Lake (yellow) perch 910 kcal/kg; edible portion 60%	Per ton		546,000	9
Protein in yellow perch 19.3%	115.8 kg protein/ ton		891.9	
kcal input/kcal output			189.2	
kcal output/labor hour			181.4	

[a] The growing of this fish is being developed in an intensive, land-saving manner by the U.S. Sea Grant Program of NOAA at the University of Wisconsin. In view of its pilot nature, substantial reduction in production inputs is to be expected.[10] All kcal values adjusted to 1 ton of fish/year.

[b] The fish reared is a carnivore and it is notable that 1 ton of perch is, even now grown on 0.2 ha of surface.

[c] Alternative uses for both land and water would be for maize growing with between 7 and 8 metric tons to be produced per ha and water equivalent on this land.

[d] Values estimated.

[e] Energy costs of production estimated from costs of growing adult fish; probably too low.

[f] Compound not specified, estimated from Kcal values for herbicides.[12]

[g] Kcal value from those supplied for trout and catfish chow;[13] calorimetric values only, compounding not considered. The value is therefore low.

[h] In intensive aquaculture where no flowing water is available naturally, high-energy inputs into water renewal are inevitable and present one of the likely future obstacles to such culture methods.

Table 4
POND POLYCULTURE IN ISRAEL[a]

Item	Amount	kcal	Ref.
	Input		
Fixed			
Pond construction	Moving 3000 m^3 of soil	610,000	7
Pond inlet	100 m, 20 cm dia. (4100 kg)	2,150,000	
Steel pipe[c]			
Pond outlet[c]	20 m, 35 cm dia. (35 kg)	3,500	
Asbestos-cement pipe			
Cement base[c] (Monk)	40 kg	3,000	
Machinery[a] (used on 100 ha for 10 years)	Jeep, tractors, etc., tank cars (22,800 kg of steel)[b]	705,200	7
Nets[d] (used on 100 ha for 5 years)	200 kg nylon	16,000	
Operating			
Labor	27 days/yr		
Machinery operation[a,e]	Fuel for jeeps, trucks, tractors, aerators, pumping	21,744,000	7
Fertilizer[f]	600 1 liquid ammonia (494 kg N_2)	7,200,000	
	600 kg superphosphate	1,800,000	
Herbicide[g]	About 2 kg	99,000	
Feed	Sorghum, 4.14 tons	9,108,000	7, 14
	Pellets, 25% crude protein, 3.38 tons	6,216,000	
Seed production[a,h]	Prorated from grow-out figures	15,000	
Total		**49,670,000**	
	Output		
Production total[i]	4150 kg	4,772,500	9
Common carp	(65.5%)		
Silver carp	(15.7%)		
Tilapia	(15.1%)		
Mullet	(3.7%)		
kcal input[j]/kcal output		10.4	
kcal in[j]/g protein out (unprocessed)		64.7	

[a] The data here condensed are in part based on a review of Israeli Fischulture.[11]
[b] From a survey of 80 farms calculated per ha of pond surface with 20 years for depreciation of farm.
[c] After Batty and Keller, Energy Requirements for Irrigation.
[d] Values for plastic twine assumed equivalent to those for PVC pipe (c).
[e] Hepher notes that pumping is not employed on all farms and that (given the water problems of Israel) about one third of the water is recycled.
[f] Chicken manure is used at 4 tons/ha/year at no energy cost except for hauling it, which is accounted for under machinery operation. Hepher adds that the use of organic manures (chicken, cattle), is now being increased to 10 tons/ha while chemical fertilizers are reduced. Kcal inputs are thus reduced, demonstrating how fish polyculture can be made more efficient by nutrient recycling.[8]
[g] Compound not specified; estimated.[12]
[h] The fish stocked in grow-out ponds are spawned and reared to fingerling size in smaller installations; energy inputs for fingerlings are here prorated from grow-out ponds and based on weight stocked.
[i] Rationale for species combination is that plant and animal growth in the ponds should be used most efficiently in addition to the extraneous feed supplied. Thus common carp are bottom feeders, silver carp, and tilapia feed mainly on floating algae and zooplankton while mullet graze on small attached algae.
[j] Considering that the data used here are a mean of 80 farms and that use of pumping is variable, that greater or lesser use of is made of organic fertilizer, etc., the input/output ratios for individual farms are likely to vary by about 25% up or down from those set down here.

REFERENCES

1. **Bardach, J. E.,** Report of Study Team 3, Aquatic food sources in supporting papers: world food and nutrition study, National Academy of Sciences, Washington, D.C., 1977, 253.
2. **Rawitscher, M. and Mayer, J.,** Nutritional outputs and energy inputs in seafoods, *Science,* 198, 261, 1977.
3. **Korringa, P.,** *Farming Marine Organisms Low In the Food Chain,* Elsevier Scientific, Amsterdam, 1976, chap. 1.
4. **Smith, R. R.,** Nutrient utilization by fish, in *Proceedings of the First International Symposium on Feed Composition, Animal Nutrient Requirements, and Computerization of Diets,* University of Utah, Logan, 1976.
5. **Scura, E.,** Aquatic Farms, personal communication.
6. **Batty, J. C. and Keller, J.,** Energy Inputs to Irrigation, Irrigation and Drainage Division, American Society of Civil Engineers, Dec., 1975.
7. **Pimentel, D.,** personal communication.
8. **Lockeretz, W.,** Energy Inputs for Nitrogen, Phosphorus, and Potash Fertilizers, mimeograph, Washington University, St. Louis.
9. **Watt, B. K. and Merrill, A.,** Composition of Foods, U.S. Department of Agriculture, Washington, D.C., 1963.
10. **Soderberg, R. W.,** University of Wisconsin, personal communication.
11. **Sarig, S.,** *Bamidgeh,* 29(3), 75, 1977.
12. **Pimentel, D.,** The Energy Crisis: Its Impact on Agriculture, *Sci. Tech.,* 1976.
13. **Halver, J.,** University of Washington, Seattle, personal communication.
14. **Pimentel, D.,** Energy Attributable to Seed, Mimeograph, Agricultural Economics, Cornell University.

Energy Inputs and Outputs for Marine Fishery Production

THE ENERGY REQUIREMENT FOR INSHORE AND OFFSHORE FISHING CRAFTS — THE CASE OF THE NORTHEAST FISHERY

Stephen P. Rochereau

Like any food production system, fisheries require a certain amount of energy input (kcal) in the form of capital equipment, fuel, and labor to obtain the protein energy yield in fish. The energy efficiency of a fishery system can thus be quantified and calculated as kcal protein output/kcal input.

The energy input in the case of fisheries includes the energy requirements of the working capital (e.g., the fishing crafts), the energy requirements of the implied infrastructure required to build the fishing crafts (from ore mining operations to the production of the finished crafts and products), and the energy requirements to produce the necessary goods and supplies for operating the fishing crafts.

The energy requirements presented in Tables 1 and 2 are the inputs for the heaviest types of steel fishing vessels, such as trawlers, combination crafts, and seiners that are characterized by a high level of automation. As shown in Tables 1 and 2 and Figure 1, the energy requirements hold a nonlinear relation with respect to vessel size, e.g., the gross registered tonnage (GRT) of a fishing craft. GRT represents the total measured cubic content of all permanently closed space, with minor exceptions from measurement by law.

Before determining the energy output of a fishing system to assess its efficiency, careful attention should be given to the structure of the fleet (e.g., the frequency distribution of the GRT index within the fleet), as well as its prevailing fishing technology because of the nonlinearity of energy requirements with respect to size. For the purpose of illustration, the total energy requirement profile of the Northeast fishing fleet is presented in Figure 2, while Figure 1 presented the energy requirement breakdown for the smallest, mode, mean, and largest fishing crafts currently operated in the fleet.

To highlight the differences in efficiency that may occur in a regional fleet, the Northeast fishery was divided into two subsystems, the inshore fishing fleet and the offshore fishing fleet employing large vessels above the 110 GRT size. The inshore fishery required only an input of 1.03 kcal per kcal of total food energy output, whereas the offshore fishery required an input of 3.90 kcal per kcal of total food energy output (Table 3), thus indicating that the smaller fishing units were more than three times more efficient than the offshore crafts.

The overall efficiency of the Northeast fishery in producing fish protein for human consumption averaged 4.1 kcal of energy per kcal of human food protein output. The inshore fishery required only 2.2 kcal per kcal of fish protein, whereas the offshore fishery required 9.6 kcal per kcal of fish protein output (Table 3). In part, the lower energy ratios for the inshore fish species are due to the greater productivity of the inshore region. The inshore pelagic species, which are primarily zooplankton feeders, are on the average 54% more efficient in storing energy (including protein and fat) for the same weight of usable biomass than most demersal fish, which are primarily carnivorous on other fish. The efficiency of the Northeast fishery is evident when compared with other fishery systems. For the U.S. overall, Hirst reported about 27 kcal of energy input per kcal of fish protein output,[3] while Leach in the United Kingdom reported about 20 kcal of energy input per kcal of fish protein output.[5] For steel stern trawlers from Scotland, Edwardson reported about 21 kcal of energy input per kcal fish protein output.[4] A major reason for the excellent efficiency of the Northeast fishing system is that about 93% of the vessels in the fleet are relatively small (5-150 gross registered tonnage).

Table 1
ANNUAL ENERGY INPUTS FOR FISHING SHIPBUILDING ACCORDING TO TYPICAL VESSEL SIZE IN GRT (GJ/YEAR)[a]

GRT	7	15	25	40	100	125	150	163	238	275	300	330
Main propulsion machinery	1.5	2.23	4.2	8.46	18.61	28.66	37.24	46.5	55.85	66	75.14	79.54
Hull and structures	12.13	23.18	50.6	131.5	294.21	458.6	619.24	764.4	910.33	1056.64	1192.11	1323.63
Outfitting and hull engineering	.95	2	8.8	25.84	98.2	210.41	378.75	582.11	757.46	981.87	1234.3	1495.85
Auxiliary machinery	3	5.96	12.8	30.55	65.76	111.08	165.57	231.19	308.47	410	509.25	619.68
Fishing deck equipment	1.9	3.62	8.34	21.61	35.79	58.95	84.9	111.62	137.38	166.62	196	225.66
Non-ferrous metallic elements	.82	1.66	3.95	9.74	26.5	49.57	81.25	119	154.57	200.37	246	291.4
Plastics, woodwork, and miscellaneous	2	9.56	14.28	19	45.87	68.8	91.75	121	145.28	169.5	193.7	217.9
Electronics and electric systems	3.25	7.8	15.63	31.3	62.55	93.83	125	156.38	187.66	219	260	300
Building total	**26**	**56**	**118.6**	**278**	**647.5**	**1080**	**1583.7**	**2132.2**	**2657**	**3270**	**3906.5**	**4553.3**

[a] 1 GJ = 2.4×10^5 kcal

Table 2
ANNUAL ENERGY INPUTS FOR FISHING OPERATIONS IN THE NORTHEAST AND TOTAL ANNUAL ENERGY INPUTS ACCORDING TO TYPICAL VESSEL SIZES IN THE NORTHEAST FLEET (GJ/Y)[a]

GRT	7	15	25	90	100	125	150	163	238	275	300	330
Fuel and lube	99	114	127.4	147.5	224.6	336.9	452.6	623.6	972.3	1064.5	1188.5	1252.2
Ice	2.5	4.1	7.3	13.9	36.6	59.5	82.2	116.3	164.6	212	270	330
Food and supplies	9.6	18.6	26	35.6	74	129	174	247	340	435	476	554
Fishing gear	35	52	70	110.3	221	271.4	360	654.5	880	1066.8	1353	1527
Structures and machinery	39.5	54.5	74.8	91	132	186.5	288.6	408	512	616.7	718.8	821
Electronics and electric systems	1	1.2	2.5	6	12	17.5	27.4	36	43.8	64.6	81.5	111.7
Fixed charges	2.4	4.8	13.7	24	67	92	233	317	350	390	437	480
Operation total (A)	189	249.2	321.7	428.3	767.2	1092.8	1617.8	2402.4	3262.7	3849.6	4524.8	5075.9
Building total (B) from Table 1	26	56	118.6	278	647.5	1080	1583.7	2132.2	2657	3270	3906.5	4553.3
Capital equipment total (C)	21.5	30.5	44	70.7	141.3	217.2	320	453.4	592	712	843	963
Total input Σ (A, B, C)	**236.5**	**335.7**	**484.3**	**777**	**1556**	**2390**	**3521.5**	**4988**	**6511.7**	**7831.6**	**9274.3**	**10592.2**

[a] 1 GJ = 2.4×10^5 kcal

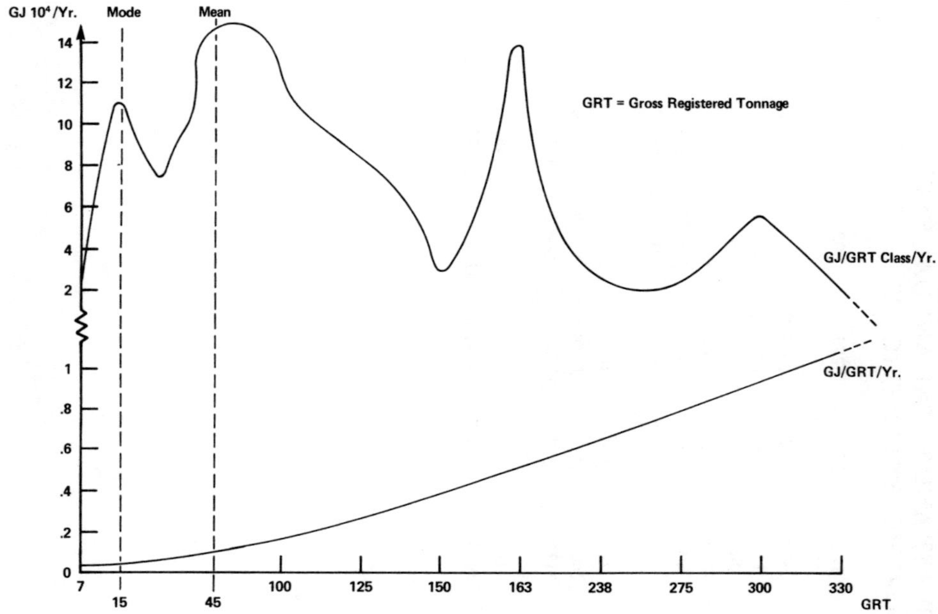

FIGURE 1. Annual energy inputs for various sizes of vessels fishing in the Northeast region.

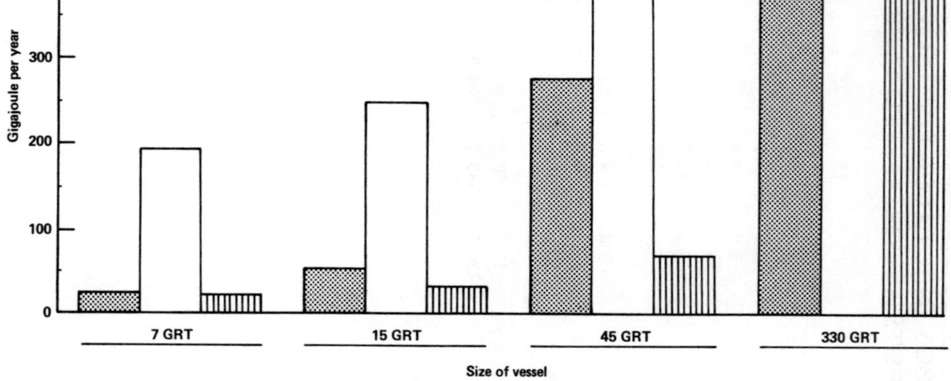

FIGURE 2. Annual energy profile of the Northeast fishing fleet (reference year 1972).

Table 3
CURRENT ENERGY EFFORT SITUATIONS IN THE NORTHEAST FISHING ZONE[a]

Fishing systems	kcal input/kcal of total food energy output[b]	kcal input/kcal human food[c] protein output
Average for total system	1.65	4.10
Inshore sub-system[d]	1.03	2.18
Offshore sub-system[e]	3.90	9.55

[a] Fishing energy requirements estimated for year 1974; energy output averaged over the 1971—1974 period to compensate for the (annual) seasonal variations in yield outputs due both to different fishing pressure levels and fluctuations between the various outputs of seafood products.[7,8]
[b] Including total harvest of fish including trash fish.
[c] Protein fish production for human food only.
[d] Including fishing units ≤110 GRT capacity.
[e] Including the rest of the fishing fleet ≥110 GRT capacity per fishing unit.

Table 4
ENERGY USE IN THE HARVEST OF SEVERAL TYPES OF SEAFOOD[6]

Seafood	Fossil energy (kcal input/kcal protein output)
Sardines (canned)	0.77
Atlantic ocean perch (redfish)	1.85
Salmon (canned pink)	5.56
Cod (fresh fillet)	6.10
Blue crab (fresh steamed)	6.90 (with shell) 3.32 (without shell)
Flounder (fresh fillet)	7.87
Haddock (fresh fillet)	11.03
Halibut (frozen steaks)	13.76
Tuna (canned)	16.63 (liquid + solids — oil) 14.38 (liquid + solids — water)
Salmon (fresh king)	25.92
Lobster (live)	49.85
Shrimp (fresh)	103.31

REFERENCES

1. **Rochereau, S. P.,** Energy Analysis and Coastal Shelf Resource Management: Nuclear Power Generation vs. Sea Food Protein Production in the Northeast Region of the U.S., Ph.D. thesis, Cornell University, 1976.
2. **Rochereau, S. P. and Pimentel, D.,** Energy tradeoffs between northeast fish production and coastal power reactors, *Energy,* 3, 575, 1978.
3. **Hirst, E.,** Food — related energy requirements, *Science,* 184(4134), 134, 1974.
4. **Edwardson, W.,** Energy Analysis and the Fishing Industry, a Report of the Energy Analysis Unit, University of Strachclyde, Glasgow, U.K., 1975.
5. **Leach, G.,** The energy cost of food production, in *The Man-Food Equation,* Bourne, A., Ed., Academic Press, New York, 1976.

6. **Rawitscher, M. and Mayer, J.**, Nutritional outputs and energy inputs in seafoods, *Science,* 198, 261, 1977.
7. Statistical Digest, No. 54—66, U.S. Department of Commerce, Washington, D.C., 1963—1975.
8. Current Fishery Statistics, No. 6400, U.S. Department of Commerce, U.S. Government Printing Office, Washington, D.C., March, 1974.

Energy Inputs and Outputs for Forestry Production

ENERGY PRODUCTION AND CONSUMPTION IN WOOD HARVEST

Douglas B. Monteith

About 370 million m³ (222 million metric tons) of wood were consumed in the U.S. in 1974. Most of this consumption was harvested from the 202 million ha of commercial forest land representing 22% of the national land area. This consumption took place in the following forms and corresponding volume ratios:

Sawlogs	48%
Pulpwood	35%
Veneer logs	9%
Fuel wood	4%
Miscellaneous	3%*

At the present time, very little wood is harvested exclusively for energy production. Although much of the manufacturing residue is used by forest industry in producing heat and some electricity, and modest amounts of fuel wood are harvested for burning in stoves and fireplaces, a vast untapped energy resource[3] is presently going largely unutilized. While much of this stockpiled solar energy may not be price competitive today, anticipated future price increases for competitive fuel stocks and environmental constraints may well make wood fuel an attractive alternative to nonrenewable energy sources.

There is, in addition, ample historical precedent for the use of wood as a significant fuel. At the turn of the century, Americans consumed 1.8 m³ of fuel wood per capita versus current per capita consumption of only 0.07 m³. Wood is also attractive as an extremely versatile material that requires less energy to produce and manufacture than nearly any other raw material. Although the tables in this volume do not include the energy requirements of product manufacture, the following estimates (adapted from Reference 13) provide some insight into the comparative advantage wood enjoys over several major alternative materials.

Material	Energy consumption in production (kcal × 10⁶/ton)
Steel	10.06
Aluminum	62.36
Cement	2.08
Plastic	2.27
Lumber	0.56

The energy balance tables are designed to facilitate the estimation of total energy production and consumption for a wide range of forest conditions and type of harvest. Energy requirements are based upon "typical" technology in use throughout the U.S. in 1977. This technology includes the chainsaw and mechanized (rubber-tired skidder or crawler tractor) transport of the trees from stump to landing. Those wishing to evaluate the energy balance of new technology (such as whole tree chipping or highly mechanized operations such as tree harvestor felling and bunching) will have to estimate the energy embodied in the equipment and develop records of fuel consumption. For mixed stands, the user will have to prorate between the corresponding hardwood stand and softwood stand tables to calculate the energy output. Standing volume is

* Includes cooperage logs, poles, piling, fence posts, hewn ties, round mine timbers, box bolts, chemical wood, shingle bolts, and other miscellaneous items.

defined as the aboveground biomass volume of wood material, exclusive of leaves and twigs.

The tables should be viewed as providing a reasonable approximation of the energy balance likely to be achieved in any particular harvest operation. Variables such as topography, technology employed, operating efficiencies, intensity of biomass utilization, and extent of unuseable forest residue will affect the actual ratios experienced.

Table 14 is provided to illustrate the use of the tables for calculating net energy production over a full rotation. Such a net energy figure represents the solar energy captured in useable form by a hectare of forest land, net of the energy costs of harvest employing 1970s technology.

Table 1
ENERGY BALANCE FOR THE FINAL HARVEST[a] OF SAWLOGS AND VENEER IN HARDWOOD[b] STANDS UNDER EVEN-AGED MANAGEMENT

		Standing volume							
		100 m^3/ha		200 m^3/ha		300 m^3/ha		400 m^3/ha	
Item	Units	Quantity/ha	kcal/ha	Quantity/ha	kcal/ha	Quantity/ha	kcal/ha	Quantity/ha	kcal/ha
Input									
Labor	hr[c]	34	—	77	—	100	—	113	—
Machinery[d]	kg[e]	11	198×10^3	31	558×10^3	54	972×10^3	73	$1,314 \times 10^3$
Tools[f]	kg[g]	0.19	3.42×10^3	0.55	9.90×10^3	0.94	16.92×10^3	1.28	23.04×10^3
Fuel[h]	ℓ[i]	167	1.69×10^6	511	5.1×10^6	840	8.49×10^6	1,143	11.6×10^6
Total			**1.89×10^6**		**5.74×10^6**		**9.48×10^6**		**12.94×10^6**
Output[j]									
Softwood	kg	1,300	5.5×10^6	3,690	15.5×10^6	6,380	26.8×10^6	8,680	36.4×10^6
Hardwood	kg	16,740	77×10^6	47,420	218×10^6	82,020	377×10^6	111,590	513×10^6
Total	**kg**	**18,040**	**82.5×10^6**	**51,110**	**234×10^6**	**88,400**	**404×10^6**	**120,270**	**550×10^6**
Output/ input			43.7		40.8		42.6		42.5
kcal output/ labor hour			2.43×10^6		3.04×10^6		4.04×10^6		4.87×10^6

[a] Final harvest under even-aged management is the removal of volume based upon the following percentages of standing volume present in m^3/ha.[9,12] Harvest types are 100, 200, 300, and 400, respectively. Sawlogs and veneer: 25, 39, 40, 41; pulpwood only: 49, 70, 80, 82; and integrated: 57, 81, 93, 95.

[b] Assumes stands are composed of 90% stems by volume of hardwood and the remaining 10% of volume is softwood.

[c] Based upon 1.76 hr/1000 kg for softwood harvest volume and 1.89 hr/1000 kg for hardwood.[2,11]

[d] Includes rubber-tired skidder, crawler tractor, or forwarder used in moving trees from the stump to a landing where the roundwood is loaded onto trucks.

[e] Assumes 6400-hr life, 3250 kg/hr production,[4,12] and 6300 kcal/kg energy embodied.[5,7]

[f] Includes chainsaws and hand tools.

[g] Based upon 6700 kcal/kg energy embodied[7] and a useful life of 1200 hr for a chainsaw and 300 hr for a saw chain.[4]

[h] Fuel used by chainsaws and machinery in harvest operation.

[i] From American Pulpwood Association, Fuel Survey, 15 pp., 1975. 10,109 kcal input per liter of fuel.[6,8]

[j] Calculated at the rates of 4600 kcal/kg for hardwood and 4200 kcal/kg for softwood.

Table 2
ENERGY BALANCE FOR THE FINAL HARVEST[a] OF SAWLOGS AND VENEER IN SOFTWOOD[b] STANDS UNDER EVEN-AGED MANAGEMENT

		Standing volume							
		100 m^3/ha		200 m^3/ha		300 m^3/ha		400 m^3/ha	
Item	Units	Quantity/ha	kcal/ha	Quantity/ha	kcal/ha	Quantity/ha	kcal/ha	Quantity/ha	kcal/ha
Input									
Labor	hr[c]	24	—	46	—	71	—	98	—
Machinery[d]	kg[e]	8	144×10^3	23	414×10^3	41	738×10^3	55	990×10^3
Tools[f]	kg[g]	0.14	2.52×10^3	0.41	7.38×10^3	0.71	12.78×10^3	0.97	17.46×10^3
Fuel[h]	ℓ[i]	126	1.27×10^6	384	3.89×10^6	632	6.39×10^6	860	8.69×10^6
Total			**1.42×10^6**		**4.31×10^6**		**7.14×10^6**		**39.7×10^6**
Output[j]									
Softwood	kg	11,710	49.2×10^6	33,180	139×10^6	57,390	241×10^6	78,080	328×10^6
Hardwood	kg	1,860	8.6×10^6	5,270	24.2×10^6	9,110	41.9×10^6	12,400	57.0×10^6
Total		**13,570**	**57.7×10^6**	**38,450**	**164×10^6**	**66,500**	**283×10^6**	**90,480**	**385×10^6**
Output/ input			40.6		38.1		39.6		39.7
kcal output/ labor hour			2.40×10^6		3.57×10^6		3.99×10^6		3.93×10^6

[a] Final harvest under even-aged management is the removal of volume based upon the following percentages of standing volume present in m^3/ha.[9,12] Harvest types are 100, 200, 300, and 400, respectively. Sawlogs and veneer: 25, 39, 40, 41; pulpwood only: 49, 70, 80, 82; and integrated: 57, 81, 93, 95.

[b] Assumes stands are composed of 90% stems by volume of softwood and the remaining 10% of volume is hardwood.

[c] Based upon 1.76 hr/1000 kg for softwood harvest volume and 1.89 hr/1000 kg for hardwoods.[2,11]

[d] Includes rubber-tired skidder, crawler tractor, or forwarder used in moving trees from the stump to a landing where the roundwood is loaded onto trucks.

[e] Assumes 6400-hr life, 3250 kg/hr production,[4,12] and 6300 kcal/kg energy embodied.[5,7]

[f] Includes chainsaws and hand tools.

[g] Based upon 6700 kcal/kg energy embodied[7] and a useful life of 1200 hr for a chainsaw and 300 hr for a saw chain.[4]

[h] Fuel used by chainsaws and machinery in harvest operation.

[i] From American Pulpwood Association, Fuel Survey, 15 pp., 1975. 10,109 kcal input per liter of fuel.[6,8]

[j] Calculated at the rates of 4600 kcal/kg for hardwood and 4200 kcal/kg for softwood.

Table 3

ENERGY BALANCE FOR THE HARVEST[a] OF SAWLOGS AND VENEER IN HARDWOOD[b] STANDS UNDER ALL-AGED MANAGEMENT AND INTERMEDIATE HARVESTS UNDER EVEN-AGED MANAGEMENT

		Standing volume							
		100 m³/ha		200 m³/ha		300 m³/ha		400 m³/ha	
Item	Units	Quantity/ha	kcal/ha	Quantity/ha	kcal/ha	Quantity/ha	kcal/ha	Quantity/ha	kcal/ha
Input									
Labor	hr[c]	17	—	38	—	50	—	56	—
Machinery[d]	kg[e]	5	90.0×10^3	15	270.0×10^3	27	486×10^3	36	648×10^3
Tools[f]	kg[g]	0.10	1.80×10^3	0.27	4.86×10^3	0.47	8.46×10^3	0.64	11.52×10^3
Fuel[h]	ℓ[i]	91	0.92×10^6	279	2.82×10^6	458	4.62×10^6	623	6.29×10^6
Total			**1.01×10^6**		**3.10×10^6**		**5.11×10^6**		**6.95×10^6**
Output[j]									
Softwood	kg	317	1.33×10^6	900	3.78×10^6	1,560	6.55×10^6	2,120	8.90×10^6
Hardwood	kg	8,620	39.6×10^6	24,420	112×10^6	42,230	194×10^6	57,460	264×10^6
Total	**kg**	**8,937**	**41.0×10^6**	**25,320**	**116×10^6**	**43,790**	**201×10^6**	**59,580**	**273×10^6**
Output/ input			40.6		37.4		39.3		39.3
kcal output/ labor hour			2.4×10^6		3.0×10^6		4.02×10^6		4.88×10^6

[a] Harvest under all-aged management is the selective removal of individual stems so as to maintain a reverse J-shaped distribution of trees in each age class. Volumes are based upon the following percentages of standing volume removal in m³/ha.[9,12] Harvest types are 100, 200, 300, and 400, respectively. Sawlogs and veneer: 12, 17, 20, 20; pulpwood only: 18, 26, 29, 30; and integrated: 24, 34, 39, 40.

[b] Assumes stands are composed of 90% stems by volume of hardwood and the remaining 10% of volume is softwood.

[c] Based upon 1.76 hr/1000 kg for softwood volume harvest and 1.89 hr/1000 kg for hardwoods.[2,11]

[d] Includes rubber-tired skidder, crawler tractor, or forwarder used in moving trees from the stump to a landing where the roundwood is loaded onto trucks.

[e] Assumes 6400-hr life, 3250 kg/hr production,[4,12] and 6300 kcal/kg energy embodied.[5,7]

[f] Includes chainsaws and hand tools.

[g] Based upon 6700 kcal/kg energy embodied[7] and a useful life of 1200 hr for a chainsaw and 300 hr for a saw chain.[4]

[h] Fuel used by chainsaws and machinery in harvest operation.

[i] From American Pulpwood Association, Fuel Survey, 15 pp., 1975. 10,109 kcal input per liter of fuel.[6,8]

[j] Calculated at the rates of 4600 kcal/kg for hardwood and 4200 kcal/kg for softwood.

Table 4
ENERGY BALANCE FOR THE HARVEST[a] OF SAWLOGS AND VENEER IN SOFTWOOD[b] STANDS UNDER ALL-AGED MANAGEMENT AND INTERMEDIATE HARVESTS UNDER EVEN-AGED MANAGEMENT

		Standing volume							
		100 m^3/ha		200 m^3/ha		300 m^3/ha		400 m^3/ha	
Item	Units	Quantity/ha	kcal/ha	Quantity/ha	kcal/ha	Quantity/ha	kcal/ha	Quantity/ha	kcal/ha
Input									
Labor	hr[c]	11	—	26	—	34	—	38	—
Machinery[d]	kg[e]	4	72.0×10^3	11	198.0×10^3	19	342×10^3	26	468×10^3
Tools[f]	kg[g]	0.07	1.26×10^3	0.20	3.60×10^3	0.34	6.12×10^3	0.46	8.28×10^3
Fuel[h]	ℓ[i]	66	0.67×10^6	202	2.04×10^6	332	3.36×10^6	452	4.56×10^6
Total			**0.74×10^6**		**2.24×10^6**		**3.71×10^6**		**5.04×10^6**
Output[j]									
Softwood	kg	6.030	25.3×10^6	17,090	71.8×10^6	29,550	124×10^6	40,200	169×10^6
Hardwood	kg	450	2.07×10^6	1,290	5.93×10^6	2,220	10.2×10^6	3,020	13.9×10^6
Total		**6480**	**27.4×10^6**	**18,380**	**77.7×10^6**	**31,770**	**134×10^6**	**43,220**	**183×10^6**
Output/ input			37.0		34.7		36.1		36.3
kcal output/ labor hour			2.49×10^6		2.99×10^6		3.94×10^6		4.82×10^6

[a] Harvest under all-aged management is the selective removal of individual stems so as to maintain a reverse J-shaped distribution of trees in each age class. Volumes are based upon the following percentages of standing volume removal in m^3/ha.[9,12] Harvest types are 100, 200, 300, and 400, respectively. Sawlogs and veneer: 12, 17, 20, 20; pulpwood only: 18, 26, 29, 30; and integrated: 24, 34, 39, 40.

[b] Assumes stands are composed of 90% stems by volume of softwood and the remaining 10% of volume is hardwood.

[c] Based upon 1.76 hr/1000 kg for softwood harvest volume and 1.89 hr/1000 kg for hardwoods.[2,11]

[d] Includes rubber-tired skidder, crawler tractor, or forwarder used in moving trees from the stump to a landing where the roundwood is loaded onto trucks.

[e] Assumes 6400-hr life, 3250 kg/hr production,[4,12] and 6300 kcal/kg energy embodied.[5,7]

[f] Includes chainsaws and hand tools.

[g] Based upon 6700 kcal/kg energy embodied[7] and a useful life of 1200 hr for a chainsaw and 300 hr for a saw chain.[4]

[h] Fuel used by chainsaws and machinery in harvest operation.

[i] From American Pulpwood Association, Fuel Supply, 15 pp., 1975. 10,109 kcal input per liter of fuel.[6,8]

[j] Calculated at the rates of 4600 kcal/kg for hardwood and 4200 kcal/kg for softwood.

Table 5
ENERGY BALANCE FOR THE FINAL HARVEST[a] OF PULPWOOD IN HARDWOOD[b] STANDS UNDER EVEN-AGED MANAGEMENT

		Standing volume							
		100 m^3/ha		200 m^3/ha		300 m^3/ha		400 m^3/ha	
Item	Units	Quantity/ha	kcal/ha	Quantity/ha	kcal/ha	Quantity/ha	kcal/ha	Quantity/ha	kcal/ha
Input									
Labor	hr[c]	68	—	154	—	199	—	226	—
Machinery[d]	kg[e]	22	39.6×10^3	62	$1{,}116 \times 10^3$	108	$1{,}944 \times 10^3$	147	$2{,}646 \times 10^3$
Tools[f]	kg[g]	0.39	7.02×10^3	1.09	19.62×10^3	1.89	34.02×10^3	2.57	46.26×10^3
Fuel[h]	ℓ[i]	334	3.37×10^6	1,022	10.3×10^6	1,679	17.0×10^6	2,285	23.1×10^6
Total			**3.77×10^6**		**11.44×10^6**		**18.98×10^6**		**25.79×10^6**
Output[j]									
Softwood	kg	2,600	10.9×10^6	7,380	31.0×10^6	12,750	53.6×10^6	17,350	72.9×10^6
Hardwood	kg	33,480	154×10^6	94,850	436×10^6	164,030	735×10^6	223,170	1027×10^6
Total		**36,080**	**165×10^6**	**102,230**	**467×10^6**	**176,780**	**808×10^6**	**240,520**	**1099×10^6**
Output/input			43.8		40.8		42.6		42.6
kcal output/labor hour			2.43×10^6		3.03×10^6		4.06×10^6		4.86×10^6

[a] Final harvest under even-aged management is the removal of volume based upon the following percentages of standing volume present in m^3/ha.[9,12] Harvest types are 100, 200, 300, and 400, respectively. Sawlogs and veneer: 25, 39, 40, 41; pulpwood only: 49, 70, 80, 82; and integrated: 57, 81, 93, 95.

[b] Assumes stands are composed of 90% stems by volume of hardwood and the remaining 10% of volume is softwood.

[c] Based upon 1.76 hr/1000 kg for softwood harvest volume and 1.89 hr/1000 kg for hardwoods.[2,11]

[d] Includes rubber-tired skidder, crawler tractor, or forwarder used in moving trees from the stump to a landing where the roundwood is loaded onto trucks.

[e] Assumes 6400-hr life, 3250 kg/hr production,[4,12] and 6300 kcal/kg energy embodied.[5,7]

[f] Includes chainsaws and hand tools.

[g] Based upon 6700 kcal/kg energy embodied[7] and a useful life of 1200 hr for a chainsaw and 300 hr for a saw chain.[4]

[h] Fuel used by chainsaws and machinery in harvest operation.

[i] From American Pulpwood Assocition, Fuel Supply, 15pp., 1975. 10,109 kcal input per liter of fuel.[6,8]

[j] Calculated at the rates of 4600 kcal/kg for hardwood and 4200 kcal/kg for softwood.

Table 6
ENERGY BALANCE FOR THE FINAL HARVEST[a] PULPWOOD IN SOFTWOOD[b] STANDS UNDER EVEN-AGED MANAGEMENT

		Standing volume							
		100 m³/ha		200 m³/ha		300 m³/ha		400 m³/ha	
Item	Units	Quantity/ha	kcal/ha	Quantity/ha	kcal/ha	Quantity/ha	kcal/ha	Quantity/ha	kcal/ha
					Input				
Labor	hr[c]	48	—	110	—	142	—	161	—
Machinery[d]	kg[e]	17	306×10^3	47	846×10^3	81	$1,458 \times 10^3$	110	$1,980 \times 10^3$
Tools[f]	kg[g]	0.29	5.22×10^3	0.82	14.76×10^3	1.42	25.56×10^3	1.93	34.74×10^3
Fuel[h]	ℓ[i]	251	2.54×10^6	769	7.77×10^6	1,264	12.8×10^6	1,719	17.4×10^6
Total			**2.85×10^6**		**8.63×10^6**		**14.28×10^6**		**19.42×10^6**
					Output[j]				
Softwood	kg	23,420	98.4×10^6	66,370	279×10^6	114,780	482×10^6	156,160	656×10^6
Hardwood	kg	3,720	17.1×10^6	10,540	48.5×10^6	18,230	83.9×10^6	24,800	114×10^6
Total		**27,140**	**115×10^6**	**76,910**	**327×10^6**	**133,010**	**566×10^6**	**180,960**	**770×10^6**
Output/input			40.4		37.9		39.6		39.7
kcal output/labor hour			2.40×10^6		2.97×10^6		3.99×10^6		4.78×10^6

[a] Final harvest under even-aged management is the removal of volume based upon the following percentages of standing volume present in m³/ha.[9,12] Harvest types are 100, 200, 300, and 400, respectively. Sawlogs and veneer: 25, 39, 40, 41; pulpwood only: 49, 70, 80, 82; and integrated: 57, 81, 93, 95.

[b] Assumes stands are composed of 90% stems by volume of softwood and the remaining 10% volume is hardwood.

[c] Based upon 1.76 hr/1000 kg for softwood harvest volume and 1.89 hr/1000 kg for hardwoods.[2,11]

[d] Includes rubber-tired skidder, crawler tractor, or forwarder used in moving trees from the stump to a landing where the roundwood is loaded onto trucks.

[e] Assumes 6400-hr life, 3250 kg/hr production,[4,12] and 6300 kcal/kg energy embodied.[5,7]

[f] Includes chainsaws and hand tools.

[g] Based upon 6700 kcal/kg energy embodied[7] and a useful life of 1200 hr for a chainsaw and 300 hr for a saw chain.[4]

[h] Fuel used by chainsaws and machinery in harvest operation.

[i] From American Pulpwood Association, Fuel Supply, 15 pp., 1975. 10,109 kcal input per liter of fuel.[6,8]

[j] Calculated at the rates of 4600 kcal/kg for hardwood and 4200 kcal/kg for softwood.

Table 7

ENERGY BALANCE FOR THE HARVEST[a] OF PULPWOOD IN HARDWOOD[b] STANDS UNDER ALL-AGED MANAGEMENT AND INTERMEDIATE HARVESTS UNDER EVEN-AGED MANAGEMENT

		Standing volume							
		100 m^3/ha		200 m^3/ha		300 m^3/ha		400 m^3/ha	
Item	**Units**	**Quantity/ha**	**kcal/ha**	**Quantity/ha**	**kcal/ha**	**Quantity/ha**	**kcal/ha**	**Quantity/ha**	**kcal/ha**
				Input					
Labor	hr[c]	25	—	57	—	74	—	84	—
Machinery[d]	kg[e]	8	144.0×10^3	23	414.0×10^3	40	720.0×10^3	55	990.0×10^3
Tools[f]	kg[g]	0.14	2.52×10^3	0.41	7.38×10^3	0.70	12.60×10^3	0.95	17.10×10^3
Fuel[h]	ℓ[i]	137	1.38×10^6	418	4.22×10^6	686	6.94×10^6	934	9.44×10^6
Total			$\mathbf{1.53 \times 10^6}$		$\mathbf{4.64 \times 10^6}$		$\mathbf{7.67 \times 10^6}$		$\mathbf{10.45 \times 10^6}$
				Output[j]					
Softwood	kg	480	2.02×10^6	1,350	5.67×10^6	2,340	9.83×10^6	3,170	13.3×10^6
Hardwood	kg	12,930	59.5×10^6	36,630	168×10^6	63,340	291×10^6	86,190	396×10^6
Total		**13,410**	$\mathbf{61.5 \times 10^6}$	**37,980**	$\mathbf{174 \times 10^6}$	**65,680**	$\mathbf{301 \times 10^6}$	**89,360**	$\mathbf{410 \times 10^6}$
Output/input			40.2		37.5		39.2		39.2
kcal output/labor hour			2.46×10^6		3.05×10^6		4.07×10^6		4.88×10^6

[a] Harvest under all-aged management is the selective removal of individual stems so as to maintain a reverse J-shaped distribution of trees in each age class. Volumes are based upon the following percentages of standing volume removal in m^3/ha.[9,12] Harvest types are 100, 200, 300, and 400, respectively. Sawlogs and veneer: 12, 17, 20, 20; pulpwood only: 18, 26, 29, 30; and integrated: 24, 34, 39, 40.

[b] Assumes stands are composed of 90% stems by volume of hardwood and the remaining 10% of volume is softwood.

[c] Based upon 1.76 hr/1000 kg for softwood harvest volume and 1.89 hr/1000 kg for hardwoods.[2,11]

[d] Includes rubber-tired skidder, crawler tractor, or forwarder used in moving trees from the stump to a landing where the roundwood is loaded onto trucks.

[e] Assumes 6400-hr life, 3250 kg/hr production,[4,12] and 6300 kcal/kg energy embodied.[5,7]

[f] Includes chainsaws and hand tools.

[g] Based upon 6700 kcal/kg energy embodied[7] and a useful life of 1200 hr for a chainsaw and 300 hr for a saw chain.[4]

[h] Fuel used by chainsaws and machinery in harvest operation.

[i] From American Pulpwood Association, Fuel Supply, 15 pp., 1975. 10,109 kcal input per liter of fuel.[6,8]

[j] Calculated at the rates of 4600 kcal/kg for hardwood and 4200 kcal/kg for softwood.

Table 8
ENERGY BALANCE FOR THE HARVEST[a] OF PULPWOOD IN SOFTWOOD[b] STANDS UNDER ALL-AGED MANAGEMENT AND INTERMEDIATE HARVESTS UNDER EVEN-AGED MANAGEMENT

		Standing volume							
		100 m^3/ha		200 m^3/ha		300 m^3/ha		400 m^3/ha	
Item	Units	Quantity/ha	kcal/ha	Quantity/ha	kcal/ha	Quantity/ha	kcal/ha	Quantity/ha	kcal/ha
				Input					
Labor	hr[c]	17	—	39	—	51	—	57	—
Machinery[d]	kg[e]	6	108.0×10^3	17	306.0×10^3	29	522.0×10^3	40	720.0×10^3
Tools[f]	kg[g]	0.10	1.80×10^3	0.29	5.22×10^3	0.51	9.18×10^3	0.69	12.42×10^6
Fuel[h]	ℓ[i]	99	1.00×10^6	303	3.06×10^6	498	5.03×10^6	678	6.85×10^6
Total			**1.11×10^6**		**34.4×10^6**		**36.2×10^6**		**36.1×10^6**
				Output[j]					
Softwood	kg	9,050	38.0×10^6	25,640	108×10^6	44,320	186×10^6	60,310	253×10^6
Hardwood	kg	680	3.13×10^6	1,930	8.88×10^6	3,330	15.3×10^6	4,540	20.9×10^6
Total		**9,730**	**41.1×10^6**	**27,570**	**116×10^6**	**47,650**	**201×10^6**	**64,850**	**274×10^6**
Output/input			37.0		34.4		36.2		36.1
kcal output/labor hour			2.41×10^6		2.97×10^6		3.94×10^6		4.81×10^6

[a] Harvest under all-aged management is the selective removal of individual stems so as to maintain a reverse J-shaped distribution of trees in each age class. Volumes are based upon the following percentages of standing volume removal in m^3/ha.[9,12] Harvest types are 100, 200, 300, and 400, respectively. Sawlogs and veneer: 12, 17, 20, 20; pulpwood only: 18, 26, 29, 30; and integrated: 24, 34, 39, 40.

[b] Assumes stands are composed of 90% stems by volume of softwood and the remaining 10% of volume is hardwood.

[c] Based upon 1.76 hr/1000 kg for softwood harvest volume and 1.89 hr/1000 kg for hardwoods.[2,11]

[d] Includes rubber-tired skidder, crawler tractor, or forwarder used in moving trees from the stump to a landing where the roundwood is loaded onto trucks.

[e] Assumes 6400-hr life, 3250 kg/hr production,[4,12] and 6300 kcal/kg energy embodied.[5,7]

[f] Includes chainsaws and hand tools.

[g] Based upon 6700 kcal/kg energy embodied[7] and a useful life of 1200 hr for a chainsaw and 300 hr for a saw chain.[4]

[h] Fuel used by chainsaws and machinery in harvest operation.

[i] From American Pulpwood Association, Fuel Supply, 15 pp., 1975. 10,109 kcal input per liter of fuel.[6,8]

[j] Calculated at the rates of 4600 kcal/kg for hardwood and 4200 kcal/kg for softwood.

Table 9
ENERGY BALANCE FOR INTEGRATED[a] FINAL HARVEST[b] OF HARDWOOD[c] STANDS UNDER EVEN-AGED MANAGEMENT

		Standing volume							
		100 m^3/ha		200 m^3/ha		300 m^3/ha		400 m^3/ha	
Item	Units	Quantity/ha	kcal/ha	Quantity/ha	kcal/ha	Quantity/ha	kcal/ha	Quantity/ha	kcal/ha
Input									
Labor	hr[d]	79	—	178	—	231	—	262	—
Machinery[e]	kg[f]	26	468×10^3	72	$1,296 \times 10^3$	125	$2,250 \times 10^3$	170	$3,060 \times 10^6$
Tools[g]	kg[h]	0.45	8.10×10^3	1.26	22.60×10^3	2.19	39.42×10^3	2.97	53.46×10^3
Fuel[i]	ℓ[j]	387	3.91×10^6	1,184	12.0×10^6	1,946	19.7×10^6	2,647	26.8×10^6
Total			**4.39×10^6**		**13.32×10^6**		**21.99×10^6**		**29.91×10^6**
Output[k]									
Softwood	kg	3,020	12.7×10^6	8,540	35.9×10^6	14,780	62.1×10^6	20,100	84.4×10^6
Hardwood	kg	38,780	178×10^6	109,880	505×10^6	190,040	874×10^6	258,550	$1,189 \times 10^6$
Total	**kg**	**41,800**	**191×10^6**	**118,420**	**541×10^6**	**204,820**	**936×10^6**	**278,650**	**$1,274 \times 10^6$**
Output/input			43.5		40.6		42.6		42.6
kcal output/labor hour			2.42×10^6		3.04×10^6		4.05×10^6		4.86×10^6

[a] The simultaneous harvest of sawlogs, veneer, pulpwood, and fuel wood.

[b] Final harvest under even-aged management is the removal of volume based upon the following percentages of standing volume present in m^3/ha.[9,12] Harvest types are 100, 200, 300, and 400, respectively. Sawlogs and veneer: 25, 39, 40, 41; pulpwood only: 49, 70, 80, 82; and integrated: 57, 81, 93, 95.

[c] Assumes stands are composed of 90% stems by volume of hardwood and the remaining 10% of volume is softwood.

[d] Based upon 1.76 hr/1000 kg for softwood harvest volume and 1.89 hr/1000 kg for hardwoods.[2,11]

[e] Includes rubber-tired skidder, crawler tractor, or forwarder used in moving trees from the stump to a landing where the roundwood is loaded onto trucks.

[f] Assumes 6400-hr life, 3250 kg/hr production,[4,12] and 6300 kcal/kg energy embodied.[5,7]

[g] Includes chainsaws and hand tools.

[h] Based upon 6700 kcal/kg energy embodied[7] and a useful life of 1200 hr for a chainsaw and 300 hr for a saw chain.[4]

[i] Fuel used by chainsaws and machinery in harvest operation.

[j] From American Pulpwood Association, Fuel Supply, 15 pp., 1975. 10,109 kcal input per liter of fuel.[6,8]

[k] Calculated at the rates of 4600 kcal/kg for hardwood and 4200 kcal/kg for softwood.

Table 10
ENERGY BALANCE FOR INTEGRATED[a] FINAL HARVEST[b] OF SOFTWOOD[c] STANDS UNDER EVEN-AGED MANAGEMENT

		Standing volume							
		100 m³/ha		200 m³/ha		300 m³/ha		400 m³/ha	
Item	Units	Quantity/ha	kcal/ha	Quantity/ha	kcal/ha	Quantity/ha	kcal/ha	Quantity/ha	kcal/ha
Input									
Labor	hr[d]	56	—	127	—	164	—	186	—
Machinery[e]	kg[f]	19	342×10^3	54	972×10^3	94	$1,692 \times 10^3$	128	$2,304 \times 10^3$
Tools[g]	kg[h]	0.34	6.12×10^3	0.95	17.10×10^3	1.64	29.52×10^3	2.24	40.32×10^3
Fuel[i]	ℓ[j]	291	2.94×10^6	891	9.01×10^6	1,464	14.8×10^6	1,992	20.1×10^6
Total			**3.29×10^6**		**10.00×10^6**		**16.52×10^6**		**22.41×10^6**
Output[k]									
Softwood	kg	27,140	114×10^6	76,890	323×10^6	132,940	558×10^6	180,920	760×10^6
Hardwood	kg	4,310	19.8×10^6	12,210	56.2×10^6	21,120	97.2×10^6	28,730	132×10^6
Total	**kg**	**31,450**	**134×10^6**	**89,100**	**379×10^6**	**154,060**	**656×10^6**	**209,650**	**892×10^6**
Output/input			40.7		37.9		39.7		39.8
kcal output/labor hour			2.39×10^6		2.98×10^6		4.00×10^6		4.80×10^6

[a] The simultaneous harvest of sawlogs, veneer, pulpwood, and fuel wood.
[b] Final harvest under even-aged management is the removal of volume based upon the following percentages of standing volume present in m³/ha.[9,12] Harvest types are 100, 200, 300, and 400, respectively. Sawlogs and veneer: 25, 39, 40, 41; pulpwood only: 49, 70, 80, 82; and integrated: 57, 81, 93, 95.
[c] Assumes stands are composed of 90% stems by volume of softwood and the remaining 10% of volume is hardwood.
[d] Based upon 1.76 hr/1000 kg for softwood harvest volume and 1.89 hr/1000 kg for hardwoods.[2,11]
[e] Includes rubber-tired skidder, crawler tractor, or forwarder used in moving trees from the stump to a landing where the roundwood is loaded onto trucks.
[f] Assumes 6400-hr life, 3250 kg/hr production,[4,12] and 6300 kcal/kg energy embodied.[5,7]
[g] Includes chainsaws and hand tools.
[h] Based upon 6700 kcal/kg energy embodied[7] and a useful life of 1200 hr for a chainsaw and 300 hr for a saw chain.[4]
[i] Fuel used by chainsaws and machinery in harvest operation.
[j] From American Pulpwood Association, Fuel Supply, 15 pp., 1975. 10,109 kcal input per liter of fuel.[6,8]
[k] Calculated at the rates of 4600 kcal/kg for hardwood and 4200 kcal/kg for softwood.

Table 11
ENERGY BALANCE FOR INTEGRATED[a] HARVEST[b] OF HARDWOOD[c] STANDS UNDER ALL-AGED MANAGEMENT AND INTERMEDIATE HARVEST UNDER EVEN-AGED MANAGEMENT

Item	Units	Standing volume: 100 m^3/ha Quantity/ha	100 m^3/ha kcal/ha	200 m^3/ha Quantity/ha	200 m^3/ha kcal/ha	300 m^3/ha Quantity/ha	300 m^3/ha kcal/ha	400 m^3/ha Quantity/ha	400 m^3/ha kcal/ha
Input									
Labor	hr[d]	34	—	76	—	99	—	112	—
Machinery[e]	kg[f]	11	198×10^3	31	558×10^3	53	954×10^3	72	$1,296 \times 10^3$
Tools[g]	kg[h]	0.19	3.42×10^3	0.54	9.72×10^3	0.93	16.74×10^3	1.26	2.27×10^3
Fuel[i]	ℓ[j]	182	1.84×10^6	557	5.63×10^6	915	9.25×10^6	1,237	12.5×10^6
Total			**2.04×10^6**		**6.20×10^6**		**10.22×10^6**		**13.80×10^6**
Output[k]									
Softwood	kg	630	2.65×10^6	1,800	7.56×10^6	3,120	13.1×10^6	4,230	17.8×10^6
Hardwood	kg	17,240	79.3×10^6	48,840	225×10^6	84,460	388×10^6	114,150	525×10^6
Total	**kg**	**17,870**	**82.0×10^6**	**50,640**	**232×10^6**	**87,580**	**402×10^6**	**118,380**	**543×10^6**
Output/in put			40.2		37.4		39.3		39.3
kcal output/labor hour			2.41×10^6		3.05×10^6		4.06×10^6		4.84×10^6

[a] The simultaneous harvest of sawlogs, veneer, pulpwood, and fuel wood.
[b] Harvest under all-aged management is the selective removal of individual stems so as to maintain a reverse J-shaped distribution of trees in each age class. Volumes are based upon the following percentages of standing volume removal in m^3/ha.[9,12] Harvest types are 100, 200, 300, and, 400, respectively. Sawlogs and veneer: 12, 17, 20, 20; pulpwood only: 18, 26, 29, 30; and integrated: 24, 34, 39, 40.
[c] Assumes stands are composed of 90% stems by volume of hardwood and the remaining 10% of volume is softwood.
[d] Based upon 1.76 hr/1000 kg for softwood harvest volume and 1.89 hr/1000 kg for hardwoods.[2,11]
[e] Includes rubber-tired skidder, crawler tractor, or forwarder used in moving trees from the stump to a landing where the roundwood is loaded onto trucks.
[f] Assumes 6400-hr life, 3250 kg/hr production,[4,12] and 6300 kcal/kg energy embodied.[5,7]
[g] Includes chainsaws and hand tools.
[h] Based upon 6700 kcal/kg energy embodied[7] and a useful life of 1200 hr for a chainsaw and 300 hr for a saw chain.[4]
[i] Fuel used by chainsaws and machinery in harvest operation.
[j] From American Pulpwood Association, Fuel Supply, 15 pp., 1975. 10,109 kcal input per liter of fuel.[6,8]
[k] Calculated at the rates of 4600 kcal/kg for hardwood and 4200 kcal/kg for softwood.

Table 12
ENERGY BALANCE FOR INTEGRATED[a] HARVEST[b] OF SOFTWOOD[c] STANDS UNDER ALL-AGED MANAGEMENT AND INTERMEDIATE HARVEST UNDER EVEN-AGED MANAGEMENT

		Standing volume							
		100 m³/ha		200 m³/ha		300 m³/ha		400 m³/ha	
Item	Units	Quantity/ha	kcal/ha	Quantity/ha	kcal/ha	Quantity/ha	kcal/ha	Quantity/ha	kcal/ha
Input									
Labor	hr[d]	23	—	52	—	67	—	76	—
Machinery[e]	kg[f]	8	144×10^3	22	396×10^3	39	702×10^3	53	954×10^3
Tools[g]	kg[h]	0.14	2.52×10^3	0.39	7.02×10^3	0.68	12.24×10^3	0.92	16.56×10^3
Fuel[i]	ℓ[j]	132	1.33×10^6	404	4.09×10^6	664	6.71×10^6	904	9.13×10^6
Total			**1.48×10^6**		**4.49×10^6**		**7.42×10^6**		**10.10×10^6**
Output[k]									
Softwood	kg	12,060	50.6×10^6	34,180	144×10^6	59,100	248×10^6	80,410	338×10^6
Hardwood	kg	910	4.19×10^6	2,570	11.8×10^6	4,440	20.4×10^6	6,050	27.8×10^6
Total	**kg**	**12,970**	**54.8×10^6**	**36,750**	**155×10^6**	**63,540**	**269×10^6**	**86,460**	**366×10^6**
Output/input			37.0		34.5		36.3		36.2
kcal output/labor hour			2.38×10^6		2.98×10^6		4.01×10^6		4.81×10^6

[a] The simultaneous harvest of sawlogs, veneer, pulpwood, and fuel wood.
[b] Harvest under all-aged management is the selective removal of individual stems so as to maintain a reverse J-shaped distribution of trees in each age class. Volumes are based upon the following percentages of standing volume removal in m³/ha.[9,12] Harvest types are 100, 200, 300, and 400, respectively. Sawlogs and veneer: 12, 17, 20, 20; pulpwood only: 18, 26, 29, 30; and integrated: 24, 34, 39, 40.
[c] Assumes stands are composed of 90% stems by volume of softwood and the remaining 10% of volume is hardwood.
[d] Based upon 1.76 hr/1000 kg for softwood harvest volume and 1.89 hr/1000 kg for hardwoods.[2,11]
[e] Includes rubber-tired skidder, crawler tractor, or forwarder used in moving trees from the stump to a landing where the roundwood is loaded onto trucks.
[f] Assumes 6400-hr life, 3250 kg/hr production,[4,12] and 6300 kcal/kg energy embodied.[5,7]
[g] Includes chainsaws and hand tools.
[h] Based upon 6700 kcal/kg energy embodied[7] and a useful life of 1200 hr for a chainsaw and 300 hr for a saw chain.[4]
[i] Fuel used by chainsaws and machinery in harvest operation.
[j] From American Pulpwood Association, Fuel Supply, 15 pp., 1975. 10,109 kcal input per liter of fuel.[6,8]
[k] Calculated at the rates of 4600 kcal/kg for hardwood and 4200 kcal/kg for softwood.

Table 13
CONVERSION FACTORS

2,200 lb/metric ton
50 lb/ft^3 hardwood (756 kg/M^3)
35 lb/ft^3 softwood (529 kg/M^3)
35.34 ft^3/M^3
405 ha/1000 acres
14.31 ft^3/acre = 1 m^3/ha
4 BTU/1 kcal

Table 14
EXAMPLE OF ENERGY TABLE USE

Treatment	Year	Standing volume	Table no.	Input/hectare: Labor (hr)	Machinery (kg)	Tool (kg)	Fuel (ℓ)	Total (10^6/kcal)	Output/hectare: Softwood (kg)	Hardwood (kg)	Total (10^6/kcal)
Regeneration	0	0	—								
Intermediate cut (pulpwood)	30	100	7	25	8	0.14	137	1.53	480	12,930	61.5
Intermediate cut (integrated)	50	200	11	76	31	0.54	557	6.20	1,800	48,840	232
Intermediate cut (integrated)	65	300	11	99	53	0.93	915	10.22	3,120	84,460	402
Intermediate cut (sawlogs)	80	300	3	50	27	0.47	458	5.11	1,560	42,230	201
Final harvest (integrated)	100	300	9	231	125	2.19	1946	21.99	14,780	190,040	936
Total				**481**	**244**	**4.27**	**4013**	**45.05**	**21,740**	**378,500**	**1832.5**

Situation: A 50-ha hardwood area has been regenerated by clearcutting, and the owner wishes to determine the energy balance and input requirements implicit in allowing the stand to grow to maturity.

Net energy production = 1790×10^6 kcal.
= 17.9×10^6 kcal/year.
Wood production = 400 kg/year.

REFERENCES

1. Fuel Survey, American Pulpwood Association, Washington, D.C., 1975.
2. **Boyd, C. W., Koch, P., McKean, H. B., Morschauser, C. R., Breston, S. B., and Wangnard, F. F.,** Wood for structural and architectural purposes, *Wood Fiber,* 8(1), 3, 1976.
3. **Burwell, C. C.,** Solar biomass energy: an overview of U.S. potential, *Science,* 199(10), 1041, 1978.
4. **Carow, J.,** personal communication.
5. *Caterpillar Performance Handbook,* 7th ed., Caterpillar Tractor Company, Peoria, Ill., 1976.
6. **Cervinka, V.,** Fuel and energy efficiency in Handbook of Energy Utilization in Agriculture, Pimentel, O., Ed., CRC Press, Boca Raton, Fla., 1980.
7. **Doering, O. C., Considine, T. J., and Harling, C. E.,** Accounting for Tillage Equipment and Other Machinery in Agriculture Energy Analysis, NSF/RA-770128, Agricultural Experiment Station, Purdue University, Lafayette, Ind., 1977.
8. **Dykstra, D. P.,** in *Loggers Handbook,* Vol. 37, Pacific Logging Congress, Portland, Ore., 1977, 55.
9. **Filip, S. M.,** Harvesting Costs and Returns Under 4 Cutting Methods in Mature Beech-Birch-Maple Stands in New England, Forest Service Research Paper NE-87, U.S. Department of Agriculture, Upper Darby, Pa., 1967.
10. The Nation's Renewable Resources — An Assessment, 1975, Forest Service, U.S. Department of Agriculture, Washington, D.C., 1976.
11. **Gabriel, W. J.,** Machines and Systems Suitable for Logging Small Woodlots in the Northeast, AFRI Research Rep. 26, State University of New York, College of Environmental Science and Forestry, 1976.
12. **Nyland, R. D., Craul, P. J., Behrend, D. F., Echelberger, H. E., Gabriel, W. J., Nissen, R. C., Jr., Uebler, R., and Zarnetske, J.,** Logging and Its Effects in Northern Hardwoods, AFRI Research Rep. No. 31, State University of New York, College of Environmental Science and Forestry, Applied Forestry Research Institute, Syracuse, N.Y., 1976.
13. **Penner, S. S. and Icerman, L.,** *Energy,* Vol. 1, Addison-Wesley, Reading, Mass., 1974.
14. *Historical Atlas of the United States GPO,* U.S. Department of Commerce, Washington, D.C., 1975.

Index

INDEX

A

B

C

D

E

F

G

H

I

J

K

L

M

N

O

P

Q

R

S

V

W

Y

Z